13th Five-Year National Key Publication Project

Book Series on Theory and Technology of Intelligent Manufacturing and Robotics
Editors-in-Chief: Han Ding & Ronglei Sun

Bangchun Wen·Xianli Huang·
Yinong Li·Yimin Zhang

Vibration Utilization Engineering

图书在版编目(CIP)数据

振动利用工程＝Vibration Utilization Engineering：英文/闻邦椿等著. —武汉：华中科技大学出版社，2022.11
（智能制造与机器人理论及技术研究丛书）
ISBN 978-7-5680-8765-0

Ⅰ.①振… Ⅱ.①闻… Ⅲ.①振动—应用—研究—英文 Ⅳ.①TB535

中国版本图书馆 CIP 数据核字(2022)第 201713 号

Sales in the Chinese Mainland Only
本书仅限在中国大陆地区发行销售

振动利用工程　　　　　　　　　　　闻邦椿　黄显利　李以农　张义民　著
ZHENDONG LIYONG GONGCHENG

策划编辑：俞道凯　张少奇
责任编辑：罗　雪
责任监印：周治超
出版发行：华中科技大学出版社(中国·武汉)　　电话：(027)81321913
　　　　　武汉市东湖新技术开发区华工科技园　　邮编：430223
录　　排：武汉三月禾文化传播有限公司
印　　刷：湖北新华印务有限公司
开　　本：710mm×1000mm　1/16
印　　张：23.5
字　　数：605 千字
版　　次：2022 年 11 月第 1 版第 1 次印刷
定　　价：198.00 元

本书若有印装质量问题，请向出版社营销中心调换
全国免费服务热线：400-6679-118　竭诚为您服务
版权所有　侵权必究

Website: http://press.hust.edu.cn
Book Title: Vibration Utilization Engineering

Copyright @ 2022 by Huazhong University of Science & Technology Press. All rights reserved. No part of this publication may be reproduced, stored in a database or retrieval system, or transmitted in any form or by any electronic, mechanical, photocopy, or other recording means, without the prior written permission of the publisher.

Contact address: No. 6 Huagongyuan Rd, Huagong Tech Park, Donghu High-tech Development Zone, Wuhan 430223, Hubei Province, P. R. China.
Phone/fax: 8627-81339688; **E-mail**: service@hustp.com

Disclaimer

This book is for educational and reference purposes only. The authors, editors, publishers and any other parties involved in the publication of this work do not guarantee that the information contained herein is in any respect accurate or complete. It is the responsibility of the readers to understand and adhere to local laws and regulations concerning the practice of these techniques and methods. The authors, editors and publishers disclaim all responsibility for any liability, loss, injury, or damage incurred as a consequence, directly or indirectly, of the use and application of any of the contents of this book.

First published: 2022
ISBN: 978-7-5680-8765-0

Cataloguing in publication data: A catalogue record for this book is available from the CIP-Database China.

Printed in the People's Republic of China

Book Series on Theory and Technology of Intelligent Manufacturing and Robotics

Consultative Group of Experts

Chairman Youlun Xiong, Huazhong University of Science and Technology, Wuhan, China

Members

Bingheng Lu, Xi'an Jiaotong University, Xi'an, China

Xueyu Ruan, Shanghai Jiao Tong University, Shanghai, China

Jianwei Zhang, Universität Hamburg, Hamburg, Germany

Jianying Zhu, Nanjing University of Aeronautics and Astronautics, Nanjing, China

Zhuangde Jiang, Xi'an Jiaotong University, Xi'an, China

Di Zhu, Nanjing University of Aeronautics and Astronautics, Nanjing, China

Huayong Yang, Zhejiang University, Hangzhou, China

Xinyu Shao, Huazhong University of Science and Technology, Wuhan, China

Zhongqin Lin, Shanghai Jiao Tong University, Shanghai, China

Jianrong Tan, Zhejiang University, Hangzhou, China

Advisory Panel

Chairman Kok-Meng Lee, Georgia Institute of Technology, Atlanta, GA, USA

Members

Haibin Yu, Shenyang Institute of Automation, Chinese Academy of Sciences, Shenyang, China

Tianmiao Wang, Beihang University, Beijing, China

Zhongxue Gan, Fudan University, Shanghai, China

Xiangyang Zhu, Shanghai Jiao Tong University, Shanghai, China

Lining Sun, Soochow University, Suzhou, China

Guilin Yang, Ningbo Institute of Industrial Technology, Ningbo, China

Guang Meng, Shanghai Academy of Spaceflight Technology, Shanghai, China

Tian Huang, Tianjin University, Tianjin, China

Feiyue Wang, Institute of Automation, Chinese Academy of Sciences, Beijing, China

Zhouping Yin, Huazhong University of Science and Technology, Wuhan, China

Tielin Shi, Huazhong University of Science and Technology, Wuhan, China

Hong Liu, Harbin Institute of Technology, Harbin, China

Bin Li, Huazhong University of Science and Technology, Wuhan, China

Dan Zhang, Beijing Jiaotong University, Beijing, China

Zhongping Jiang, New York University, NY, USA

Minghui Huang, Central South University, Changsha, China

Editorial Committee

Chairmen Han Ding, Huazhong University of Science and Technology, Wuhan, China
Ronglei Sun, Huazhong University of Science and Technology, Wuhan, China

Members

Cheng'en Wang, Shanghai Jiao Tong University, Shanghai, China

Yusheng Shi, Huazhong University of Science and Technology, Wuhan, China

Shudong Sun, Northwestern Polytechnical University, Xi'an, China

Dinghua Zhang, Northwestern Polytechnical University, Xi'an, China

Dapeng Fan, National University of Defense Technology, Changsha, China

Bo Tao, Huazhong University of Science and Technology, Wuhan, China

Yongcheng Lin, Central South University, Changsha, China

Zhenhua Xiong, Shanghai Jiao Tong University, Shanghai, China

Yongchun Fang, Nankai University, Tianjin, China

Hong Qiao, Institute of Automation, Chinese Academy of Sciences, Beijing, China

Zhijiang Du, Harbin Institute of Technology, Harbin, China

Xianmin Zhang, South China University of Technology, Guangzhou, China

Xinjian Gu, Zhejiang University, Hangzhou, China

Jianda Han, Nankai University, Tianjin, China

Gang Xiong, Institute of Automation, Chinese Academy of Sciences, Beijing, China

About the Authors

▶ **Bangchun Wen** is an academician of Chinese Academy of Sciences, chairman of the China Institute of Vibration Engineering. Professor Wen is the member of China Committee of IFToMM, the International Rotor Dynamics Technical Committee, and the Asia-Pacific Vibration Conference Steering Committee. Professor Wen was awarded the title of National Young and Middle-aged Experts with Outstanding Contributions of the first group in 1984. He systematically researched and developed the new discipline"Vibration Engineering"which combines vibration with mechanism. He also did a lot of research in rotor dynamics, and theory and application of mechanical systems nonlinear vibration, vibration diagnosis of mechanical fault, electromechanical integration, as well as the theory of engineering machinery. He issued more than 700 papers, among of which more than 260 have been indexed in SCI, EI and ISTP and has published and edited chiefly over 10 monographs. He has hosted 4 times of international academic conference, and has been the editor-in-chief of 4 collections of papers of international symposium. He has completed dozens of major research projects of national and horizontal level, and gained 2 international prizes, 4 National Invention Award and Science and Technology Progress Award and 8 national patents. There are a number of research achievements which have reached the international advanced level, making significant economic and social benefits.

▶ **Xianli Huang** graduated from Dalian University of Technology. He was a lecturer from Northeastern University before he went to Michigan State University in USA. During his teaching and doing research in Northeastern University, Dr. Huang worked under supervision of Professor Wen and Professor Wen is his mentor. He got his Ph.D. Degree in Mechanical Engineering from Michigan State University. Then he worked as an engineer for Product Development in Ford Motor Company. During his 18 years of experience in Ford Motor Company he did vehicle NVH design, development and experiments. Then he went to China and worked in heavy duty truck, bus, electric vehicle design and development. His major work has been a long time NVH expert. He published two books on Truck and Electric Vehicle NVH in Chinese.

About the Authors

▶ **Yinong Li** is currently a Professor of the State Key Lab. of Mechanical Transmission and the College of Mechanical and Vehicle Engineering, Chongqing University. He received the B.Tech. degree from Chang'an University, Xi'an, China, in 1983 and M.S. degree from Chongqing University, Chongqing, China, in 1993, and Ph.D. degree from Northeastern University, Shenyang, China, in 1999, respectively. He is now a director of the Society of Vibration Engineering of China, the vice-president of a council of China Vibration Utilization Engineering Special Committee, the executive director of China Vibration and Noise Control Special Committee, a member of the editor board of Journal of Vibration Engineering (Chinese).

He has authored more than 180 papers and is the holder of 20 patents in China. He won the second prize of China Science and Technology Progress Award, and 3 first prizes of other technology awards. His research interests include active/semi-active vibration and noise control of mechanical system, nonlinear vibration of mechanical transmission system, and automated vehicle system dynamics and control.

▶ **Professor Yimin Zhang** held a Ph.D. degree and then became a professor and advisor for Ph.D. students. He was rewarded as a Chang Jiang Scholars professor in 2004, a member in Mechanical Engineering for Science Subject Review Group of State Council Degree Committee (6th and 7th). Starting from 2008 he became an expert in Mechanical Engineering and Communication and Carry Engineering science subject review group for National Nature Science Foundation Committee. He was the leader of Chang Jiang Scholars and Innovation Team Development Initiative in 2009. He was also the National Excellent Scientific and Technology Worker in 2014. He became a youth science and technology expert for China Mechanical Industry in 1995 and leading personnel for China Mechanical Industry Science and Technology Innovation in 2016.

Professor Zhang published about 700 papers and more 500 of them are included in SCI and EI retrieval systems and cited by researchers in more 20 countries from US, BT, Germany, Canada, France, Italy and China etc., and domestic citations are more than 5000. He published more than 10 monographs and teaching materials. He has more than 50 patents and software copyrights.

Preface

In recent years, "Intelligent manufacturing+tri-co robots" are particularly eye-catching, presenting the characteristics of the era of the perception of things,the interconnect of things,the intelligence of thing. Intelligent manufacturing and tri-co robots industry will be the strategic emerging industry with priority development,it is also a huge engine for "Made in China 2049". It's remarkable that the large-scale tri-co robots industry formed by smart cars ,drones and underwater robots will be a strategic area of countries to compete in the next 30 years, and have influence on economic development, social progress, and war forms. The related manufacturing sciences and robotics are comprehensive disciplines that link and cover material sciences, information sciences, and life sciences. Like other engineering sciences and technical sciences, tri-co robots industry also will be the big science that provides a way to understand and transform the world. In the mid-20th century,the publication of Cybernetics and Engineering Cybernetics created a new era of engineering sciences. Since the 21st century, the manufacturing sciences, robotics and artificial intelligence and other fields have been extremely active and far-reaching,they are the sources of the innovation of "Intelligent manufacturing+ tri-co robots".

Huazhong University of Science and Technology Press follows the trend of the times, aiming at the technological frontiers of intelligent manufacturing and robots, organizes and plans this series of Intelligent Manufacturing and Robot Theory & Technology Research. The series covers a wide range of topics,experts and professors are warmly welcome to write books from different perspectives,different aspects, and different fields.The key points of the topics include but are not limited to:the links of intelligent manufacturing,such as research, development, design, processing, molding and assembly, etc;the fields of intelligent manufacturing,such as intelligent control, intelligent sensing, intelligent equipment, intelligent systems, intelligent logistics and intelligent automation, etc;development and application of robots,such as industrial robots, service robots, extreme robots, land-sea-air robots, bionics/artificial/robots, soft robots and micro-nano robots;artificial intelligence, cognitive science, big data, cloud manufacturing, Internet of things and Internet,etc.

This series of books will become a platform for academic exchange and cooperation between experts and scholars in related fields, a zone where young scientists thrive, and an international arena for scientists to display their research results.Huazhong University of Science and Technology Press will cooperate with international academic publishing organizations such as Springer Publishing House to publish and distribute this series of books. Also, the company has established close relationship with relevant international academic conferences and journals, creating a good environment to enhance the academic level and practical value,expand the international influence of the series.

In recent years, people from all walks of life, university teachers and students, experts, scientists and technicians in various fields are more and more enthusiastic about intelligent manufacturing and robotics.This series of books will become the link between experts, scholars, university teachers, students and technicians, enhance the connection between authors, editors and readers, speed up the process of discovering, imparting , increasing and updating knowledge, contribute to economic construction, social progress, and scientific and technological development.

Finally,I sincerely thank the authors, editors and readers who have contributed to this series of books,for adding,gathering,and exerting positive energy for innovation-driven development, thank the relevant personnel of Huazhong University of Science and Technology Press for their hard work in the process of organizing and scheming of the series of books.

Professor of Huazhong University of Science and Technology
Academician of Chinese Academy of Sciences

Youlun Xiong
September, 2017

Foreword

Vibration, or oscillation, is a periodic movement in time of a system around a certain equilibrium position. The system consists of at least one element for storing kinetic energy and one for storing potential energy. The vibration is a process of energy-type exchanges between the kinetic and potential energies. The equilibrium position is the position in which the potential energy becomes zero. The periodicity of the movement is described in frequency which is the measurement of the numbers of times of the repeated events that occurred in a unit time. Vibration is an omnipresent type of dynamic behavior of a system and exists in many forms, such as sound and music.

Human beings first recognized the vibration phenomenon by entertaining themselves with percussion, string, and plate music instruments in ancient times. Just like the time when the lever principle was discovered is much later than that of its real utilization in human history, the time when the vibration and acoustics theory on the principles of the music instruments were discovered is much later than the time when the music instruments were made and played. The earliest music instrument unearthed in Henan Province, China, is a bone flute, which can be traced back about 6000–5000 years B.C., while the earliest string vibration frequency and music acoustic theory for a string instrument ever written and published in Chinese history is around 700 B.C. in "Guanzi" by Guan Zhong (?–645 B.C.). In his music-scale algorithm, if the length of the basic string for the major tone is 1, the string lengths of the next scales are either added 1/3 (4/3) of the basic length or subtracted 1/3 (or 2/3), other music scales are determined by this 1/3 length rule, and he obtained five tones by this algorithm: C, D, E, G, A. It's amazing that more than 100 years later the Greek philosopher and mathematician Pythagoras (580–500 B.C.) discovered independently the very similar theory for seven tones: The Pythagorean Scale.

Human beings first learned the vibration and acoustics of the plates and shells also from music instrument manufacturers. "Kaogongji" or "Artificer's Record" 500 B.C. in Chinese history, recorded the Bian-Qing (sound bian-ching). It is a set of percussion, made of high-quality stones such as jade, with several fixed music scales. The "Kaogongji" specifically described how to adjust the percussion music scale: "if music scale is higher, filing the surfaces of the plate, if the scale is lower, then filing the edges of the plate". These technological processes, even though not giving exactly the quantitative relation between the natural frequencies of the plate and its geometrical parameters, are indeed in accordance qualitatively with the contemporary vibration and acoustic principles of the plate and shells: filing the surfaces of the plate makes the plate thinner and natural frequencies will be decreased while filing the edges of the plate makes the plate relatively thicker thus the natural frequencies will be increased.

One of the most important dynamic properties for the vibrations of a system is the inherit frequency or natural frequency. The vibrating movements of a system occur by external forces or internal self-excitement. When the frequencies of the excitation are the same or near the system's natural frequencies, the responses of the systems will become larger and larger, which is called the resonance. The resonance has a variety of applications, such as radio, television, and music. The large magnitude vibrations, such as large magnitude earthquakes, tsunamis, could bring harmful, devastative, and even catastrophic damages to properties and human lives. People tried to predict the earthquake by vibrating devices. In 132 A.D. Zhang Heng (78–139 A.D.), a Chinese mathematician, an astronomer, and a geographer, invented the vibration utilization device: Seismograph. The shape of the device looks like a goblet. Eight (8) exquisitely casting dragons, upside down, attached to the body of the device. The eight dragons are mounted in the North, South, East, West, Northeast, Southeast, Northwest, and Southwest directions, respectively, representing the earthquake directions. There is a ball in each dragon's mouth. There is a vivid toad, mounted separately to the body, under each dragon. It is said that when an earthquake occurs the dragon in the earthquake direction will release its ball in its mouth into the toad's mouth as a prediction indication. The exact mechanism inside was not well documented. It was recorded in history books that this device had successfully predicted an earthquake about 600 miles away in 138 A.D. just 1 year before he died.

In modern days, tens of thousands of types of vibrating machines and instruments have been successfully used to accomplish a variety of technological procedures in the fields of mining, metallurgical industry, coal mining, petro-chemistry industry, machcical industry ,electronic industry , hydraulics and civil engineering construction , food and grain processing ,biological industry and process of human everyday life .The technologies involved in this science branch are so closely associated to agricultural and industrial production that it can create great economical and social benefits, and can provide great convenience and excellent service for people 's life. It becomes an inevitable means and a necessary mechanism for human production activities and life processes .A study of "Vibration Utilization Engineering "as a new branch of science has been gradually formed and developed in the latter part of the twentieth century to cope with the increasing demands of the vibration utilization.

The literature on vibration utilization is scattered over all the magazines, books, journals, and conference proceedings in different academic fields and different science branches. It would be difficult to look into and collect the information and references on different vibrating machines. It would be informative and convenient to have an encyclopedia type of a book to cover as many types of vibrating machines and as many aspects of vibration utilization as possible and provide multi-sources and cross-references for scientists and engineers to use. This is part of our intention in writing this book. The authors attempt to summarize the scientific research works on the vibration utilization over more than 30 years. The book also includes the following original and creative results led by the first author and Professor Bangchun Wen and his team colleagues:

1. The authors constructed a theoretical framework for "Vibration Utilization Engineering", this terminology is first born in domestic and international areas.
2. In the technological theory and technology creatives, the authors introduced the creative results obtained on technological theories and practical applications in the Vibration Utilization Engineering such as probability ISO-thickness screening theory; material sliding and throwing theory on different surfaces; and screening process theory and their applications in the vibration machine processes.
3. In mechanism creativities, the authors proposed a variety of new vibrating mechanisms, such as exciter-eccentric type of self-synchronically vibrating mechanism, special forms of non-linear inertial resonant type of vibrating mechanism, etc., patent approved.
4. In the nonlinear dynamic theory creativities, the authors summarized the systematic studies and experiments conducted on vibration and wave utilization technology and equipment working theories and derived theory basis on many branches of the Vibration Utilization Engineering, such as vibration synchronization and application of the controlled synchronization theory, application of the nonlinear vibration, etc. For example, the dynamic theory of a variety of vibrating machines and equipments, the equivalent mass and damping of vibrating machine systems, the theory and computing methods of the second vibration isolation of the vibrating machines and equipments, the vibrating synchronization theory and methods on duel- or multi-motor-driven vibrating machines, dynamic analysis, and dynamics parameter computation methods on nonlinear vibrating machines, the dynamic design methods on vibrating machines and their main components, many of them are first published in this book.
5. In the design theory and method creativities, the authors proposed the systematic vibrating machine dynamic design theory and method, especially those for the non-linear vibrating machines and the comprehensive design methods which contain the dynamic optimization, intelligent optimization and visualization.
6. In the engineering applications, the authors have applied their theoretical results to engineering for over 30 years. For example, the inertia-resonance-probability screens, new mechanism type of vibrating cooling machines and new mechanism type of vibrating crushers, etc. These machines are rewarded by the National Invention, National Technology Progress, etc.

This book contains seven chapters. Chapter 1 introduces the formation and development of the Vibration Utilization Engineering; Chap. 2 devotes to some of the important research results in the vibration and wave-energy utilization in some technological processes; Chap. 3 describes the theories on the technological process of the vibration utilization technology and equipment; Chaps. 4 and 5 discuss the vibration utilizations of the linear, pseudo-linear, and nonlinear systems; Chap. 6 presents the utilization of the wave and wave-energy; and Chap. 7 briefly illustrates the vibration phenomena and utilizations in nature and human societies.

The authors include Bangchun Wen, Xianli Huang, Yinong Li, and Yimin Zhang. During the editing it absorbed the research results by the team members: they are Profs. Lizhang Guan, Guozhong Zhang, Liyi Ren, Shengqing Ji, Shuying Liu, Chengxiu Wen, Zhishan Duan, Jie Liu, Tianxia Zhang, Xun He, Haiquan Zeng, Peimin Xu, Yannian Rei, Shirong Yan, Huiqun Yuan, Qingkai Han; Associate Professors Jian Fan, Jingyang Qi, Chunyun Zhao, Fenglan Wang, Suying Shu, Hongguang Li, Wali Xiong, Shide Peng; Senior Engineers Xiangyang Lin, Mingfei Luo, Yongxi Liu, Zhaomin Gong, Hong Zhang, Tianning Xu, Qinghua Kong, Huajun Wang, and Naiqing Ma; Doctors He Li, Hongliang Yao, Zihe Liu, Haiyan Wei. We also received help from Academicians Wenhu Huang, Shuzi Yang, Jinji Gao; Professors Yushu Chen, Zhaochang Zheng, Haiyan Hu, and Dianzhong Wang; Doctoral students Hong Chen, Xiaowei Zhang, Xiaopeng Li, Xueping Song, Li Wang, Tao Yu, Wei Sun, Hui Ma, and Juequan Mao. Thanks also go to Electronics and Machine Institute of Northeastern University, the Group of Xuzhuo Engineering Machines, the Capital Iron and Steel Corp, Luoyang Mine Machine Design and Research Institute, Anshan Mine Machine Limited, Henan Weimeng Vibrating Machine Limited, Chaoyang Vibrating Machine Factory, Haian Vibrating Machine Factory, and Zhongxiang Machine and Electrics Corp for their assistance.

It is noted that some of the research results in this book come out of the projects funded by the National Nature Science Foundation (Projects # 59475005, 50075015, 59075175, and 59875010), two Doctoral Student Fund Projects, and other scientific and research projects.

Shenyang, China	Bangchun Wen
Liaocheng, China	Xianli Huang
Chongqing, China	Yinong Li
Shenyang, China	Yimin Zhang

Contents

1 **Formation and Development of Vibration Utilization Engineering** 1
   1.1 Introduction 1
   1.2 Vibrating Machines and Instruments and Application of Its Related Technology and Development 3
   1.3 Applications and Developments of Nonlinear Vibration Utilization Technology 7
   1.4 Applications and Developments of Wave Motion and Wave Energy Utilization Technology 10
   1.5 Applications of Electrics, Magnetic and Light Oscillators in Engineering Technology 11
   1.6 Vibrating Phenomena and Utilization in Natures 16
   1.7 Vibrating Phenomena, Patterns and Utilization in Natures 17
   1.8 Vibrating Phenomena, Patterns and Utilization in Human Society 17
   1.9 Vista 18

2 **Some Important Results in Vibration and Wave Utilization Engineering Fields** 21
   2.1 Utilization of Vibrating Conveyance Technology 22
   2.2 Applications of Vibrating Screening Technology 24
   2.3 Applications of Vibrating Centrifugal Hydro-Extraction and Screening Technology 27
   2.4 Applications of Vibrating Crush and Milling Technology 29
   2.5 Applications of Vibrating Rolling and Forming Technology 31
   2.6 Applications of Vibrating Tamping Technology 33
   2.7 Applications of Vibrating Ramming Technology 34
   2.8 Applications of Vibration Diagnostics Technology 35
   2.9 Applications of Synchronous Vibrating Theory 37

	2.10	Applications of Resonance Theory	38
		2.10.1 The General Utilization of the Resonance	38
		2.10.2 Application of the Nuclear Magnetic Resonance	39
	2.11	Applications of Hysteresis System	40
	2.12	Applications of Impact Principles	41
	2.13	Applications of Slow-Changing Parameter Systems	42
	2.14	Applications of Chaos Theory	42
	2.15	Applications of Piecewise Inertial Force	44
	2.16	Applications of Piecewise Restoring Force	45
	2.17	Utilization of Water Wave and Wind Wave	46
	2.18	Applications of Tense or Elastic Waves	47
	2.19	Utilization of Supersonic Theory and Technology	47
		2.19.1 The Application of the Supersonic Motor	48
		2.19.2 Significance and Function in Medical Diagnostics of B-Ultrasound	49
	2.20	Applications of Optical Fiber and Laser Technology	49
		2.20.1 Application of the Optical Fiber Technology	49
		2.20.2 Application of Laser Technology	50
	2.21	Utilizations of Ray Waves	50
	2.22	Utilization of Oscillation Theory and Technology	51
	2.23	Utilization of Vibrating Phenomena and Patterns in Meteorology	53
	2.24	Utilization of Vibrating Phenomena and Patterns in Social Economy	53
	2.25	Utilizations of Vibrating Principles in Biology Engineering and Medical Equipments	54
3	**Theory of Vibration Utilization Technology and Equipment Technological Process**		**57**
	3.1	Theory and Technological Parameter Computation of Material Movement on Line Vibration Machine	57
		3.1.1 Theory of Sliding Movement of Materials	58
		3.1.2 Theory of Material Throwing Movement	69
		3.1.3 Selections of Material Movement State and Kinematics Parameters	76
		3.1.4 Calculation of Real Conveying Speed and Productivity	80
		3.1.5 Examples	86
	3.2	Theory and Technological Parameter Computation of Circular and Ellipse Vibration Machine	89
		3.2.1 Displacement, Velocity and Acceleration of Vibrating Bed	89
		3.2.2 Theory of Material Sliding Movements	91
		3.2.3 Theory of Material Throwing Movements	96

Contents

- 3.3 Basic Characteristics of Material Movement in Non-harmonic Vibration Machines 102
  - 3.3.1 Initial Conditions for Positive and Negative Sliding Movements 102
  - 3.3.2 Stopping Conditions for Positive and Negative Sliding Movements 104
  - 3.3.3 Calculations of Averaged Material Velocity 104
- 3.4 Theory on Material Movement in Vibrating Centrifugal Hydroextractor 105
  - 3.4.1 Basic Characteristics of Material Movement on Upright Vibration Hydroextractor 106
  - 3.4.2 Characteristics of Material Movement on Horizontal Vibration Hydroextractor 114
  - 3.4.3 Computation of Kinematics and Technological Parameters of Vibration Centrifugal Hydroextractor 115
- 3.5 Probability Theory on Material Screening Process 119
  - 3.5.1 Probability of Screening for Material Particle Per Jump 119
  - 3.5.2 Falling Incline Angle and Number of Jumps of Materials on Screen Length 123
  - 3.5.3 Calculation of Probability of Material Going Through Screens for a General Vibration Screen 124
  - 3.5.4 Calculation of Probability of Material Going Through Screens for a Multi-screen Vibrating Screen 126
- 3.6 Classification of Screening Method and Probability Thick-Layer Screening Methods 130
  - 3.6.1 Screening Methods 130
  - 3.6.2 Screening Methods for Probability Thick Layer Screens 133
- 3.7 Dynamic Theory of Vibrating Machine Technological Processes 139

4 Linear and Pseudo Linear Vibration 143
- 4.1 Dynamics of Non-resonant Vibrating Machines of Planer Single-Axis Inertial Type 143
- 4.2 Dynamics of Non-resonant Vibrating Machines of Spatial Single-Axis Inertial Type 150
- 4.3 Dynamics of Non-resonant Vibration Machines of Double-Axis Inertial Type 153
  - 4.3.1 Dynamics of Non-resonant Vibrating Machines of Planer Double-Axis Inertial Type 153
  - 4.3.2 Dynamics of Non-resonant Vibration Machines of Spatial Double-Axis Inertial Type 157

4.4	Dynamics of Non-resonant Vibration Machines of Multi-axis Inertial Type		158
	4.4.1	General Pattern of Planer Movement	159
	4.4.2	Values of Displacement, Velocity and Acceleration Curves and Differential Coefficients When θ_2 is Equal to $\Pi/2$	160
4.5	Dynamics of Inertial Near-Resonant Type of Vibration Machines		163
	4.5.1	Dynamics of Single Body Near-Resonant Vibration Machines	163
	4.5.2	Dynamics of Double Body Near-Resonant Vibration Machines	165
4.6	Dynamics of Single Body Elastic Connecting Rod Type of Near Resonance Vibration Machines		168
4.7	Dynamics of Double Body Elastic Connecting Rod Type of Near Resonance Vibration Machines		171
	4.7.1	Balanced Type of Vibration Machines with Double Body Elastically Connecting Rod	171
	4.7.2	Non-balance Double Body Type of Elastically Connecting Rod Vibration Machines	173
4.8	Multi-body Elastic-Connecting Rod Type of Near-Resonant Vibration Machines		177
4.9	Dynamics of Electric–Magnetic Resonant Type of Vibrating Machines with Harmonic Electric–Magnetic Force		180
	4.9.1	Basic Categories of Electric–Magnetic Forces of Electric–Magnetic Vibration Machines	180
	4.9.2	Dynamics of Electric–Magnetic Type of Vibrating Machines with Harmonic Electric–Magnetic Force	180
	4.9.3	Amplitudes and Phase Angle Differentials of One-Half-Period Rectification EMTVM	184
	4.9.4	Amplitudes and Phase Angle Differentials of One-Half-Period Plus One-Period Rectification EMTVM	185
4.10	Dynamics of Electric–Magnetic Type of Near-Resonant Vibration Machines with Non-Harmonic Electric–Magnetic Force		186
	4.10.1	Relationships Between Electric–Magnetic Force and Amplitudes of Controlled One-Half-Period Rectification EMTVM	187
	4.10.2	Relationships Between Electric–Magnetic Force and Amplitudes of the Decreased Frequency EMTVM	190

Contents

5 Utilization of Nonlinear Vibration 195
- 5.1 Introduction 195
- 5.2 Utilization of Smooth Nonlinear Vibration Systems 201
  - 5.2.1 Measurement of Dry Friction Coefficients Between Axis and Its Bushing Using Double Pendulum 201
  - 5.2.2 Measurement of Dynamic Friction Coefficients of Rolling Bearing Using Flode Pendulum 203
  - 5.2.3 Increase the Stability of Vibrating Machines Using Hard-Smooth Nonlinear Vibrating Systems 207
- 5.3 Engineering Utilization of Piece-Wise-Linear Nonlinear Vibration Systems 210
  - 5.3.1 Hard-Symmetric Piece-Wise Linear Vibration Systems 210
  - 5.3.2 Soft-Asymmetric Piece-Wise Linear Vibration Systems 213
  - 5.3.3 Nonlinear Vibration Systems with Complex Piece-Wise Linearity 218
- 5.4 Utilization of Vibration Systems with Hysteresis Nonlinear Force 222
  - 5.4.1 Simplest Hysteresis Systems 223
  - 5.4.2 Hysteresis Systems with Gaps 226
- 5.5 Utilization of Self-excited Vibration Systems 231
- 5.6 Utilization of Nonlinear Vibration Systems with Impact 233
- 5.7 Utilization of Frequency-Entrainment Principles 236
  - 5.7.1 Synchronous Theory of Self-synchronous Vibrating Machine with Eccentric Exciter 238
  - 5.7.2 Double Frequency Synchronization of Nonlinear Self-synchronous Vibration Machines 250
- 5.8 Utilization of Nonlinear Vibration Systems with Nonlinear Inertial Force 259
  - 5.8.1 Movement Equations for Vibration Centrifugal Hydro-Extractor with Nonlinear Inertial Force 259
  - 5.8.2 Nonlinear Vibration Responses of Vibration Centrifugal Hydro-Extractor 261
  - 5.8.3 Frequency-Magnitude Characteristics of Vibration Centrifugal Hydro-Extractor 263
  - 5.8.4 Experiment Vibration Responses of Vibration Centrifugal Hydro-Extractor 264
- 5.9 Utilization of Slowly-Changing Parameter Nonlinear Systems 265
  - 5.9.1 Slowly-Changing Systems Formed in Processes of Starting and Stopping 266
  - 5.9.2 Slowly-Changing Rotor Systems Formed in Active Control Processes 267

		5.10	Utilization of Chaos	271
			5.10.1 Major Methods for Studying Chaos	271
			5.10.2 Software of Studying Chaos Problems	273
			5.10.3 Application Examples of Chaos	275

6 Utilization of Wave and Wave Energy ... 285
- 6.1 Utilization of Tidal Energy ... 285
- 6.2 Utilization of Sea Wave Energy ... 287
- 6.3 Utilization of Stress Wave in Vibrating Oil Exploration ... 288
  - 6.3.1 Mechanism and Working Principles of Controllable Super-Low Frequency Vibration Exciters ... 289
  - 6.3.2 Effect of Stress Wave on Oil Layers ... 290
  - 6.3.3 Experiment Results and Analysis ... 299
  - 6.3.4 Elastic Stress Wave Propagation When a Controllable Vibration Source is Working ... 305

7 Utilization of Vibrating Phenomena and Patterns in Nature and Society ... 309
- 7.1 Utilization of Vibration Phenomena and Patterns in Meteorology ... 309
- 7.2 Periodical Vibration and Utilization of the Tide ... 316
- 7.3 Vibration Patterns and Utilization in Other Natural Phenomena ... 318
  - 7.3.1 Periodical Phenomenon of Tree Year-Rings ... 318
  - 7.3.2 Bee's Communications Using Vibrations ... 319
- 7.4 Utilization of Vibration Phenomena and Patterns in Some Economy Systems ... 320
  - 7.4.1 Fluctuation and Nonlinear Characteristics in Social Economy Systems ... 320
  - 7.4.2 Growth and Decline Period in Social Economy Development Process ... 324
  - 7.4.3 Active Role of Macro-adjustment in Preventing Big Economy Fluctuations ... 325
- 7.5 Utilization of Vibration Phenomena and Patterns in Stock Market ... 326
  - 7.5.1 Stock Fluctuation is One of Typical Types of Economy Change Form in Social Economy Fields ... 326
  - 7.5.2 Stock Market Characteristics and General Patterns of Oscillation ... 328
  - 7.5.3 Some Principles in Stock Operations ... 332
- 7.6 Obey the General Rules in the Stock Operations ... 332
- 7.7 The Entering Point and Withdrawing Points in the Stock Operations ... 334

	7.8	Utilization of Vibration Phenomena and Pattern in Human Body	335
		7.8.1 Vibration is a Basic Existing Form of Many Human Organs	335
		7.8.2 Some Diseases Make Abnormal Fluctuations (Vibration) in Human Organs Physical Parameters	336
		7.8.3 Medical Devices and Equipment Based on Vibration Principles	341
		7.8.4 Artificial Organs and Devices Using Vibration Principles	342
	7.9	Prospect	343
References			345

Chapter 1
Formation and Development of Vibration Utilization Engineering

1.1 Introduction

Utilization of vibration and wave is a type of the most valued applicable technology developed during the latter part of the twentieth century and still in rapid development state now. Since the technology is closely associated with industrial and agricultural production, can create huge social and economic benefits for our society and provide excellent service for human beings, it becomes a necessary mechanism in human production and life. The scientists in the world put a lot of efforts to study the technology in this field, the authors and their teams spent more than 30 years in the investigation and research in this field and achieved a series of research results and hence brought about the formation and development of the branch.

One of the indications for the rapid development of the technology is the variety of the vibrating devices, equipment and machines and the diversity in the fields of their application and utilization. In the industries, such as mining, metallurgy, petrochemistry, mechanical engineering, electricity, hydraulics, civil engineering, construction, building material, railway, highway transportation, light industry, food processing and farmland cultivation, tens of thousands of vibrating devices, equipment and machines have been used to accomplish many technological processes such as material feeding, conveyance, screening, material distribution, drying, cooling, dewatering, selecting, crashing, grinding, shaping, burnishing, shakeout, ramming, rolling, paving, drilling, loading, pile sinking, pile pulling, extracting oil, cleaning, bundling, aging, cutting, checking pile, detecting, exploring, testing, diagnosing etc. These machines include the vibrating feeder, vibrating conveyor, vibrating shaping machine, vibrating screen, vibrating centrifugal dehydrator, vibrating drier, vibrating cooler, vibrating freezer, vibrating breaker, vibrating polisher, vibrating grinding mill, vibrating burnisher, vibrating roller, vibrating paver, vibration earthener, vibration sinking piling machine and a variety type of shakers and shockers, etc.

Utilization technology of the vibration and wave is based in a wide range of vibration principles. In addition to linear and nonlinear vibration theories, the linear and nonlinear stress wave are used for detection and geology exploration. In petroleum exploitation, the elastic waves generated by vibrations are used to increase the oil output. In the Ocean Engineering, the energy of the tidal waves can be used to generate electricity. In the medical equipment, the supersonic wave is used to diagnose and cure diseases. In the other medical settings, the color ultrasound, computed tomography (CT) and magnetic resonance imaging (MRI) are the medical applications of the vibrations. The application of the light-fiber and laser is the typical application example of the vibrations.

In the perspective of the electronics and communications, the oscillating circuits in televisions and radios, door bells, telephone mechanical and electronic watches, clocks, light-fiber communications, recorder, just to name a few of them, they all work effectively due to the vibration principles.

In retrospect of the past history, we can see that the utilization of vibration and waves has been made some new reforms in some scientific fields and even brought about new revolutions in some fields and industries. For example, the study and research on self-synchronization theory have successfully pushed the appropriate inertial vibrating machines and mechanical industries into significant changes. The successful introduction of the controllable electric–magnetic feeders in some industrial enterprises has improved their production automations. The utilization of the Vibrating Roller and Spreaders in road constructions has improved the highway construction quality and enlarged the highway usage life; the quartz oscillators have made a revolution in the Watch and Bell Industries; the successful creation of the supersonic motors has made the compact, small power and lower rotation motors a reality; the color ultrasound and CT machines have made the medical detection and diagnostics technology a revolution; the successful study of light-fiber has promoted a revolution in the communication technology. We can see from above examples that the utilization technology of the vibration and waves has very important effects on human's life and production activities.

Generally speaking, in the social and economic life, the growth and attenuation of the human population, the periodicity of the insect calamity, the ups and downs of the stock markets, and increase and decrease of the growth speeds in social economy development process can be all classified, in some sense, as different forms of vibrations and oscillations. In nature and the universe there exist vibrations everywhere: the Moon's phase, the tide's in and out, the tree's annual rings, etc. It is no doubt that studying on the vibration and wave phenomena, finding out the intrinsic patterns, and making use of the patterns effectively can produce social and economic benefits and bring benefits to mankind.

The vibration and waves can be categorized roughly as two types: linear and nonlinear systems of vibrations, waves (sound wave, light wave) and electric–magnetic oscillations. We can classify the utilization of vibration and waves into the linear and nonlinear vibration utilization, wave motion and energy utilization and the electric–magnetic oscillation utilization in engineering, the phenomena, patterns and utilization of the vibrations in nature and human social society (see Fig. 1.1). The vibration utilization, from very low frequency (tidal waves) to very high frequency (ultrasound), can all be effectively utilized (see Fig. 1.2).

1.2 Vibrating Machines and Instruments and Application …

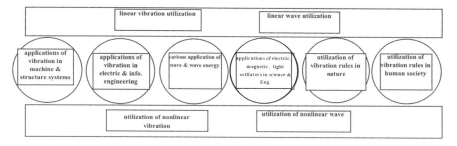

Fig. 1.1 Linear and nonlinear vibration utilization fields

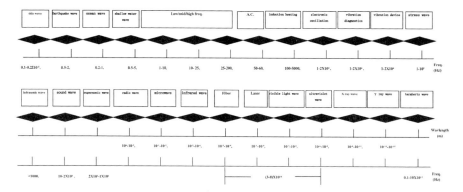

Fig. 1.2 Vibration and wave distribution by frequency or wavelength. *Note* Waves are arranged from left to right not totally by its magnitude or wavelength

1.2 Vibrating Machines and Instruments and Application of Its Related Technology and Development

Linear and nonlinear vibration utilization technology are accomplished via vibration mechanics or vibrating instruments which could generate the vibration. The vibrating machines or vibration instruments as a kind of special equipment or device have been widely used in industry and agriculture productions. Table 1.1 shows some major usages for vibrating machines and instruments.

New vibration utilization technologies emerged as science and technology are developing. Here are some examples:

(1) Vibrating Drying Technology

Drying is a complex technological process in industry production. The technology is newly developed in recent years in vibration utilization engineering. The vibrating vulcanizing machine is developed on the basis of a general vulcanizing machine.

Table 1.1 Vibration machines and equipment categorized by their usages

Type	Usage	Machines or device
Conveyer/Feeder types	Conveyance, feeder, part arrangement	A variety of bulkhead vibrators, EM vibration feeder, inertial resonance feeder, vibration hopper conveyance such as EM vibration conveyor, inertial resonance conveyor, elastic connecting rod type of vibration conveyor
Screen/Dryer types	Screen, selection, dry, cooling, hydro-extraction	EM vibration screen, inertial vibration screen, resonance screen, rotating vibration screen, probability screen, vibration dryer, vibration centrifugal hydro-extractor, hydro-extraction screen, vibration ore selector, table
Crush/Cleaning type	Powder polish, crush, sand washing, ice-crushing, cleaning, dust-removing	Vibration polisher, coarse crusher, inertial vibration crusher, vibration sand washer, vibration loader, vibration shovel, pneumatic shovel, rock drill
Shaping/Compacting type	Shaping, solidifing, rolling	Vibration shaper, vibration densifier
Ramming/Pile driving type	Rolling of road, spread, pile driving, ramming, excavating, loading, rock drilling	Vibration roller, oscillation roller, spreader, vibration pile driver, attaching type of vibrator, insertion type of vibrator, rammer
Test/Display type	Excitation, test, display, etc.	A variety of exciters, vibration tables, vibration simulators, dynamic balance test machines, fatigue test machines, mechanical type of vibration measuring device, a variety of vibration motors
Monitoring/Measuring types	Monitoring, diagnostics	A variety of monitoring and diagnostic device and equipment
Others	Ageing, cutting, massaging, bundling, well cementation	A variety of exciters

1.2 Vibrating Machines and Instruments and Application ...

Table 1.2 Utilization of nonlinear vibration characteristics

Sequence number	1	2	3	4	5	6	7	8	9	10	11
Types of nonlinear vibration characteristics	Utilization of piece-wise inertial force system	Utilization of smooth nonlinear system	Utilization of piecewise-linear nonlinear system	Utilization of nonlinear hysteresis system	Utilization of self-excitation vibration system	Utilization of nonlinear vibration with impact	Utilization of slow-changing parameter system	Utilization of frequency entrainment principle	Utilization of nonlinear elastic system	Utilization of bifurcations	Utilization of Chaos
Application examples	Vibration conveyers, vibration centrifuges	Friction pendulum, flode pendulum, electric–magnetic feeder	Vibration feeder, vibration centrifugal hydro extractor, inertial resonance screen	Vibration roller, spreader, pile driver	Pneumatic or hydraulic Rock drill, pneumatic puncher, oscillator	Rammer, inertial vibration puncher, vibration shaper	Working process of many vibration machines	A variety of self-synchronous vibration machines	Utilization of longitudinal and transverse elastic vibration systems	Frog-rammer applications of hyperbolic movement state	Chaos encryption system, chaos mixer

(2) Vibrating Crusher

Crushing of materials is a widely used technological process in the mining industry. Most of the ore materials need to be crushed and grinded. The breaking means of a traditional crusher has its limitations. For example, when the materials to be broken have compressive strength of more than 2×10^8 Pa the breaking process will consume large energy, the ore and materials are difficult to break and crush or the materials are over-broken. The machines are very complex. The development of vibrating breaking and crushing technology then overcomes these shortcomings. Cylindrical inertial vibrating crusher is used to break the ore and other mineral materials by the centrifugal forces generated by an eccentric mass and the extruding and impacting forces. The crushing rate by the vibrating crusher is much larger than that of traditional cylindrical inertial crusher and can be adjusted in a large range. It has bright future in its applications in finer breaking of the materials.

(3) Vibrating Spreader and Vibrating Roller

Vibrating spreader and vibrating roller are the key equipment in the road construction and are the typical application of the vibration technology. A vibrating spreader first spreads materials onto the whole width of the road, and then uses the exciter of the rolling mechanism to roll the road surface. The vibration system is the key device to determine the effects of spreading the materials and the density and quality of the spreading and rolling.

There are vibrating motors mounted on the connecting plate in the vibrating rollers. The motors drive two eccentric shafts to rotate in opposite directions at high speeds. The eccentric lumps on the shafts generate the centrifuge forces to drive the vibrating rollers into the forced vibrations to roll the road surface. The amplitude adjusters mounted on the eccentric drafts are used to adjust the vibration amplitudes. The introduction of the vibration can increase the density of the rolled surface from 90% to above 95% and improve the quality and usage life of the roads.

(4) Vibrating Forming and Technology

The vibrating forming on the metal and porous materials can greatly reduce the energy consumed and improve the quality of the parts formed compared to the static force forming. The test data show that introducing the vibration in the processes of metal plastic processing can reduce the energy consuming, increase efficiency, and improve the processed part quality.

(5) Vibrating Ageing

Vibrating ageing can reduce the internal residual stresses of metal parts to some extent, and stabilize the sizes and shapes of the parts after processing. The vibrating ageing is to exert the periodical forces on the parts and force the parts vibrate in the frequency range of their resonance. The periodical dynamic forces make metal atoms in the internal structure generate displacement and crystal lace dislocations and thus relax the stresses. The vibrating ageing equipment generally includes an exciting device, a vibration measuring unit and a control device for dynamic stress. Compared to heat treatment, the vibrating ageing has advantages of easy operation, less transportation, short period of production and consuming less energy and is a new technology of saving energy.

(6) Vibrating Diagnostics and Vibration Measurement Technology

Diagnosing the faults by analyzing the different characteristics of the vibration signals from the machines and structures is a new technology developed in recent decade. It is widely used to analyze and diagnose the vibration signals by the Fuzzy Theory, Gray Theory or Neural Network.

(7) Vibrating Equipment and Device for Exercise and Sports

Vibrating massager belongs to this type. Many work-out and sports equipments are made use of the vibration technique and become an indispensable sport device.

(8) Mechanical Medical Equipment, Instrument and Device

The artificial hearts and heart pacer are all the medical devices using vibration principles.

1.3 Applications and Developments of Nonlinear Vibration Utilization Technology

The vibration machines and devices mentioned above can be classified into two types: linear or pseudo-linear (approximately linear) and nonlinear according to their linearity and nonlinearity. Some of the nonlinear vibrating machines and equipment using the nonlinearity characteristics in the systems for better efficiency are applied intentionally, and others not intentionally but made use of the nonlinearity naturally existed inherently (see Table 1.2 above).

(1) Utilization of the Vibrating Systems with Piece-Wise Friction or Impact and Piece-Wise Inertial Forces

In the vibrating machines, in order to have relative movement, i.e. the relative sliding or jumping, between machine body and materials, the piece-wise friction force or mass inertial force (sliding movement), or impact force and piece-wise mass inertial (throwing movement) must be generated in the vibrating machines. These are the necessary conditions to keep the vibrating machines working normally. This type of machines is the vibrating feeder, conveyer, screen, centrifugal dehydrator and cooler, etc.

(2) Utilization of Smooth Nonlinear Vibration System

Friction pendulum belongs to smooth nonlinear vibration system in its normally working neighborhood. The friction pendulum can be used to measure the friction coefficient between axial and axial bushing. There are two methods to do it: the first one is to use the decay value of the every vibration period of the double pendulum to directly calculate the friction coefficient; the second is to use the working principle of the Flode pendulum to measure and calculate the friction coefficient. The latter is more accurate than the former.

For some vibrating machines working in resonance frequency range, such as electric–magnetic vibration feeder, near-resonant vibrating conveyer and resonant screen, their amplitudes are not stable. To overcome such shortcomings, the smooth harden type of nonlinear restoring force vibration machines may be used.

It is pointed out that the two fixed ends of the main resonant leaf springs in the electric–magnetic vibration feeder can be made into the curved shapes so that as the vibration amplitude increases the working length of the leaf springs will be shortened, therefore their stiffness will increase as a result which in turns stabilize the vibration amplitudes for these kinds of machines.

Besides, the harden type of smooth nonlinear vibration systems can also be used to the vibration isolation of vehicles. Vehicle loads change from time to time in order to keep its resonant frequency changing slowly when its loads are changing, the isolation springs can be made into the harden type of smooth nonlinear system.

(3) Utilization of Piece-Wise-Linear Nonlinear Vibration Systems

Piece-wise-linear nonlinear systems have been applied to nonlinear vibrating conveyors, nonlinear vibrating screens, nonlinear vibrating centrifugal dehydrators, vibrating and vibrating centrifugal tables. The piece-wise-linear nonlinear vibration systems are generally classified as three categories:

i.   Total symmetric and non-symmetric harden type of piece-wise-linear nonlinear vibration systems.
ii. Symmetric and non-symmetric soften type of piece-wise-linear nonlinear vibration systems.
iii. Compound or complex type of piece-wise-linear nonlinear vibration systems.

Those types of the vibrating machines have many advantages: they are in the process of constant promotion and are expected to be developed further.

(4) Utilization of Nonlinear Hysteric Systems

The vibration systems with elastic–plastic deformation belong to this category, such as vibrating shape machine, vibrating roller, vibrating pulling pile machine. Their vibration systems belong to the nonlinear vibration systems with hysteric restoring forces. In order for vibrating-shape machines to work effectively the plastic deformation is necessary and must increase the area of the Hysteresis loop as much as possible. This kind of vibrating machines exists widely in the engineering. The categories of this type of vibrating machines are as follows:

i. The nonlinear vibration systems with parallelogram Hysteresis restoring forces;
ii. The nonlinear vibration systems with closed Hysteresis loop restoring forces;
iii. The nonlinear vibration systems with asymmetric restoring forces;
iv. The nonlinear vibration systems with gap Hysteresis restoring forces.

In our studies we have conducted deep researches on this type of nonlinear systems.

(5) Utilization of Self-excitation Nonlinear Vibration Systems

In engineering the self-exciting vibration systems have been widely used. For examples the pneumatic and hydraulic rock drill and stone crusher in mining industry, air pick in coal industry, air shovel in a casting shop, steam hammer in a forging shop, non-piston pneumatic eliminator in coal depressing plant. The working process of the steam engines, reciprocal oil cylinder or a variety of mechanisms composed of parts driven by pistons, is also self-excited. In radio communications and instrument industry the electronic oscillators in radio and television circuits, the wave generators in a variety of instrumental display devices, switch type of temperature adjustors in constant temperature containers, watches and clocks in everyday life and the string music instruments are all self-excited. Heart beating is a self-excited as well.

(6) Utilization of Impact Nonlinear Vibration Systems

The technological utilization accomplished by impact is frog rammer, impact pile machine, vibrating sand remover and vibrating drills with impact. The impact type of vibrating machine is a special example of the non-linear vibrating machines. By theory computation and experiment verification, the instant acceleration generated under the impact situation is several, even hundreds of times compared to the maximum acceleration generated by a general linear vibrating machine. Making use of such a big impact force, we can accomplish many works, such as ramming road surface, removing sands from forged parts and crushing rocks.

(7) Utilization of Slowly-changing Parameter Vibration Systems

In engineering, many working processes belong to slow-changing parameter systems. For examples the working process of the controls on vibrations of the rotating machines is a kind of the slow-changing parameter systems. The speed of the slow-changing process will affect directly on the vibrating amplitudes controlled. Therefore we may make use of the speed of the slow-changing control process to achieve the optimal control effects. For example, for a rotor system with slow-changing parameters, the startup and braking process, the exciting frequency slow-changing will make the resonant curve generate a shift and affect its stability, rotor stiffness slow-changing will make the unstable resonant curve shift. For the roller system of slow-changing support stiffness, if we choose a proper stiffness slow-changing speed, we could effectively constrain the vibration passing the resonant region. From this example we can see that the slow-changing parameter systems can be made use of in industry.

(8) Utilization of Frequency Entrainment Principles

In engineering the frequency entrainment phenomena have been used widely. The realization for this principle is that two induction motors, parallelly installed in the same vibrating system, drive two eccentric rotor exciters separately. Nowadays there are millions of such self-synchronous vibrating machines based on this principle. Test shows that the two exciting motors operate separately and their revolutions per minute (RPM) are 962 rpm and 940 rpm respectively. Their revolutions are all 950 rpm while they operate together. This is called frequency entrainment. In its engineering utilizations, for example, the two induction motors drive two eccentric rotor exciters which are installed to the same vibrating system.

(9) Utilization of Bifurcation Solution

Some of the bifurcation solutions for nonlinear equation can be utilized and others can not. Therefore the study on the solution has an engineering significance. For example the effective working zone for a frog rammer is the maximum impact speed zone of its periodic movements, i.e. the bifurcation solution of the periodic movements. The movements of the materials in the vibrating conveyers have many bifurcation solutions, such as equal-periodical jump, double periodical jump, triple periodical jump and n-periodical jump and non-periodical jump. Selecting the most ideal, or the movement forms which have the optimized technological index and other parameters, is the most valuable and most important research topics.

(10) Engineering Utilization of Chaos

Chaos has a wide range of utilization. For example, cancel the vibration of the boring shaft in a boring machine by the chaos movement of the steel balls in the vibration cavity in an impact vibration absorber; accelerate the mix of the materials by chaos movement of the multi materials in a vibrating mixer; some people even have tried to speed up the mix of materials by chaos. The chaos has the unique unpredictable characteristics and thus has been successfully utilized to encrypt the confidential communication systems.

In addition, some harmful instable vibrations may occur in some nonlinear systems. In these cases, we should try to transit the unfavorable vibration to favorable chaos movements. This is naturally a way of utilizing the chaos movement. However is there another more effective measure than chaos movements? It is a worth-study question.

1.4 Applications and Developments of Wave Motion and Wave Energy Utilization Technology

Utilization of wave motion and wave energy is a new direction in vibration utilization in engineering fields in the recent decade. Scientists in the world did a lot of research on it and many results have been applied. The subject is still under study

1.5 Applications of Electrics, Magnetic and Light Oscillators ...

and development (see Table 1.3).

Table 1.3 Classification and usage of wave and wave energy

Type of waves	Usage	Device and equipment
Utilization of sea wave and tides (water wave and wave energy)	Sea wave and tidal electric generation, tidal transportation	Tidal electricity generator, tidal power station, tidal transportation device and ships
Utilization of elastic wave (stress wave)	Vibration oil extraction, vibration exploration, column test, structure health diagnostics	Vibration oil extraction equipment, geology exploration vehicle, vibration exciter for column test
Utilization of subsonic and sound wave	Equipment and structure diagnostics	Subsonic wave generator
Utilization of ultrasound	Ultrasound motor, ultrasound water-fuel mixing, ultrasound medical treatment (color ultrasound, ultrasound calculus smashing, ultrasound diagnostics	Ultrasound motor, ultrasound fuel-water-mixer, ultrasound medical equipment, color ultrasound, ultrasound calculus crusher, ultrasound washer
Utilization of ultraviolet wave and microwave	Ultraviolet disinfection, microwave heating, microwave communication	Ultraviolet wave generator, microwave oven, microwave communication equipment
Utilization of visible light wave	Heating, laser technology, fiber-optic technology	Water-heater, laser machining, fiber-optic communication
Utilization of infrared wave	Heating, imaging, medical treatment, military reconnaissance	Heater, thermal-imaging, infrared medical equipment, infrared military reconnaissance and monitoring equipment
Utilization of X-ray wave, γ-ray wave and β-ray wave	Medical treatment, diagnostics	X-ray machine, medical CT, industry CT, β ray medical equipment and other monitoring and diagnostics devices

1.5 Applications of Electrics, Magnetic and Light Oscillators in Engineering Technology

The rapid development and application of the big information capacity satellites and satellite live broadcasting lead to the more and more light-electric device and instruments based on the vibration principles being used. For examples, the quartz oscillators in the watches and clocks, light-fiber vibration sensors, electronic timing devices, and the oscillators used in the communication systems and ultrasound devices as shown in Table 1.4.

A. Electronic Oscillator

The application of the electronic oscillators to watch industry starts a very important revolution in the watch industry. Until now most of the mechanical watches are replaced by the electronic watches.

B. Ringtone Circuit and Electronic Music

The electronic ringtone circuits, which are small in volume, light in weight and pleasant in sound, have been used in cell phones. The core element of the ringtone circuit is the specialized integrated circuit composed of a rectifying circuit and an

Table 1.4 Type and usage of the oscillators

Type of oscillators	Usage	Device
Electronic oscillator	Electronic watch, electronic wall clock, electronic music, etc.	Electronic watch, electronic clock, electronic instruments, radio, TV, walkie-talkie, signal generator
EM oscillator	Telephone, piezoelectric ceramic oscillation circuit, etc.	Telephone, ringtone, magnetic deformation ultrasound generator, piezoelectric ceramic oscillation circuit, etc.
Laser oscillator	Laser oscillation circuit, etc.	Laser oscillator, etc.

oscillating circuit. The ring input signal is rectified by the rectifying circuit into a direct-current (DC) voltage about 10 V. The DC voltage was fed into the oscillating circuit, which outputs signals in two frequencies. The output signals, modulated by an ultra-low frequency oscillator, are fed into an amplifier. The amplified signals then are sent to a piezoelectric ceramic or to a transformer to match impedance, to broadcast the ringtone.

The electronic music is a sound created by oscillators of different frequencies and then output via some rhythms and different frequencies by a computer.

C. Piezoelectric Ceramics Electric–Magnetic Oscillator

The USA produced an oscillator with simple structure and great harmonic characteristics. Its basic structure and principle are that for a piezoelectric ceramics plate, whose two surfaces are covered with electrodes and are polarized along the thickness of it, its one surface electrode is partitioned into two parts. Such piezoelectric ceramics will be telescopic vibrating in 180° phase differential along oblique symmetric thickness. This kind of vibration pattern could cancel the redundant parasitic vibration signals therefore it could obtain high vibration quality factor.

The piezoelectric ceramic transformer in liquid crystal display of color television is a real example of vibration utilization in the electronic engineering.

D. Laser Oscillator

The vibration has been widely used in the light-electronic devices. The Okinawa Electric Industry Inc., Japan, produced a laser system of short wavelength using arranged semi-conduct chips. The stable oscillating characters are obtained by making the semi-conduct laser device and the second harmonic cell (SHC), which reduces the wavelength to half, into one body. The new system has the notch structure between the laser cell and the SHC. The notch makes the lights resonate repeatedly. A little of lights leaked in the resonant state is sent back to the SHC by the concave and convex reflection of the SHC. The laser magnified by the light oscillation is used to reduce the laser wavelength through the SHC to its half of the wavelength. Since the distance between the concavity and convexity of the SHC determines the oscillation wavelength of the laser, the same wavelength can be used to make the oscillation irrelevant to the ambient temperatures.

(1) Sea Wave Electricity Generation

Oceans cover about 71% of the surface of the Earth and contain 97% of water in the world. The inexhaustible water resource is the new hope of the new energy for the twenty-first century. The oceanographers estimate that the energy of the ocean wave would be as high as 90 trillion kW·h which is 500 times of the total electricity generation in the world.

Scientists noticed the sea wave electricity generation as early as in twentieth century. Scientists in the USA, Japan and Great Britain designed the sea wave electricity generation stations in the 90s in the last century and the efficiency of the station reached about 60%. The successful wave electricity generation unfolds the future application and development of the utilization of the sea wave energy for the human beings for the new century. Best of all, electricity generation by sea wave energy has no pollution, no negative biological effect on environment. This is one of the reasons people put a great expectation on it.

Sea wave electricity generation station uses a pneumatic turbine which converts the energy, generated by guiding the sea wave into a narrow grotto, into electricity. The station uses a high efficient collecting and converting device. A computer controls everything, from wave-guide, electricity generator to electricity transmission. The first wave electricity station in China was installed in Pearl Delta, Guangdong Province. The station's pneumatic chamber is 3m×4m and two generators, 3kW·h and 5kW·h, have been installed. The station has a trumpet-shaped opening on the bank. The basic principle of generating electricity is that the pneumatic chamber converts the sea wave energy into the reciprocal movement of the air which is used to drive the generators to generate electricity.

(2) New Technology for Oil Drilling by Super Low Frequency Controllable Vibration Source

The effect of the natural earthquakes and artificial earth surface vibration on the output of the gas and crude oil was brought to people's attention in the research on earthquake prediction. An oil well in Illinois, the US, is about 1,500 ft deep. When trains pass over on the surface the oil pressures in the well fluctuate and it has some effect on the oil output. The oil output increased 100%. The gas and oil outputs in Shengli, Daguang and Liaohe oil fields in China, before and after The Haicheng Earthquake in 1975 and The Tangshan Earthquake in 1976, had been increased obviously. Scientists note that compress and expansion waves of the Earth surface are the source of the micro additional pressures on the oil layers. It can increase the pressure to drive the oil and improve the fluidity of the oil. Both theory and simulation tests in labs show that vibration can increase the recovery ratio of the oil.

The authors and our team participated in a project on the technology and achieved a great result. The controllable vibration excitation source used in the study is the surface vibration device. The device is a powerful double-rotor eccentric type of exciter. It consists of an alternative speed-adjustable motor, a gearbox and two eccentrics which rotate relatively. A distributor and a controller adjust the frequencies and excitation forces of the vibrations. The symmetry in the structure makes the centrifuge forces cancelled each other in the horizontal direction and the centrifuge forces superimposed each other in the vertical direction. The centrifuge forces are in the magnitude of above 100 kN, the mass of the bass is approximately equal to the resultant force of the centrifuge forces. The working frequencies are in the range of 4–40 Hz. The vibrating device is put on the surface of the earth near the oil well. The whole device is impacting the earth surface just like a huge hammer beating the surface. The vibration of the device exerts an enormous impact force to the earth surface and generates a huge elastic waves. The waves propagate into the layers of the earth. The low frequency waves have long wavelength and have large ability of penetration into the depth of the earth and reach the oil layer as deep as several hundreds even thousands miters. They form a wave field in the oil layers and disturb the oil layers. The oil field under the continuous action of the wave field can generate the vibrating effects: the viscosity and condensation point decrease, the permeating flow accelerates moisture content decreases, and therefore the outputs are increased.

Vibration technique compared to other output increase measures has many unique advantages. Vibration technique on the oil well and near by excitation does not need the operations both in the well and outside well, saving a lot of unnecessary costs; just a little vibration could benefit many oil wells; and vibration has no pollution whatsoever; it is a technology with less investment, quick results, high efficiency, simple and easy operations. It is worth promoting.

(3) Ultrasonic Water-Fuel Emulsion Combustion Technology

In 70s last century the oil prices were up and crude oil supplies were short. Current high fuel price and shortage in supply are just the repeat of the history. So people were looking for ways to increase the fuel combustion efficiency. It has a special realistically significance in the high oil prices. One of the ideas is to add the additives to fuels. One of the additives is naturally the water. The water-fuel emulsion is a process in which water and other liquids are thoroughly mixed in a way in which water molecules are completely incorporated into those of the liquids. However liquids such as water and fuel are very hard to be mixed in a normal condition. The mix can be made possible when non-mixing liquids are processed by the cavitation processes in the ultrasound fields. It uses a fluid-dynamic ultrasound emulsifier to conduct the emulsification process of the watered-fuel. Fluidics device and a monitor, ultrasound device, moisture teller form a system of technological processes. The combustion system accelerates the combustion speed of the emulsified fuel due to its finer vaporization, it is advantageous to the complete combustion, increase flame temperature, and the stability and balance of the flame, increase the combustion efficiency and strength the heat transfer effects and at the same time, decrease the detrimental gases and harmful gas contents in its exhaust gas, which greatly reduces the pollution to the atmosphere. This technology has been used in food process, agricultures, transportation, etc., and generated a great social and economic benefits.

1.5 Applications of Electrics, Magnetic and Light Oscillators …

(4) Ultrasound Wave Vibrating Cutting Technology

The ultrasound vibrating metal cutting (USVMC) is the new technology very rapidly developed in recent years and there are the surprising effects in many application fields. The USVMC technology is the pulse metal cutting way of ultrasound frequencies. The cutting forces on the horizontal direction in the boring spindle are formed by the traditional cutting forces to the high-frequency pulse cutting forces.

In one cycle time T of the USVMC cutting process, the cutting time is t_c and ($T - t_c$) is not in cutting. This makes the cutting lubricant oil come into the cutting area sufficiently to get sufficient cooling and lubricating. The cutter tool heat radiation condition gets improved and cutting temperatures are decreased. In USVMC process the cutting tools touch with the cut materials in a discontinuous manner like the traditional cutting process. In addition, the cutting tools of the USVMC will push the cutting fluid to vibrate in an ultrasound manner. The advantage is that the cutting oils do not adhere to the cutting tools and the cut materials due to its ultrasound characteristics. The other advantage for the USVMC is that the cutting surface roughness caused by the different materials of the cutting tools and cut materials can be reduced to its minimum. The USVMC technology can be also applied to the precise cutting of the special parts.

(5) Ultrasound Application in Medical Science

The ultrasound is discovered to use for diagnostics check on human health in 1940s and rapidly developed since then. Nowadays it is widely used to diagnose a variety of human body diseases. The healthy and unhealthy organs and tissues of human bodies have different absorption characteristics and reflection patterns to the ultrasound. It is this characteristic that is used to diagnose the diseases. In the ultrasound medical device, an ultrasound is generated. The ultrasound is sent to a human body. The ultrasounds are reflected and refracted back when they meet human body organs and tissues. The reflected and refracted ultrasound signals are recorded, analyzed and displayed in wave shapes, curves and images. These analysis results provide doctors and physicians the evidence on the healthy situations of the organs and tissues. B-Ultrasound is another medical treatment device using the ultrasound technology.

(6) Ultrasound Motors

The piezoelectric ultrasound motors are simply called the ultrasound motors. It is made use of the anti-piezoelectricity effect of the piezoelectric material to generate the micro-mechanical vibration in the elastic body (the stator) and convert the micro-mechanical vibration into macro-rotational (or rectilinear) movement in one direction in the rotor. The vibration frequencies in the stator are above 20 kHz, the motor is called the piezoelectric ultrasound motor.

The ultrasound motors have the following characteristics:

i. *Simple and compact in structure, large torque/mass ratio (about 30 times that of a traditional motor).*

ii. *Low speed and big torque, no gear box for speed reduction and can be used for direct drive;*
iii. *Small inertia for rotating parts and quick response time (millisecond magnitude).*
iv. *Self-lock upon electricity when electricity supply gets sudden shut off and instant stop.*
v. *Accurate controllability of rotating speed and position.*
vi. *No electrical-magnetic field, no external magnetic inference and good electrical-magnetic compatibility.*
vii. *Low noise operation.*
viii. *Working in some harsh conditions.*
ix. *Can easily be made rectilinear motor.*
x. *Can be made into different shapes: round, square, hollow and rod, etc.*

Due to its so many unique characteristics, it has been utilized in the cameros, watches, robots, automobiles, aerospace, precise positioning instruments and micro-machining systems. It is not difficult to predict that it will be utilized much more in many of the fields in the industry such as man-made satellites, missiles, rockets, magnetic-lift trains etc.

The ultrasound motor technology is a very young branch of science in its second decades of life. Its theory, design, manufacturing technology and material are not matured and perfected yet. How to resolve this problem is the task for researchers in the ultrasound motor fields in the world in the century.

(7) Vibration by Continuous Reflection and Refraction of Light in Fiber Optics

Making use of fiber optics in information transmission is a revolution in communication fields. The continuous reflection process of light in fiber optics is a typical vibration process. Its successful research will have a big leap in the human communication cause.

(8) Vibration Heat Image Technology

In infrared heat image non-destruction detection technology, the vibration heat image technology has been applied to the non-destruction detection of the composite materials. The basic principle of the vibrating heat image system is to detect the structures of the materials under the mechanical vibration excitations.

1.6 Vibrating Phenomena and Utilization in Natures

People could use the statistical data from annual precipitation to find its periodical patterns and predict the precipitation for a year.

Tide is a periodical vibration. Its rise and fall and amplitudes are dependent upon the relative positions between the moon and the earth, and the relative positions between the sun and the earth as well. People can predict the tide rise and fall according to the moon segments. At the same time people can compute the effects of the sun's gravitation according to the relative positions between the moon and the sun. This is very useful for the navigation and the time arrangements for port entry and exit of the ships.

The density of the year-rings of the trees is caused by the periodical changes of the weathers. In generalized views, it is a vibrating phenomenon. The vibrating characteristics have been applied in archeology and geology exploration. The sunflowers rotate with the sun's rise and fall daily. This is a special type of the vibrations.

1.7 Vibrating Phenomena, Patterns and Utilization in Natures

A periodical pattern can be obtained by statistical annual precipitation in a region. This is a vibrating pattern of the precipitation. It can be used to estimate the precipitation in a year in that region.

The tide is a periodical vibration. Its rises and falls depend not only on the relative position of the earth to the moon but also on the relative position between the moon and the sun. People can calculate the rise and fall of the tides by the phase change of the moon, and also calculate the scales of the tides by the relative positions between the moon and the sun. This information is very useful for navigation and the time arrangements for ships to enter or leave the ports.

The uneven density of the annual rings in the trees is caused by the periodical changes of the climate. In a broad perspective it is a vibrating phenomenon. The vibrating characteristics have been utilized in the study of the archaeology and geology. Sun flowers follow the sunrise and sunset and swing from east to west. This is a special pattern of vibration as well.

1.8 Vibrating Phenomena, Patterns and Utilization in Human Society

People can estimate the rises and falls of a certain stock by the outside and inside affecting factors, i.e. mastering the vibrating patterns of the stock change process.

Naturally, economists could estimate the increase or decrease trends of a country by its domestic and international economic and social factors, i.e. the vibrating patterns in the economy development processes, and further propose some corresponding measures to reduce the losses of the national economy due to some factors.

The occurrences of the economy and finance crises are of a form of vibration phenomena. Extending the high economy growth periods and shortening the cycle of the economy and finance crises are effective measures to treat the vibration phenomena.

In ecology science, the speeds of the human population growth in history have their ups and downs. This vibrating phenomenon can be described by the corresponding vibration theory. In current situations some country has effectively used this vibration pattern to reduce the speed of its population growth.

In the life science fields, especially the utilizations of vibration and waves are everywhere in the organs of the human body itself and in the medical devices and equipment. The lung's inhale and exhale, the heart's beating, the vocal cords' and the eardrum's vibrations, the pulse's throbs, and the blood circulations are the effective utilizations of the vibrations. The medical devices and equipment of the Electroencephalogram and electrocardiograms, B-ultrasound, medical CT, X-ray, etc., are also the effective utilization of the vibrations.

1.9 Vista

The developments and evolutions of the utilization of the vibrations enrich the contents of the branch of the vibration utilization engineering. The vibration has been used to change the world, improve the life and create the values. This is an important change and it brings out a beautiful future for the vibration utilization engineering (VUE).

In order to promote the development of VUE, the following topics can be investigated to bring about the further development of the branch and benefit the world:

(1) Micro and Macro Study on Vibration Utilization Engineering from the Perspective of System Engineering

The techniques of the vibration utilization spread in different fields and different science branches, it is difficult to study such a broad and extensive scientific and technology problem systematically and comprehensively. The authors and our team have preliminarily established the theoretical framework of the vibration utilization engineering by conducting the investigation in depth on some branches and summarizing and inducting on other branches over thirty years. It is necessary to investigate further and in depth the basic theories, the associated development and applications of the branch in the system engineering perspective. By multi-disciplinary integration the associated theories and technology levels on VUE can be strengthened and we can make the contents covered by VUE really an indispensable means and a necessary mechanism in the human manufacture activities and life process.

(2) Conduct Vigorously the Study and Development on Real Application of Vibration Utilization Engineering

The study and development on the practical applications of the vibration utilization engineering should be enhanced and the extents of the applications should be spread to a wider domains and ranges. The study range should not be limited to the engineering scopes but the fields of the social science and people's life.

(3) Conduct Studies on Basic Principle and Working Mechanism of Vibration Utilization Engineering

On the basis of the study of the practical application of the vibration utilization engineering, the nonlinear vibration and wave theory, including the theoretical solutions, and numerical methods, the stability, bifurcation and chaos of the nonlinear vibration systems should be enhanced.

(4) Strengthen Combination of Vibration Technology and Information Technology

We should enhance the combination of the vibration technology with information technology, i.e. the multi-media technology, integrated-circuit technology, light-fiber technology, internet technology, and artificial intelligence technology and make the vibration utilization engineering a new and high technology industry on the basis of intelligence and a new industry with high technology.

In the time of knowledge economy, the vibration utilization engineering technology should play an active role in the process of increasing people's material life and culture life.

Chapter 2
Some Important Results in Vibration and Wave Utilization Engineering Fields

People often have relative knowledges and experiences on the harmful vibrations, because those harmful vibrations bring danger and bad effects on them. For examples: earthquakes bring huge losses to people and properties, their resonance and secondary harmonic resonances cause the damages of mechanic equipment, bridges and planes; vehicle vibrations make passengers uncomfortable; environment noises make people uneasy, etc. However for useful vibrations people get used to. If they don't analyze and understand it, it is hard for them to recognize the importance of the vibration utilization and its great contribution to the world.

Great achievement by Vibration Utilization Engineering in history has made a great contribution to human society development, and has brought the important changes for science and technology and production industry time and time again, we even could say that it brought science technology and industry revolutions in many fields. We might as well illustrate it by some typical examples: applications of synchronization theory make the self-synchronous vibration machines appear in many vibration utilization fields, at the same time as a result of it, many industry branches for manufacturing such as vibration motors and machines emerge; in highway construction, applications of vibration rammers make it possible to ensure the high density required by highway road surfaces which cost as much as 20~40 million RMB/km; as we all know that quartz oscillators promote a revolution in watch industry which replace the complex mechanical watches in some degree; applications of ultrasound (B- ultrasound) and medical CT make a revolution in medical detection technology; it is the great significance that Magnetic Resonance Image (MRI)technology was rewarded as Nobel prize. Optical fiber technology is a key revolution in communication technology.

We will introduce some of the important achievements of theories and technologies of Vibration Utilization Engineering in the following aspects:

(1) From the perspectives of technological process of the VUE, there are vibrating conveyer technology, vibrating screening technology, vibrating dewatering technology, vibrating crushing technology, vibrating compacting ramming

technology, vibrating ramming technology, vibrating sinking pile technology and vibrating diagnostics technology etc.

(2) From the perspectives of application of vibration principles, there are synchronous theory, vibration screen theory, vibrating hysteresis theory, vibration impact theory, slow-changing theory, chaos theory, piece-wise inertial and restoring theory, etc.

(3) From the perspectives of wave and wave energy utilization, there are some applications of water wave and wind wave, sound wave and ultrasound wave, optical fiber and laser, and a variety of ray waves.

(4) From perspectives of oscillating principles, there is some utilization of quartz oscillator, electronic-magnetic oscillator, and laser oscillator.

(5) The utilization of vibration principles in nature, social economy and biology fields has its important significance.

Studying those intrinsic vibration patterns and making use of them effectively are the bounden responsibilities for science and technology staffs.

In summary the purpose of study and research on theory and technology of vibration utilization should be to provide the theory and technology support for the science and technology and the national economic development and benefit the human society.

2.1 Utilization of Vibrating Conveyance Technology

The vibrating conveyance technology is the one that moves the material from one position to another position by vibrating machines. The powder materials will be dust flying up during the conveyance process and hence being harmful to operators. The working body in the vibrating conveyance can be sealed off so that the dust of the powder materials will not be harmful to the operators.

The vibrating feeder is a special form of the vibrating conveyance engineering. The automatic controls of the vibrating amplitude adjusting and feeding in the electric–magnetic vibrating feeders can be easily realized by the Silicon Controlled Rectifier (SCR). This feeder is light in weight, small in volume, cost-effective and easy to adjust output. The automatic control in the production processes can be realized very easily.

Figure 2.1 shows the 20 m long vibrating conveyor of the spring-vibration-isolator balance type by the authors for an iron ore mine in Beijing. The main resonant springs are installed between the upper and lower troughs in the direction of the vibration. The exciter of the elastic-connecting rod type is installed between the upper and lower troughs and continuously drives the troughs to relatively vibrate and move the materials in the trough body forward and thus accomplishing the mission of material conveying. The two troughs are articulated by swing rods, its middle fulcrum is supported by a prop with rubber articulation.

The prop is fixed on the lower support. In order to enhance the vibration isolation effects, the vibration isolation springs are installed under the lower trough props.

2.1 Utilization of Vibrating Conveyors Technology

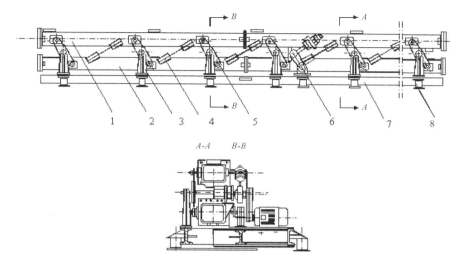

Fig. 2.1 The 20 m long vibrating conveyor of the spring-vibration-isolator balance type. 1-upper trough; 2-lower trough; 3-hinge support frame; 4-shear rubber spring; 5-hinge; 6-elastic connecting rod type of exciter; 7-supporting frame; 8-vibration isolation spring

The long distance vibrating conveyor has the following advantages: excellent vibration isolation; trough bodies are sealed off thus having no harm to operators and environments.

In the 1970s and 1980s, it is driven by an electric–magnetic exciter. The impulsive currents adjusted by an SCR flow through the coil of the electromagnet which changes the periodical electric forces generated by the electromagnet, thus making the relative vibrations between the connecting forks fixed with the trough and the electromagnet. Due to the resonant leaf springs, installed between the trough (including the connecting fork) and the electromagnet, the system is working in its near resonant conditions, the working body generates a certain magnitude of vibrations in its near resonant vibration, and thus changing the material amount out of the electric–magnetic feeder (Fig. 2.2).

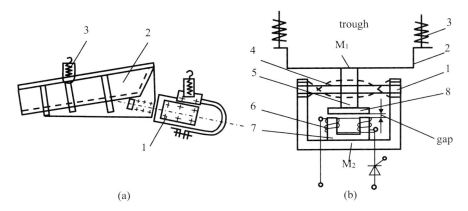

Fig. 2.2 The schematic of the electric–magnetic vibrating feeder developed by authors in the **a** Schematic, **b** Electric–magnetic exciter, 1-exciter; 2-working mechanics; 3-elastic element; 4-leaf spring; 5-connecting fork; 6-coil ;7-core; 8-armature

2.2 Applications of Vibrating Screening Technology

The vibrating screens have been widely used to screen a variety of porous materials in metallurgy, coal, electricity, food processing, and agriculture industries. The screening of the porous materials is to serve the purpose of sorting the porous materials by their sizes. The purposes of sorting and grading are:

1. Use the material with different sizes to different usages. For example, the sorted coal can increase the use efficiency.
2. Due to the technological requirements, the materials must be graded. For example, before the ore and fuel material are sent into a furnace, they must be screened for grading. The powder fine materials must be separated from the ores to avoid the ventilation to be affected by the excessive powder.

There are a variety of screens and resonant screens and many screening methods: general screening, thin-layer screening, probability screening, thick-layer screening and probability iso-thick screening etc. Figure 2.3 shows a large self-auto-synchronous screen by authors. Many iron-steel companies in China have used this type of screen. This type of screen includes 2500 mm × 7500 mm, 2500 mm × 8500 mm, 3000 mm × 9000 mm and their weights are in the range of 30–50 tons.

For this type of large screens the height is dramatically reduced, it is lightweight and compact due to their uses of the internationally original self-auto-synchronous vibrating mechanism with exciter-eccentric rotation installation; the embedded exciter is installed on the top of the body to make the body stiffness and strength bigger; due to the secondary vibration isolation system, the screens have excellent vibration-isolation characteristics and can be installed on the top of the buildings without causing the building vibration.

2.2 Applications of Vibrating Screening Technology

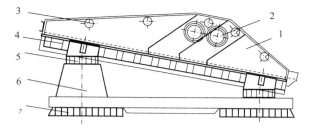

Fig. 2.3 Exciter-deflection type of cold sintering ore vibration screen. 1-vibration body; 2-deflection type of self-synchronous exciter ;3-screen upper frame circular connecting beam;4-screen lower frame rectangular connecting beam ;5-primary vibration isolation spring ;6-vibration isolation frame; 7-secondary vibration isolation spring

Figure 2.4 is the inertial resonant probability screen developed by authors. The vibrating screen has a new vibrating mechanism in which the double bodies, composed of screen box and balancing body, are working in near resonant conditions. The resonant shear rubber springs are installed between the screen box and the balancing bodies to make the system work in the near resonant conditions. There are three layers for screening surface 4. The machine has large output thanks to its probability screening. The whole machine is suspended by the vibration isolation springs 2. The advantages of the screen are: stable working state, no noise, good vibration isolation and high screening efficiency and high screening output which is about as 3~5 times high as a general screen.

Figure 2.5 is a nonlinear inertial resonant screen with gap rubber springs. It is composed of screen box 1, eccentric of the exciter 4, balancing body 5, a gap main resonant nonlinear spring 2, leave spring 6 and vibration isolation spring 7. The balancing body is simply supported on the screen box, the main nonlinear resonant system is composed of screen box, balancing body and nonlinear rubber spring, while the isolation system is composed of the screen box, balancing body and isolation spring.

The vibration magnitude in the main vibration direction is relatively large due to the fact that the resonant frequency of the main vibrating system is near the working frequency in the main vibration direction, while the vibration magnitude

Fig. 2.4 Inertial resonance type of probability screen. 1-inertial resonance type of exciter; 2-vibration isolation spring; 3-screen box; 4-screen surface

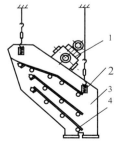

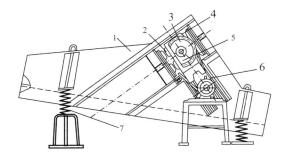

Fig. 2.5 Structure of an inertial type of resonance screen. 1-screen box; 2-main vibration spring; 3-pulley; 4-eccentric; 5-balancing body; 6-leaf-spring; 7-vibration isolation spring

perpendicular to the main vibration direction is relative small because it's far away from the resonant frequency in that direction. Therefore the trajectory of the body is a long ellipsoid, which is good for the screening process.

The difficulty in the screening technology is to screen the wet and fine material and there are some output and screening deficiency in dealing with those materials. Figure 2.6 shows a bi-directional semi-spiral vibrating screen for fine material. This is a type of non-resonant vibrating screen. The double exciters, installed under the body, drive the body vibrate in the vertical direction. There are 8 semi-spirals. The materials are fed from the upper left of the spirals into the separator, and then are guided onto the screen surfaces of the 8 spiral troughs. Owing to the vibrations the materials on the inclined screen surfaces will be moved forward to the right along the spiral troughs and be discharged from the exits eventually. The screen surface areas are large due to total of 8 spiral troughs while the volume it takes is small. Besides, the vertical vibration of the screen surfaces is good for the screening of the fine materials.

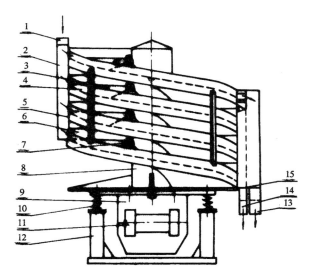

Fig. 2.6 Schematic of bi-directional semi-spiral vibration fine screen. 1-entrance; 2-feed tube; 3-semi-spiral trough; 4-screen surface; 5-screened material; 6-material on screen; 7-mid support tube; 8-shallow support tube; 9-exciter fixture; 10-vibration isolation spring; 11-excitation motor; 12-vibration isolation spring support; 13-discharger for material on screen; 14-screened discharger; 15-support plate

2.3 Applications of Vibrating Centrifugal Hydro-Extraction and Screening Technology

The vibrating centrifugal hydro-extractor is a type of vibrating machine used for wet coal and fine particle water extracting. The schematic of a horizontal vibrating centrifuge is shown in Fig. 2.7. It is a system of two bodies working in the nonlinear, near resonant state. The working body is a cone, hollow rotor with screening holes. The cone rotor rotates continuously to generate the centrifugal forces, which makes the water contained in the material in the cone rotor be thrown outside of the cone via the screening holes while the left-over material within the cone will be forced to move toward the larger end of the cone by the radial vibration of the cone body, finally the materials without water are discharged to the receiving container. The design parameter selections for rotating speed of the cone rotor, the vibration frequencies and magnitudes of the vibrating mechanism are based on the vibrating centrifuge material movement theory. In general the throwing movement of the material will not occur, only some sliding in the positive direction or, some slight negative sliding may occur.

The working principle of the vibrating mechanism for this type of vibrating centrifugal hydro-extractor is the cone body is fixed on the vibrating body 3 by a pair of bearings. The vibrating body vibrates in the horizontal direction. The vibrating body is supported by several sets of the vertical leaf-springs. The lower ends of the leaf-springs are supported on the vibrating body 3. The vibrating body 3 is installed on the springs, vibrating in the horizontal direction, on which the twin axle inertial exciter is installed. There are gap rubber springs between the two vibrating bodies with the nonlinearity of the piece-wise linearity. The main resonant frequency of the

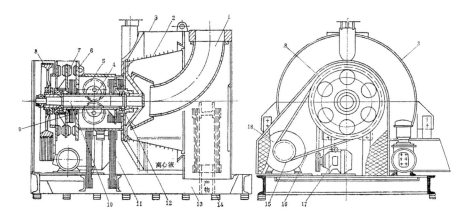

Fig. 2.7 Horizontal type of vibration centrifuges for coal or fine particle coal hydro-extraction. 1-feed tube; 2-cone screen body; 3-cylinder receiving centrifuge fluid; 4-main shaft casing tube; 5-inertial exciter; 6-gap spring impact plate on body 2; 7-gap rubber spring on body 1; 8-pulley; 9-short-plate spring; 10-mail shaft; 11-long-plate spring; 12-bearing; 13-frame; 14-vibration isolation spring; 15-pinion; 16 triangle belt; 17-excitation system motor; 18-cone rotation motor

vibrating system is usually taken the forcing frequency which is a little bit smaller than the equivalent resonant frequency of the system: $\omega = 0.85\omega_0 \sim 0.9\omega_0$.

The centrifugal vibrating hydro-extractor is complex in structure but has good water extracting effect. That is why many companies in coal industry are still using them.

Figure 2.8 shows a new type of the spring table being used for screening the fine article ores in nonferrous metal ore dressing plants (such as tungsten ore dressing plant). The exciting forces, generated by the eccentric wheel 2 rotations, are transferred to bed 13 via rod 1. The bed is connected to the soft spring 8 and the hard spring 11 by the spiral rod 12. The soft and hard springs are compressed when the machine is in motionless condition. The bed and the soft nonlinear vibrating system composed of the soft springs will vibrate by the exciting forces. The system vibrates in near resonant state. The hard spring is compressed and released from time to time, the movement of the bed will result in asymmetric acceleration. The asymmetric acceleration curves will screen well the materials with different densities on the bed.

In order to increase the capacity and retrieve efficiently the fine ores, a new vibrating centrifugal table is developed. There is no essential difference from the spring table from the point of views of the vibration mechanism. The difference lies in that the screening of the spring table happens in the gravity fields while that of the centrifugal table is accomplished by the centrifuge which is much larger so that the gravity and the screening process is greatly enhanced.

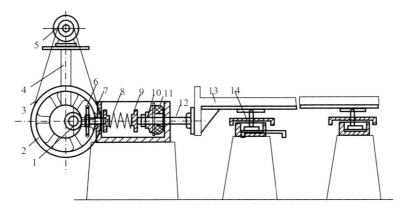

Fig. 2.8 Spring table. 1-eccentric bar; 2-eccentric wheel; 3-triangle belt; 4-frame; 5-motor; 6-handle wheel; 7-spring box; 8-metal soft spring; 9-spring base; 10-nut; 11-rubber harden spring; 12-auger; 13-bed surface; 14-supporting base

2.4 Applications of Vibrating Crush and Milling Technology

The crush of the material is the technological process to crush the mineral ores in mining enterprises. Most of raw materials mined need to be milled and/or crushed. Millions of tons of ore and stone materials need to be crushed.

The traditional crushers have their limitations: when the compressive strength limit of the ores reaches to 2×10^5 Pa, either the crushing process consumes excessive energy or the ores are difficult to crush and equipment's usage would be complex.

The development of the vibrating crush technology overcomes the shortcomings of the traditional technology. The earliest vibrating crusher was developed by the Leningrad Mine Screening Institute in the former USSR. The inertial vibrating cone crushers use the centrifugal forces generated by the eccentrics to crush ores or other materials and the exciters are used to squeeze the materials in the gaps between the inner and outer cones or with some impact and wearing to crush them.

The successful development of the inertial vibrating cone crusher is a big breakthrough in the material crushing fields. Its advantages are that its crushing ratio (above 50) is much larger than that of the traditional crushers (5~10). For the same crushing ratio the energy consumed on crushing a unit material is relatively less and can be adjusted in a big range.

The vibrating cone crusher with single inertial exciter is shown in Fig. 2.9. The outer cone and frame are supported on the rubber vibration isolation spring device. The inner cone is supported on the ring-shaped sliding chassis on the frame. The inner and outer cones can rotate around the same point (center of a sphere). The swing movement of the inner cone is driven by an inertial exciter with its sliding case installed in the symmetrical axis of the inner cone. The rotation of the non-balanced inertial exciter is driven by the U-joint via the driving device. The crushing ratio can be adjusted by changing the eccentric static moment. The inner surface of the outer cone and the outer surface of the inner cone form a crushing cavity, the huge

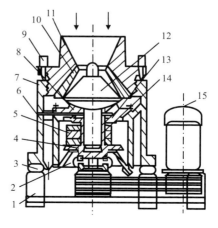

Fig. 2.9 Vibration cone crusher with single exciter. 1-rubber base; 2-univeral joint; 3-spring base; 4-belt wheel support; 5-eccentric; 6-low frame; 7-upper frame; 8-sphere shaft tile; 9-adjust-ring lock; 10-inner liner; 11-feeder; 12-inner cone; 13-axle sleeve; 14-axis; 15-motor

centrifugal forces generated by the high-speed rotation of the non-balanced rotor push the inner cone surface rolling over the material on the outer cone surface. The crushing force magnitudes of the inertial cone crushers are dependent upon the centrifugal forces on the inner cone and the mass of the outer cone etc. while those of the traditional cone crushers are dependent upon the strength of the materials and the cavity-filling rate.

The eccentrics of the traditional crushers are constants, while the eccentrics of the inertial cone crushers are dependent upon the crushing forces and the thickness of the materials. The materials to be crushed in the crushing cavity are rolled and squeezed more than 30 times in a flash of a few seconds. The materials are easily self-crushed from their weak areas with defectives and cracks. The inertial cone crushers can realize a big crushing ratio and energy-efficient. They can be started up even with a negative load. It doesn't matter even the crushing cavity contains the un-crushable materials and there is no need for an over-load protection.

The vibrating cone crusher with double inertial exciters is new type of vibrating cone crushers using vibration synchronous principles developed by authors, as shown in Fig. 2.10. The system is composed of outer cone, inner cone, two self-synchronous inertial exciters, upper and lower connecting plates, suspension device, vibration isolation springs etc. The inner and outer cone surfaces are embedded with protective plates and form a crushing cavity. The movable inner crushing cone, two exciters, and the upper and lower connecting plates are suspended on the vertical plate of the outer crushing cone via the suspension device. The motors drive the two exciters rotating in the same direction at the same speed by tire-axle connectors to generate the centrifugal forces. The movable inner crushing cone rotates with a circle trajectory around the centerline of the machine to perform the effective crushing function.

This new type of the vibrating inertial cone crusher has its originality in its structure and mechanism. It differs from the KNΠ and traditional crushers in structure. In addition to include their advantages, the new crusher has its special crushing cavity

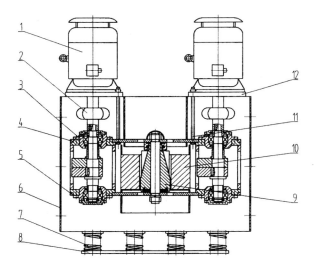

Fig. 2.10 Vibration crusher with double inertial exciters. 1-motor; 2-connector; 3-bearing; 4-exciter case; 5-exciter low base; 6-external case; 7-vibration isolation spring; 8-base; 9-inner cone; 10-outer cone; 11-exciter base; 12-motor base plate

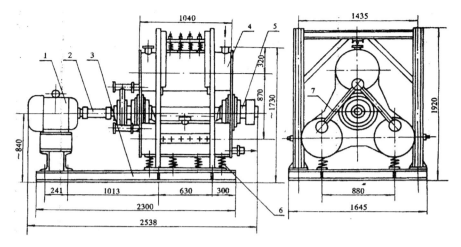

Fig. 2.11 Tri-cylinder vibration ball mill. 1-motor; 2-connecting shaft; 3-base; 4-cylinder body; 5-inertial exciter; 6-vibration isolation spring

between surfaces of the cone and the cylinder to obtain the crushing products with excellent size characters. In the crushing experiments on pebbles with low compressive strength and rocks with high compressive strength, for the feeding materials with size less than 40 mm, the size less than 5 mm for the crushed output is above 80% and 5 mm above 90%. When the crushing materials are crystal body, the flake body and powder contained in the output is rare to meet requirements of the building sands. This crusher has the following advantages: low height, convenient and cost-effective to manufacture.

Figure 2.11 shows a tri-cylinder vibrating ball mill. The motor drives the triple-cylinders to vibrate by the elastic coupler and the single axle inertial exciter. The cylinders are supported on the base through vibration isolation springs. Steel balls and materials to be milled are put into the cylinders. The cylinders are driven to vibrate in near a circle or an ellipsoid trajectory and the materials, steel balls are impacting on cylinder bodies continuously and milling thus the materials are crushed. The vibrating machine has high efficiency and can mill the material size as small as several millimeters, however its noise is severe.

2.5 Applications of Vibrating Rolling and Forming Technology

Constructing the highways, airport runways and dams etc. all needs the vibrating rollers. The density of a highway would be above 95% by using the vibrating rollers to roll the surfaces, which is higher than 90% by using a static roller. The density requirements for a highway must be over 95% to ensure the surface quality or the life expectancy of the highway would be greatly reduced. As estimated that the constructing cost would be in the range of ¥ 20,000,000~¥ 40,000,000 per kilometer, the cost would increase 1/3 if the life expectance is shorten 1/3, the economic benefit by using the vibrating rollers is very significant.

There are many types of vibrating rollers, such as the vibrating and oscillating rollers with single steel wheel and double steel wheels.

Figure 2.12 shows a typical vibrating roller. The exciter with an eccentric is installed in the vibrating wheel and is driven by a hydraulic motor. The inertial force generated by the exciter's rotation is exerted on the soil to roll it. The rolling process is accomplished while the roller is moving forwards or backwards and it takes more than once to do it. The soil rolling process is a typical nonlinear dynamics process with elastic–plastic deformation and the nonlinear system shows the characteristics of the hysteresis and the size of the hysteresis loop area has direct impact on rolling effects (Fig. 2.13).

The vibrating rollers are used to roll the road surface outside the rollers while there is another kind of rollers which roll inside the rollers—the vibrating forming. A vibrating forming machine is shown in Fig. 2.14. The motor drives a double-axle exciter to vibrate vertically by a gear synchronizer. The working table is supported on the vibration isolation springs and the springs are installed on the

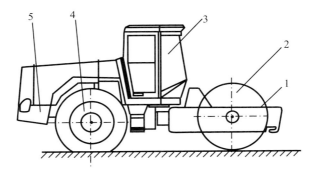

Fig. 2.12 Schematic of vibrating roller. 1-vibration wheel frame; 2-vibrating wheel; 3-driver compartment; 4-rubber tier type of rear driving wheel; 5-rear wheel supporting frame

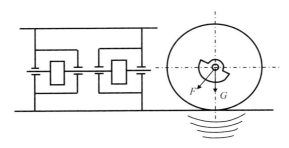

Fig. 2.13 Schematic of vibrating wheel for vibrating roller

2.6 Applications of Vibrating Tamping Technology

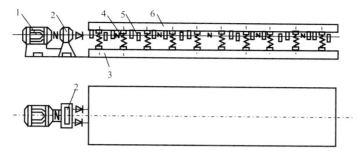

Fig. 2.14 Schematic of a vibration shaper. 1-motor; 2-gear synchronizer; 3-base; 4-supporting spring; 5-eccentric; 6-table

base. Due to the excitations from the exciter, the table vibrates vertically and on the table there are the materials, the concrete products, and/or reinforced concrete and their moulds to be formed. The gravity or other forms of pressures may be exerted when necessary.

2.6 Applications of Vibrating Tamping Technology

The vibrating tamping is used in the construction engineering process. In the concrete pouring process the concrete must be tamped to remove the bubbles, increase the density and quality. In the 50's of the twentieth century, the tamping was performed manually. In the 60's of the last century, the concrete vibrator was invented (Fig. 2.15).

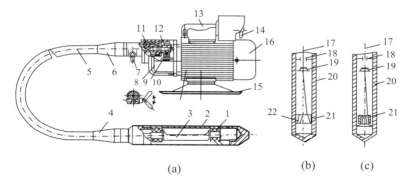

Fig. 2.15 Insertion type of rammer. **a** eccentric insertion type of rammer, **b** outer flow channel type of rammer, **c** inner flow channel type of rammer. 1, 11-bearing; 2-outer shell; 3-eccentric shaft; 4, 6-tube connector; 5-soft shaft; 7-soft-tobe lock wrench; 8-accelerator; 9-motor rotor shaft; 10-siphone device; 12-pinion; 13-handler; 14-swatch; 15-rotation disk; 16-motor; 17-shaft; 18-bearing; 19-universal joint; 20-shell body; 21-flow cone; 22-channel

The mechanism of the vibrating tamper is very simple. The motor drives a center flexible axle, which can be swung, to rotate around the inner surface of the inner cone of the outer sleeve in a galaxy movement. The diameter of the center axle is almost equal to the diameter of the inner cone, the revolution speed of the center axle around the inner cone, i.e. the rotation speed of the eccentric center axle, is the multiple of the motor's rotation speed, i.e. the rotation of the center axle's rotation, generally in the range of above 10,000 rpm. Under such high rotation of the revolution the outer cone generates 10,000 rpm vibration.

Another type of the vibrating tamper has the adherence tamping. It is adherently tamped the top of the concrete pouring parts to increase the quality. The other type of the tamper has a table on which the concrete pouring parts are put and the vibrating table is used to tamp the concrete pouring parts.

The spreader is used for the concrete road surfaces and the array of the concrete vibrators is also used as the main tamping equipment for the concrete road surfaces to increase the road surface quality.

2.7 Applications of Vibrating Ramming Technology

Figure 2.16 shows the frog rammer. The vibrating rammer is composed of rammer head, rammer frame 2, transmission axis frame 8, chassis 4 and motor 5 etc. The motor drives the shaft with an eccentric to rotate by a set of triangle belts. When the eccentric rotates to a certain angle, the head is raised by the inertial force and when the head is raised to a certain height, the whole machine is moved a frog leap forwards

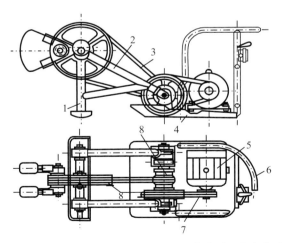

Fig. 2.16 Frog rammer. 1-rammer head; 2-rammer frame; 3, 7-triangle belt; 4-chassis; 5-motor; 6-handler; 8-transmission axis frame

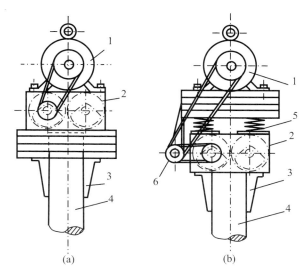

Fig. 2.17 Vibration impact forging hammer. **a** motor installed direct on body, **b** motor installed on vibration isolator base. 1-motor; 2-bi-axle inertial exciter; 3-clamper; 4-impact hammer; 5-spring; 6-mid pulley

a certain distance. Then the head begins to fall with an increasing speed to impact and roll the soil.

Figure 2.17 shows a vibrating hammer which is composed of a double-axle inertial exciter, a clamper and a impact hammer. In order to prevent an impact damage to the motor a spring is used to separate the motor and exciter as shown in Fig. 2.17b. The motor drives the exciter by a belt to rotate. To prevent the belt from stretching or shortening due to the vibration, there is an added middle belt pulley whose shaft is installed in the same plane as the exciter. In addition to the rammers mentioned above, a hydraulic exciter-driven static-and dynamic pressure vibrating rammer has been studied by the authors.

2.8 Applications of Vibration Diagnostics Technology

The safety and robustness of the equipment operations are always the important study subjects. People have been devoted to study the monitoring and diagnostics of the equipment operation. In the nineteenth century industry revolution, the diagnostics study on equipment started almost on the same days as the equipment was operated. The engineers used their human sensory organs to obtain the state information of the equipment and used their engineering judgments and experiences to diagnose the equipment faulty. Even though the diagnostic methods are rather rudimentary and simple ,they are very effective as the equipment at that time was simple and the forms of the faulty are relative obvious. As the manufacture and electronic industries develop the functions, features and structures of the equipment become more complex, the reliability and robustness theory were developed in 1960's. The diagnostics task on the equipment can be accomplished by the analysis and estimation on the material life expectance and partial detection on the equipment characteristics in advance.

The current mechanic equipment tends to be large and large (or microminiaturized), specialized, high speed, high efficiency, precision, integrated, synthesized (function), modularized, automatic, intelligent, digitalized, web-oriented, and environment-friendly. Those features demand the high requirements on design, manufacture, assemble, transportation and maintenance etc. The requirements lead to the more complexity on the equipment and working mechanisms. Their dynamic characteristics, such as coupling, randomness, suddenness and nonlinearity, are more obvious. For example, the larger the equipment is, the more cost the manufacturing will take, the more costly when the equipment fails. The higher speeds the equipment is, the larger the dynamics loads, the speedier the equipment might fail. The accident odds will be larger when the equipment becomes more complex.

According to an incomplete statistics by 11 fertilizer factories in China, the damages caused by big-unit unexpected stops would be around several hundreds of millions RMB during 1976~1985. In recent 30 years, the axle-broken accidents of the 200,000 kilowatts generating units in three power plants in China have caused damages over hundreds of millions RMB. Overseas, in 1970s', the Three-Mile nuclear power plant in the U.S.A. was damaged.

To prevent the severe accidents from happening, it is imperative for the designs, manufacture and operations on the equipment to monitor the operating states and diagnose the faulty. It is equally imperative to detect, diagnose the possible fault and take effective actions in timely manners to prevent them from happening.

The use of the monitoring and diagnostic equipment could achieve good economic benefits. As some experts estimated that the ratio of the benefit of using the diagnostic equipment to the cost would be around 10:1, some experts estimated the ratio would be 17:1.

The monitoring and diagnostics devices generally include the sampling of signals (sensing technology), signal transmitting (web and fiber-glass technology), signal processing and storage (multi-media technology), knowledge inference and decision (human-intelligence technology) and final decision and accident treatment (expert systems).

The developments of the sensing technology in 1960's made the measurements of the diagnostic signals and data much easier; the usage of computers compensates the low efficiency of human's handlings on the data processing. The intelligent monitoring and diagnostic technologies, based on aforementioned signal measurement and analysis, have been used in the military, steel, ship building and mechanical manufacture fields.

The development of the human intelligence technology, especially the applications of the expert system in the diagnostics fields, provides the strong technology supports for the intelligence diagnostics. The diagnostics process based on the signal process

2.9 Applications of Synchronous Vibrating Theory

and the numerical computation is being replaced by the one based on the knowledge processing and knowledge inference.

2.9 Applications of Synchronous Vibrating Theory

The synchronous phenomenon is a form of movement existed in nature, human life and production processes. As early as the seventeenth century, Huygnes (1629~1695) found that for the two identical wall clocks hung on a swing thin plate their pendulums can swing synchronously while they can't when the clocks ware hung on a fixed wall.

Dr. Blehman in the former USSR in 1960's first studied the self-synchronous theory of a twin-exciter vibrator. The application of the theory made the vibrator mechanism much simpler, much easier to be maintained and its reliability increased drastically. There are many synchronous vibrating machines being used in the industries: self-synchronous feeder, conveyor, vibrating screen, probability screen, vibrating cooler, vibrating drier, vibrating rammer, and vibrating crusher etc. The authors proposed the exciter-eccentric synchronous theory based on the self-synchronization theory and applied this theory to the study of the large vibrating screen for cold mines. Nowadays tens of thousands of the vibrating machines made of the vibration synchronization theory have been used in agriculture and industry.

In the engineering industries, there are many types of synchronous phenomena and problems encountered. Some mechanical equipment is required to have two or more than two components, such as rotating shaft, mechanism, rods, and cylinder pistons etc., to have the same displacements, velocity, acceleration, phase and acting forces, i.e. to require them to operate in the synchronous manners to perform their functions. In the past, to realize the synchronous operations of two exciters it used a pair of gear with gear ratio 1 to drive the eccentric exciters. The proposal of the vibration synchronous theory promotes the important change and reform in the vibration utilization engineering fields and industrial branches. From point of view of technology, the theory simplifies the vibration structures, makes the maintenance job easier and therefore brings the huge economic and social benefits.

At the same time many countries established special vibrating motor factories and their motor manufacture sectors undertook a revolutionary change. The change was triggered by the synchronization theory study and successful applications.

We have concluded the several new conclusions as follows and applied the theories to the engineering practices and achieved economical and social benefits.

1. Synchronous theory of the self-synchronous vibrators with eccentric exciters.
2. Vibrating synchronous conveyance theory.
3. Multiple-frequency synchronous theory (see Fig. 2.18).
4. Compound synchronous theory.

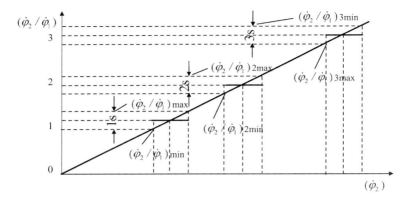

Fig. 2.18 Double frequency synchronization schematic

2.10 Applications of Resonance Theory

2.10.1 The General Utilization of the Resonance

In the vibration systems, the resonance occurs when the exciting frequencies are coincident with the natural frequencies of the systems. The nature resonance phenomena can be observed everywhere in the nature and industrial engineering. The resonance principle has been used in the industry because they have the same vibrating magnitude when excited under the resonance conditions, its exciting force or energy required is minimum.

The resonance principle has long time been used in the communication industry. The most basic principle in the wireless communication is "tuning" (adjusting the receiving frequency to match the frequency of the transmitting signal, make it in the resonant condition), and then magnify it to obtain a big volume and receive the transmitted signals. In addition to the radios, the resonance and magnifying principles have been used in many electronic instruments as well.

The resonance principle has been used in the vibrating instrument and equipment, especially in the vibrating machines. For example, the electric–magnetic vibrating feeder, inertial resonance feeder, long distance elastic-rod vibrating conveyor, elastic-connecting rod vibrating screen, inertial resonance screen, near-resonance vibrating centrifugal water-extractor, spring-table, etc.

The exciting force of the resonance vibrating machines is in general about 1/5 or less than that of the non-resonance vibration machines as shown in Fig. 2.19, the size and volume of its exciters and connecting mechanisms can be made rather small and hence its working life is extended. Taking an electric–magnetic vibrating feeder as an example, its volume and mass are about 1/2~1/5 of those of non-resonance vibrators, and energy consumed is reduced to 1/2~1/5 due to the resonance applications. This is

Fig. 2.19 The comparison of the exiting force in the case of resonance and non-resonance

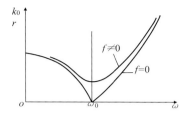

a good example to illustrate the economical benefits of the utilization of the resonance in the vibrating machines.

Some people proposed that the resonance can be used to generate a huge amount of energy by increasing the system mass and springs, then release the huge energy to accomplish the works which need the huge energy in a short time.

We can predict that the resonance theory would be developed further not only at present but in the future.

2.10.2 Application of the Nuclear Magnetic Resonance

The phenomenon of Nuclear Magnetic Resonance (NMR) was demonstrated already in 1946 and has already resulted in Nobel Prize in both Physics and Chemistry. The Nobel Prize in Physiology or Medicine 2003 had been rewarded to Paul C. Lauterbur, University of Illinois, Urbana, IL, USA and Sir Peter Mansfield, University of Nottingham, School of Physics and Astronomy Nottingham, U.K. for their discoveries concerning Magnetic Resonance Imaging (MRI). This is very exciting and important news for the vibration utilization engineering.

The NMR technology is to use the physical principle (Quantum Mechanics Principle) to analyze the structure and characteristics of the molecules without destructing the sample's internal structures by measuring the NMR spectrum characteristics. It is a non-destructive detection method. The NMR technology is to obtain the information of the nuclei from the energy changes of the nuclei in the magnetic fields. It is the extremely high-resolution analysis technology and has been used in physics, chemistry, medicine, petrol-chemistry and archaeology etc. It is also used in the oil-quality analysis, food study, medicine development, and underwater survey. MRI (Magnetic Resonance Imaging) technology is used to examine all organs of the body with detailed images of the organ and is outstanding for diagnostics, and it provides the vivid scientific evidence to doctors to determine the causes of the diseases and the follow up treatment of the disease on brain, kidney and lung etc.

The NMR is a phenomenon resonance jump of the energy grade of the nuclei in an external magnetic field. A nucleus has a positive charge and a spin movement. The nuclear spin generates a magnetic moment, called "nuclear magnetic moment". The nuclear magnetic moment is directly proportional to the orbital angular moment of the nucleus. In an external magnetic field, the nucleus associated with the nuclear magnetic moment has the corresponding potential energy. Due to the interactions between the nuclear magnetic moment and the magnetic field, a radio frequency signal of the proper frequency can induce a transition between spin states. This action places some of the spins in their higher energy state. If the radio frequency signal is then switched off, the relaxation of the spins back to the lower state generates a measurable amount of radio frequency signal at the resonant frequency associated with the spin flip. This process is called NMR. The NMR frequency is the product of the magnetic spin ratio and the magnetic induction intensity.

The NMR can be realized in the following two ways based on the principles above: 1. change the incident electronic-magnetic wave's frequencies while keeping the external magnetic field the same; or 2. adjust the magnetic field's intensity while keeping the frequency the same. The object is to put on a glass container and in between two probes; the coil is rounded on outer surfaces of the glass container, in which the radio frequency current was input via a radio frequency oscillator. Thus the coil emits an electric–magnetic wave to the object, the function of the modulation oscillator is to make the frequency of the radio frequency wave change continuously near the natural frequency of the object. When the frequency is coincident to the NMR frequency, a peak in the output of the radio frequency oscillator will occur and can be displayed in an oscilloscope and the resonance frequency can be read.

The nuclear resonance spectrum device is specially used to observe the nuclear magnetic resonance and is composed mainly of the magnet, probe, spectrum device. The function of the magnet is to generate a magnetic field in which the probe is placed between the two poles and used to detect the nuclear magnetic resonance signals; the spectrum device is used to magnify the resonance signals and display and record them. There are many NMR instrument types as the science is in process of development, more different types of NMR will emerge.

2.11 Applications of Hysteresis System

Some of the vibrating machines are used to ram to the soil, or to shape the powder materials. The vibration systems formed by the vibrator and the working object are the vibrating systems with hysteresis restoring force. The system with hysteresis restoring force is a system with both elastic–plastic deformations at the same time in the working process. The vibrating rammer is a machine that rams the soil. The working effectiveness and ramming speed are directly dependent on the size of the hysteresis loop. Obviously the larger the hysteresis loop area the bigger the speed and the higher the efficiency. From the point of view of increasing the working speed and efficiency it is benificial to increase the plastic deformation.

There are many hysteresis systems: the broken-line hysteresis (similar to parallelogram) and the curve hysteresis, etc. The following is the most special hysteresis system with a gap. The inertial cone crusher is modeled as a hysteresis system with a gap. Because the crushing cavity of the crusher, in general, has a large gap, there is a gap between the material and the crushing cavity when the cone presses the materials.

Fig. 2.20 Double hysteresis loop with a gap

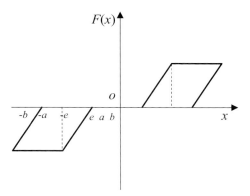

When the inner and outer cone are in the same position, the material begin to touch the crushing cavity and elastic and plastic deformations occur till the materials are crushed, and the hysteresis restoring forces occur in the systems as shown in Fig. 2.20.

2.12 Applications of Impact Principles

It is known that the impact force may be tens and even hundreds of times of the forces (i.e. harmonic vibration force) when impact time is shortened according to the momentum and impulse principles. The impact force can be computed. Under the constant impact velocity $v = \omega r$, the impact force can be computed and compared to the maximum inertial force:

$$F = mv/t = m\omega r/t$$

where F is the impact force; m is the vibrating mass; v is the impact velocity; ω is the angular velocity of the simple harmonics; r is the amplitude of the simple harmonics; t is the impact time.

It can be seen from the Table 2.1 that as the impact time decreases the impact force increases 10, 100 and even 1,000 times. The utilization of the impact principle is based on this idea and impact type of the vibrating machines is also based on

Table 2.1 Relation between impact time and impact force

Impact force	Impact time	Impact force value
F	$t = 1/\omega$	The maximum harmonics force $= 1\ m\omega^2 r$
F	$t = 0.01/\omega$	Impact force $= 10\ m\omega^2 r$
F	$t = 0.01/\omega$	Impact force $= 100\ m\omega^2 r$
F	$t = 0.001/\omega$	Impact force $= 1000\ m\omega^2 r$

the principle. The other types of the machines are the frog rammers and the impact hammers.

2.13 Applications of Slow-Changing Parameter Systems

There exist the slow-changing parameter systems in the vibration utilization engineering. All technologies that make use of the vibration are to accomplish their tasks in the changing states. The most of the changing processes have the characteristics of the slow-changing parameters, i.e. the changes of the parameters are completed in the many, tens, hundreds and even tens of thousands of cycles. This is so called the slow-changing system, or the slow-changing parameter system.

For example, the soil pressing and ramming process, material forming process, the crushing of the materials, and grinding process are all the slow-changing system. The soil intensity and crushing degree of the materials increase as the time elapses; the plastic deformation speed of the soil decreases as the soil intensity increases and the crushing difficulty of crushing materials increases as the time elapses. The fault development processes of the equipment and component exist the slow-changing characters. To decrease the amplitude passing the resonance, we tried to make use of the slow-changing characteristics, as shown in Fig. 2.21, by changing the stiffness of the supports. The test shows that when the slow-changing speed changes, the amplitude passing the resonance is different. Thus the purpose of decreasing the amplitude of the system passing the resonance can be achieved by choosing the proper slow-changing speeds.

2.14 Applications of Chaos Theory

It was found that there was a deterministic motion, so called "chaos", which is very sensitive to the initial condition in the nonlinear dynamic field in the second half of the twentieth century. The finding of the chaos was considered as the important discovery and outstanding scientific achievements after theories of the relativity and quantum mechanics.

Chaos is a special motion form and a question before the scientists and technical workers is: can this chaos theory be utilized in the industries? The answer is yes. chaos can be used in the technology and industry, in the encryption of communications and in the other technological processes.

As the rapid development of the powerful micro-computers, internet technology and multi-media technology, the computer networks increasingly become the important ways of exchanging information. The information spread, internet trades and remote control can be performed via the computer networks. The information security and encryption have become more important.

2.14 Applications of Chaos Theory

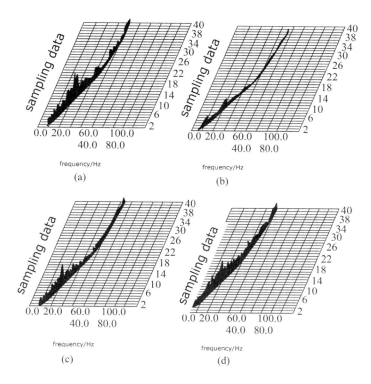

Fig. 2.21 The relationship between the slow-changing velocity k_d and amplitude. **a** $k_d = 5$, **b** $k_d = 10$, **c** $k_d = 15$, **d** $k_d = 20$

The traditional encryption has the shortcomings: the efficiency is low when the data to be encrypted and decrypted are horrendously huge. Since Matthews proposed the data encryption technology based on the chaos systems in 1989, as the chaos systems have the sensitivity to the initial conditions, and possess the white noise statistical characteristics. Chaos-based encryption technology has become the study focus in domestic and abroad. The widely used encryption chaos system nowadays is the one-dimensional discrete system (i.e. logistic mapping). It has the following advantages: simple, quick generation of chaos sequences, and high decryption efficiency. However its disadvantage is that the simple structure leads to a small space for the encryption keys. Peres and Castillo respectively decrypted the lower-dimensional encryption systems using double-structure phase space technology or neural network methods. So the low dimensional chaos encryption should be studied further.

The continous chaos systems have the complex evolution rules and high randomness. The authors proposed a method of constituting a discrete chaos system by the least square algorithm based on the continous chaos system. Its multiple initial values and structure parameters can be used to design the encryption keys to expand the encryption space dramatically and improve the encryption (decryption) real number

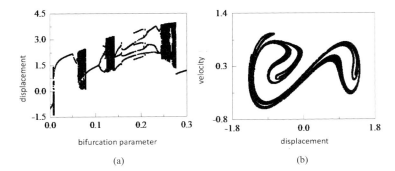

Fig. 2.22 Application of chaos in the information encryption. **a** Poincare section of a dynamic system, **b** bifurcation of a dynamic system

chaos sequence mapping as the position sequence method of the encryption and decryption. The experiment results show that the method preserves the advantages of the high efficiency on encryption and decryption of the one-dimensional discrete chaos system and further increases the anti-decryption capability of the encryption system.

Figure 2.22 shows the Poincare section and bifurcation of a nonlinear dynamic system. Using the pictures of the continous chaos system constitutes a discrete chaos system as the initial values and structure parameters for the multi-iterations and design as encryption keys. In this way the encryption and decryption efficiency and the anti-decryption capability of the sytsems have been increased.

2.15 Applications of Piecewise Inertial Force

Nonlinear principles play an important role in the study and development of the vibration utilization engineering. The reason for it is that many systems in the vibration utilization engineering are nonlinear. Besides, nonlinear systems must be used to meet some of the technological requirements in the vibrating equipment. The nonlinear inertial force terms in the vibration systems, i.e. the piecewise inertial force vibration system, are a special vibration form in the vibration utilization engineering. It is a necessary means to accomplish the material conveyance.

The vibration system, in which the inertial terms are nonlinear, is applied in the vibration utilization engineering: to transport the materials from one place to another in a vibrating conveyor, the materials must have a relative movement to the working surfaces thus a piecewise inertial force nonlinear systems appeared.

The working surfaces of the vibrating machines, in general, have the following vibrations to perform their functions: harmonic line vibrations, non-harmonic line vibrations, circle vibration, oval vibrations, etc. When vibrating machines are

2.16 Applications of Piecewise Restoring Force

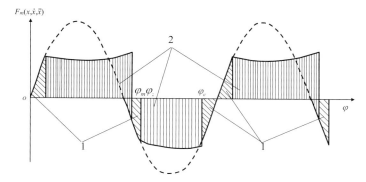

Fig. 2.23 Piecewise inertial forces in the material movement

designed by different parameters (vibration amplitudes, frequencies, vibration phase and inclining angles etc.), the materials on the working surfaces can realize the different forms of movements:

1. Relative still. The materials move with the working surfaces without relative movements.
2. Positive sliding. The materials keep contact with the working surfaces and at the same time move in the conveying direction relatively to the working surfaces.
3. Negative sliding. The materials keep contact with the working surfaces and at the same time move in the opposite direction to the conveying direction relatively to the working surfaces.
4. Throwing movement. The materials are thrown away from the working surfaces and moving forward in a parabolic trajectory.

Of all the four movements, except the movement form of the relative still, three other movements can be used to accomplish the functions of feeding, conveying, screening and cooling etc.

In the nonlinear systems of the piecewise inertial forces, there are two different ways of conveying and screening:

1. The nonlinear systems with the piecewise inertial forces and friction (Fig. 2.23).
2. The nonlinear systems with the piecewise inertial forces and throwing, and frictions.

2.16 Applications of Piecewise Restoring Force

The following vibration systems with the piecewise restoring force will be discussed:

1. Piecewise linear harden type of restoring force of nonlinear vibration systems.
2. Piecewise linear harden type of nonlinear vibration systems.
3. Piecewise linear soften type of nonlinear vibration systems.

Fig. 2.24 Resonance curves for inertial type of resonance screen

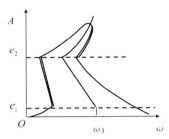

4. Complex piecewise linear type of nonlinear vibration systems.

Using the piecewise linearity in the nonlinear systems can increase the stability of the vibration amplitude of the systems, increase the technological effects of the equipments and obtain asymmetric movement trajectories.

Figure 2.24 shows a schematic of the resonance curves for a nonlinear system with multi-pieces of complex piecewise linearity. The working point can be chosen in a proper interval.

2.17 Utilization of Water Wave and Wind Wave

Water waves include the tidal waves, sea waves and shallow water waves etc. The tidal frequencies can be the lowest: there could be a big and a small tide in everyday. The tidal energy can be utilized for power generation. There is a power generation station in Jiangxia, Zhejiang Province, China. The tidal power generation must have a condition of storing the seawater, i.e. to have a bay of storing a large amount of seawater. When the tide rises the bay stores the water and then releases the water when the tide falls. The water level difference can be used for power generation.

Sea waves can be also used for power generation. A big sealed container is put on the sea surface. The bottom of the container is open and connected with the seawater. When the waves rise, the air in the container is compressed, and an anti-siphon valve is installed in a proper position. The compressed air drives the air turbine to rotate and thus generate power by a generator.

The airflow forms a wind. The wind is a form of energy. The amplitude and direction of the wind change constantly. If the horizontal axis represents the time in a Cartesian coordinate system, the vertical represents the wind direction, and if the north wind is taken as positive and the south as negative, then the wind flow would be represented by a vibration curve; the vertical can also be used to represent the wind speed, then a vibration curve can also be obtained. The wind due to the airflow is changing and vibrating constantly. The wind energy can be used to convert into electric power. The wind energy is one of the cleanest energy resources without any pollution.

The typhoons and hurricanes bring damages to human beings and properties. The energy in them could not be controlled yet from the science and technology nowadays. However the time in the future may come when the human beings may utilize them for our well-beings.

2.18 Applications of Tense or Elastic Waves

The vibration, including sound and light etc., can propagate in the matters. Vibration can be propagated in the solid matters by constantly changing of the stress and strains. The wave is called tense or elastic wave.

Using the tense wave to survey and exploration of the underground minerals or other materials is widely used in the mineral explorations.

The tense wave is used to detect the damages of piles. When a vibration signal is input at the top of a pile, the signal will propagate downward along the pile. When a crack or a fraction is defected in the pile, the signal receiver installed on top of the pile will receive a special waveform of a reflection wave due to the fraction, thus judging the pile has a fraction.

Vibration oil extraction is a classical example of effectively making use of the tense waves. The crude oil is generally deposited in depth of 3,000~5,000 m underground and is mixed with water. The crude oil pumped out from the underground contains a large amount of water. If the water contents can be decreased, the real oil output will be increased.

It is found that when an earthquake occurs near an oil field, the production of the crude oil will rapidly increase. This gives people an enlightenment that exerting the vibration to the oil field could increase the oil production about 30%. In 1920's we developed an ultra-low frequency, 80 ton excitation vibrating oil extraction equipment. This equipment has been used in a real oil extraction in an oil field and achieved a very good effect.

2.19 Utilization of Supersonic Theory and Technology

Sound wave can propagate in a variety of media. Audible frequency range for a human being is about 20~20,000 Hz. The sound whose frequency is lower than 20 Hz is called an infrasonic wave while the sound whose frequency is larger than 20 kHz is called supersonic wave. It is the wave which can propagate the sound that people could live in a mutually communicating social environment on the bases of the sound propagation. Were it not for the sound propagation that the world would have been a silent world. As matter of a fact human beings had been naturally made use of the wave principles and live in a rich and colorful environment.

The wavelength of the infrasonic wave is relatively long and not so easy to be reflected and refracted when it hits an object, not so easy to be absorbed and it propagates in a long distance. The infrasonic wave has been used not only to weather forecast, earthquake analysis and military reconnaissance but also to state monitoring of the mechanic equipment especially suitable for far field measurements. The infrasonic wave has potential applications in the biology, medicine science and agriculture etc.

The supersonic wave has strong penetration capability and well directional propagations. Its characteristics in the wave speed, attenuation and absorption in different media are different, thus is a widely used means of state monitoring and fault diagnostics of the equipment. The supersonic wave has also important applications in the production and everyday life such as oil–water mixture, cutting and machining, metal plastic machining, disease diagnostics and treatments etc.

2.19.1 The Application of the Supersonic Motor

Promotion and application of the piezo-supersonic motor are an important revolution for the compact, low-power and low speed motors are widely used in a variety of field and industry branches.

The piezo-supersonic motor is also called supersonic motor. The inverse piezoelectric effect of the piezoelectric material is used to make the micro vibration in the elastic body (stator). The micro vibration of the stator is converted into macro unidirectional rotation (or line movement) through the friction between the stator and rotator. The vibration frequencies are above 20 kHz and so the motor is called piezo-supersonic motor.

The movement mechanism, structure characters and practical experience indicate that the supersonic motor has the following advantages: compact and simple in structure, higher ratio of rotation moments to weight (is about 3~10 times that of the traditional motor); low speed and large torque, no need for reduction gear mechanism and direct drive; small inertial of the moving part (rotator), quick response (in the grade of milli-second); auto-lock when power is off and can be stopped completely; good speed and position control with high precision; no electric–magnetic field generated and no external magnetic field interference and good electric–magnetic compatibility; low noise operation; it can work in a severe environment conditions; it can easily be made to straight line supersonic motor; it has multi-shapes, such as round, square, hollow and bar etc.

Due to its characteristics above, the supersonic motors have been used in cameras, robots, automobiles, aeronautics, precision positioning device, micromachine etc.

2.19.2 Significance and Function in Medical Diagnostics of B-Ultrasound

The ultrasound technology has been developed very rapidly since it was found that the ultrasound could be used for detection and diagnostics on diseases in 1940's. Nowadays it has been widely used to diagnose the diseases for human organs. Human body's normal and abnormal tissues and organs absorb the ultrasound in different way and the reflection patterns are different. It is the principles that the ultrasound diagnostics is made use of. The ultrasound device generates ultrasound and emits into the human body and the ultrasound propagates in the human body. In its propagation when it encounters the tissues and organs it will reflect and scatter. The instrument receives the reflected and scattered waves, analyzes it and displays the signals in the forms of wave shapes, curves or images. Doctors use this information as basis for the judgments and diagnostics if the organs and tissues are healthy or not.

Unlike the ultrasound diagnostics, "inverse piezoelectric effect" principle is made use of for the ultrasound treatment and it exerts a high-frequency electric field upon the crystal thin slice and makes the latter generate the vibration of the corresponding frequency. When it is used for a specific treatment, it can be used in different manners: for example fixed touch method, the ultrasound probe is put on a fixed treatment position; moving touch method, the ultrasound probe slowly moving in a straight line around the treatment position; underwater radiation method, treatment in lukewarm water; acupoint treatment method, use a small ultrasound probe to stimulate the acupuncture point. The ultrasound method can also be combined with other methods.

2.20 Applications of Optical Fiber and Laser Technology

2.20.1 Application of the Optical Fiber Technology

The optical fiber is a conductor realizing the information transfer by the light transmitting in the quartz fibers. Since the wavelengths of a light can be adjusted, a fiber can be used to transmit multi-information. That is different from the conductor for electricity in which can only transmit one piece of information. It is low cost, high efficiency and convenient to use. It brought about an effective revolution in the communication technology.

The light itself is a wave. Its transmission in a fiberglass is a constant reflection and deflection process. It is a vibration (or wave) from the point of view of vibration theory.

The electric–magnetic wave contains a broad spectrum range, from the long wavelength of 10^5 km for the electric power transmission, to the short wavelength of 10^{-6} μm for the cosmic ray. Arranging the waves by wavelength from long to short: electric power transmission long wave, radio wave, microwave, and infrared wave (wavelength 0.76~1,000 μm), visible light wave (wavelength 0.4~0.76 μm), ultraviolet wave, X ray wave and γ ray wave etc. From the point of view of the wave and wave energy utilization, the electric–magnetic and light waves have already been used in industry, agriculture, military and every aspects of ordinary life. The electric–magnetic wave has been successfully used not only as a media for information transmission but also as a hopeful media for electric power transmission in the future. For example, the electric–magnetic wave is the indispensable transmission means for modern communications. The microwave can be used for information transmission and the microwave oven is a utensil for modern families; infrared wave has been used not only for disease treatment but also for the state monitoring and fault diagnostics of the mechanical equipments and for the reconnaissance of enemy's targets; the light from the sun has been used for heating and power generating.

2.20.2 Application of Laser Technology

The Laser Machining technology is a special machining technology of cutting, welding, heat treatment and micro-machining using of the character-pair material of the laser ray and material interaction (including metal and non-metal materials). The shape technology of the laser quick prototype is a rapidly developed branch of the laser machining technology family. The laser measurement device is a high precision measuring device using the laser wavelength as a basic unit and is the datum instrument for length measurement and has been used in the Computer Numerical Controlled (CNC) lathe and static and dynamic precision measurement of the advanced equipment.

Laser machining and measurement technology as one of the advanced manufacturing technologies play a more and more important role in the progress of machining technology reform of traditional industry, military modernization, development of auto industry and national economics development.

For traditional industries, such as electronics, automobile, steel, petroleum, ship building and aviation, laser technology can be used for their reforms; laser technology can provide new type, of laser equipment and instrument for high-tech fields such as information, material, biology, energy, space and ocean.

2.21 Utilizations of Ray Waves

X ray, β ray, γ ray are used in many branches and human's daily life. X ray has been used in the fluoroscope and medical CT. These two devices are very important medical devices used for diagnosing human body organs. Their successful development can be called a revolution in medical technology.

The fluoroscope can be used to look at the parts of human body and movement within the body. When some disease occurs in some parts and/or organs in a body, its different depth of spot mark can be displayed on an oscilloscope and doctors can judge if the parts or organs may have a disease or not. This is a very effective diagnosing means.

The medical CT is an advanced medical diagnostic and detection device and equipment for obtaining the part and organ section images by roentgen scope scanning. A rotating gantry is mounted on the CT scanner. The gantry has an X ray tube mounted on one side and an arc-shaped detector mounted on the other side. An X ray beam is emitted as the rotating frame spins the X ray tube and detector around a patient. Each time the X ray tube and detector make a 360-degree rotation and the X ray passes through the patient's body, the image of a thin section is acquired. During each rotation the detector records about 1,000 images (profiles) of the expanded X ray beam. A dedicated computer then reconstructs each profile into a two-dimensional image of the section scanned.

The difference between industry CT and medical CT is that the industry CT's main purpose is to detect defectives inside the metal parts. Since the X ray has weak penetration capability into a metal part to acquire clear-enough images, the γ ray, which is much stronger than X ray, is used in the industry CT. The other side of the coin is that γ ray is stronger than X ray many times, so it is suitable for human body detection.

β ray is also used in medical treatment to kill the harmful and abnormal cells to human body.

Those rays are used not only in medical branches but also in other departments.

2.22 Utilization of Oscillation Theory and Technology

The applications of oscillation principle to engineering technology have caused many important reforms in science technology and industry. The development and application of the electric–magnetic oscillators started communication and information technology and the development and the application of the quartz oscillator started a watch industry and timing industry revolution.

Whether the communication or information technology industry or timing industry, they all change in a great degree due to the new oscillating circuit invention.

The development of the sciences, the progress of the society and the raising living standards for people all require a more convenient, more accurate means and method for measuring time. In the end of the eighteenth century appearance of the mechanical watches hastened parturition of the big development of the modern manufacture industry and human civilization entered into the industrialized society. Along with the development of the electronic science and technology, the appearance and wide application of the quartz crystal oscillators especially in the middle of the twentieth century made the significant changes to the bell and watch manufacture industry. As the basic element for the modern communication, the adoption of the quartz oscillator dramatically increases the stability and accuracy of the signal and systems, lays a foundation for digital communication and leads the human civilization into the information and digital times.

The quartz crystal oscillator is the widely used precision element and has been developing in the direction of highly integrated, extremely low energy consumed, highly time-stabilized, high frequency accurate and multi-function derived productions. It has been used not only in the timing basis, but also in the micro-intelligent sensors, and micro-intelligent actuator. It is also a strong lever for the development of the multi-scientific branches and technology fields.

Quartz crystal oscillators, with high stability, high precision oscillating characteristics, replace the complex mechanical components, and constitutes a all-new timing tool- electronic watch. It has many advantages: novel, intuitive, compact, low cost, accurate, convenient so that it enters people's life, study and work.

The quartz crystal, concerted with CMOS circuit, constitutes a crystal oscillator with frequency 32,768 Hz. Its output signal forms a periodical impulsive signal with a period of 1 s after 2^{15} frequency-dividing downs. Then the impulsive signal is output through 3 grade-counters in terms of hours, minute and seconds and the time signals are displayed in either an LED (light emitting diodes) or an LCD (liquid crystal display) via coding circuit, driving and control circuits.

Now the electronic magnetic oscillation principles have been used widely in different branches. It is one of the important forms on which people rely in the material civilization and exists in micro and macro and universe world in different forms. It constitutes a bridge and a link for interaction among different matters.

The electric–magnetic oscillations can be categorized into electric field oscillation, magnetic filed oscillation, and mutual oscillation of the electric and magnetic fields. The electric field can be used to accelerate the speed of a particle movement; the magnetic field can be used to generate the wave movement for the particle movement and the mutual oscillation of the electric and magnetic fields can be used to electric–magnetic waves.

The feature that the electric field can accelerate the charged particle speed has its application values: in television set, the oscillation of the electric field is used to accelerate the electronics, the accelerated electronics are beamed to the screen to display the images; the high energy accelerator uses the alternative electric field to accelerate the charged particles to such a level that the particles carry an extremely high energy and high speed. These accelerated particles are used to impact a different matter component to obtain the structural characteristics and composition on the matters, analyze the secrets of the matter world and discover new patterns and the most basic elements. It can also be used to generate the nuclear chain reaction or fusion and the peaceful utilization of the nuclear energy.

The feature that oscillation of the magnetic field can make the charged particles deflexion has its application values: in television set, the oscillation of the magnetic field polarizes the high speed electronics and the electronics are beamed accurately at the corresponding spot on the screen to generate a clear image; the high energy accelerator uses the alternative electric field to polarize the high speed charged particles and generate the synchronous radiated light which forms a synchronous radiation light source. The alternative magnetic field makes the high energy electronics fluctuate and generate free electronic laser.

The core of the quartz watch, the CMOS crystal oscillator, is composed of the quartz oscillator, BW, SMOS reversing phaser, an assistance, and oscillation capacitor, etc. Its basic function is to generate a sinusoidal wave with a frequency as the time unit for a clock.

2.23 Utilization of Vibrating Phenomena and Patterns in Meteorology

The changes of the atmospheric pressure on earth surface make air flow. The flows of the air could generate winds in different directions and magnitudes. The winds can be make use of generating powers. This is one of the most cleaning energies in Nature. The hurricanes can bring a huge damages to human and properties sometimes. However they contain a huge amount of energy. As the development of science and technology, not for a long time, it will be made use of by human.

2.24 Utilization of Vibrating Phenomena and Patterns in Social Economy

There exist the vibrations (fluctuations) everywhere in social economy. We feel intuitively that all the economical phenomena are in the process of constant fluctuations. For example, the economical conditions in a country or a region are good sometimes and are bad in other times; the prices of the agriculture products rise sometimes and fall in the other times; the exchange rates, stock prices, fixed property investment, commercial product inventory and energy consumptions are all changing and fluctuating constantly as time goes. The statistical data show that the operation processes of the economical systems often fluctuate with statistical periods of different frequencies. The effective measures must be taken in the early stages to alleviate, so called "soft landing", or prevent if possible, the economical crises before the burst.

The occurrence of the financial and economical crises is a form of vibration phenomena. Taking active and effective measures to prolong the periods of the economy and finance growth and shorten the periods of the economical and financial crises is one of the effective methods to treat the vibration phenomena.

The economical fluctuations have their own patterns. The economists could infer the situations of the economical growth and development in a country or a region on the basis of the domestic and international economical and social factors, i.e. the vibration patterns in the economical development processes, thus propose the corresponding measures to reduce the losses of the social economy due to some factors.

People could predict rise and fall of a stock according the inner and outer influence factors and a general stock patterns. Mastering the vibration patterns in the stock fluctuations, people could benefit from it.

2.25 Utilizations of Vibrating Principles in Biology Engineering and Medical Equipments

In order to accurately detect and diagnose and effectively treat the diseases in human body, the scientists develop a variety of medical devices and equipments, of which many of them are using the vibration and wave principles. Here are some of them:

1. B-ultrasound. A variety of normal and abnormal tissues and organs in human body absorb, reflect and refract the ultrasound differently. The ultrasound diagnostic device generates and emits the ultrasound signal into a human body. When the signal encounters the tissues and organs it will be absorbed, reflected and scattered somehow. The device receives the signals and displays them in forms of wave patterns, curves and images. The doctors will interpret them to judge if the tissues and organs are healthy or not.
2. Ultrasound stone crush. The ultrasound can be used to crush calculus (stone) in human body, such as kidney stone, gallstone.
3. Electrocardiogram. Measuring the intensity of the current flowing through the device to detect the situation of the blood flowing through a heart.
4. Electroencephalogram (EEG). Measuring the intensity of the current flowing through the device to detect the situation of the blood flowing through a brain.
5. Fluoroscope. Can be used to look at the organs inside the human body.
6. CT. Medical CT uses the X ray to scan the human body to obtain the images of the tissue and organ sections.
7. MRI CT. MRI CT has much less damages to human body compared to fluoroscope.

The artificial human organs using vibration principles:

1. Artificial heart. When a heart becomes abnormal and can't be repaired, the artificial heart can replace the malfunctioned original biological heart. The artificial heart itself is a self-vibration exciter.
2. Cardiac pacemaker.When the pulse of the human is abnormal, such as heart beats are far less than the heart beats of a normal man, a cardiac pacemaker can be implanted into human body to help realize the normal heart beating. The pacemaker is a pulse generator. Its frequency is almost coincident to the heart beats of a normal person. It continuously triggers the heart and makes human's heart beat normally.
3. Hearing aid. The hearing aid is actually a sound amplifier. When it receives a sound from outside, it amplifies it and broadcast it to the ears.

Vibration Utilization Engineering has been widely used in many technical branches and industry departments. Its development is directlies dependent on the discoveries and applications of new principles and new technologies in vibration and wave utilization. For the future outlook new theoretical and technical inventions and discoveries will emerge in the field of vibration utilization engineering and will be widely used in a variety of industry fields and promote the science technology and industry develop further and benefit the society and human beings as a whole.

Chapter 3
Theory of Vibration Utilization Technology and Equipment Technological Process

3.1 Theory and Technological Parameter Computation of Material Movement on Line Vibration Machine

The technological processes of the vibrating machines used for material feeding, conveying, screening, cooling and dewater etc. are accomplished in general in the continuous movements of the materials along the vibrating working surfaces. The working quality of the vibrating machines is directly related to the movements of the materials. Expressing the movement theory of the materials on the vibrating working surfaces has its importance in correctly selecting the kinematics parameters and guaranteeing the effective operation of the technological process.

The working surfaces of the vibrating machines, in general, have the following vibrations to perform their functions: harmonic line vibrations, non-harmonic line vibrations, circle vibrations, oval vibrations, etc. When vibrating machines are designed by different parameters (vibration amplitudes, frequencies, vibration phase and inclining angles etc.), the materials on the working surfaces can realize the different forms of movements:

(1) Relative still. The materials move with the working surfaces without relative movements.
(2) Positive sliding. The materials keep contact with the working surfaces and at the same time move in the conveying direction relatively to the working surfaces.
(3) Negative sliding. The materials keep contact with the working surfaces and at the same time move in the opposite direction to the conveying direction relatively to the working surfaces.
(4) Throwing movement. The materials are thrown away from the working surfaces and moving forward in a parabolic trajectory.

Of all the four movements, except the movement form of the relative still, three other movements can be used to accomplish the functions of feeding, conveying, screening and cooling etc. Thus in analyzing the kinematics characteristics and electing and computing the kinematics parameters, the three working states must be considered all together. Among the three movement forms, the properties of the positive and negative sliding are similar and the principles are the same, but they are totally different from the throwing movement. The sliding and throwing movement will be discussed respectively as follows.

Besides, the different movement trajectories of the working surfaces on the vibrating machines lead to different basic formulas in computing the material movements, this section will first explain the theory of the material movement in the line vibrating machines. The circle and eclipse vibration machine theories will be dealt with in the following sections.

3.1.1 Theory of Sliding Movement of Materials

(1) Displacement, Velocity and Acceleration of the Working Surface

The line vibrating machines with harmonic movements include the vibrating feeder, vibrating conveyer, vibrating cooler and spiral vibrating feeder etc. with line movements.

For the dynamics of the line vibrating machines the displacement of the working surface of the machine can be expressed:

$$S = \lambda \sin \omega t \quad \omega t = \varphi \tag{3.1}$$

where λ is the single amplitude of the surface along the vibration direction; ω is the vibration angular frequency; t is the time and φ is the vibration phase angle.

Projecting the displacement onto the y direction (vertical to the working surface) and x direction (parallel to the working surface) obtains the displacements in y and x directions:

$$S_y = \lambda \sin \delta \sin \omega t \quad S_x = \lambda \cos \delta \sin \omega t \tag{3.2}$$

where δ is the angle between vibration direction and the working surface.

Taking the 1st and the 2nd order derivatives of Eq. (3.2) with respect to t gives the velocities v_y, v_x and accelerations a_y and a_x in y and x directions:

$$v_y = \lambda \omega \sin \delta \cos \omega t \quad v_x = \lambda \omega \cos \delta \cos \omega t \tag{3.3}$$

$$a_y = -\lambda \omega^2 \sin \delta \sin \omega t \quad a_x = -\lambda \omega^2 \cos \delta \sin \omega t \tag{3.4}$$

3.1 Theory and Technological Parameter Computation ...

Fig. 3.1 Force analysis and law of motion of vibration working surface

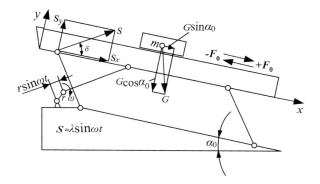

(2) Positive Sliding Index D_k and Negative Sliding Index D_q

First we study the movement of a thin layer material on the surface (the interaction between particles can be ignored in this condition). When the layer becomes thick the correction can be made according to the test.

Figure 3.1 shows the force analysis. Assume that the material moves relative to the surface, the relative displacements are Δy and Δx, the corresponding velocities, $\Delta \dot{y}$ and $\Delta \dot{x}$ and accelerations $\Delta \ddot{y}$ and $\Delta \ddot{x}$. The force balance in the x direction:

$$F = -m(a_x + \Delta \ddot{x}) + G \sin \alpha_0 \qquad (3.5)$$

While the normal pressure of the particle acting on the surface in the y direction:

$$F_n = m(a_y + \Delta \ddot{y}) + G \cos \alpha_0 \qquad (3.6)$$

where m, G are the mass and gravity of the material particle respectively; α_0 is the inclining angle of the working surface.

When the material slides it keeps touch with the surface and the normal pressure $F_n \geq 0$, relative acceleration $\Delta \ddot{y} = 0$. When it is thrown, the normal pressure $F_n = 0$, relative acceleration $\Delta \ddot{y} \neq 0$.

When the material keeps touch with the surface, the friction limit of the working surface to the material is:

$$F_0 = \mp f_0 F_n \qquad (3.7)$$

where f_0 is the static friction coefficient of the material to the working surface.

In the symbol $\pm$, "$-$" indicates the positive sliding, "$+$" denotes the negative sliding because in the positive sling the friction force is opposite to the x direction and the negative sling friction force is in the same direction as the x direction.

At the instant time when the sliding occurs, the relative acceleration of the material to the working surface $\Delta \ddot{x} = 0$. Since the throwing doesn't occur, $\Delta \ddot{y} = 0$. Thus in Eq. 3.5 sum of F and friction limit should be zero:

$$F + F_0 = 0 \qquad (3.8)$$

Substituting Eqs. (3.5) and (3.7) into (3.8) and substituting Eqs. (3.6) and (3.4) give:

$$m\omega^2 \lambda \cos \delta \sin \omega t + G \sin \alpha_0 \mp f_0(-m\omega^2 \lambda \sin \delta \sin \omega t + G \cos \alpha_0) = 0 \qquad (3.9)$$

Since $f_0 = \tan\mu_0$ (μ_0 is the friction angle), $G = mg$, substituting them into Eq. (3.9) and simplifying it obtain the positive beginning sliding phase angle φ_{k0} (simply called it a positive sliding starting angle) and the negative beginning sliding phase angle φ_{q0} (simply called it a negative sliding starting angle) as:

$$\varphi_{k0} = \arcsin \frac{1}{D_k} \quad \varphi_{q0} = \arcsin \frac{1}{D_q} \qquad (3.10)$$

The positive and negative sliding indices are respectively:

$$D_k = K \frac{\cos(\mu_0 - \delta)}{\sin(\mu_0 - \alpha_0)} \quad D_q = K \frac{\cos(\mu_0 + \delta)}{\sin(\mu_0 + \alpha_0)} \qquad (3.11)$$

where $K = \omega^2 \lambda / g$, K is the vibration strength (or mechanical index); μ_0 is the static friction angle and g is the gravity acceleration.

For most vibrating machines, $\mu_0 \pm \alpha_0 = 0°–180°$, $\mu_0 \pm \delta = -90°–90°$, $\sin(\mu_0 \pm \alpha_0)$ and $\cos(\mu_0 \pm \delta)$ are all positive. From Eqs. (3.10) and (3.11) the positive sliding beginning angle φ_{k0} is in the range of $0°–180°$ while the negative sliding; beginning angle φ_{q0} is in the range of $180°–360°$. That means that the velocity v_x curve computed from Eq. (3.3) exists only within the first half period of the vibration period. When φ is in the range of $180° - \varphi_{k0}$, the material begins positive sliding; when φ is in the range of $540° - \varphi_{q0}$, the material begins negative sliding (Fig. 3.2).

When the positive sliding index $D_k < 1$ it can be seen from Eq. (3.10) that there is no solution for φ_{k0}. So the positive sliding condition is $D_k > 1$.

When the negative sliding index $D_q < 1$ material cannot negatively slide. So the negative sliding condition is $D_q > 1$.

Since the negative sliding has no direct importance for most of the vibrating machines, it is hoped in general that the large positive sliding occurs. In common practice the positive and negative sliding indices D_k and D_q are selected first in designs. The guided values are $D_k = 2–3$, $D_q \approx 1$. Then the vibration direction angle δ must be derived from Eq. (3.11) as:

$$\frac{D_k}{D_q} = \frac{\sin(\mu_0 + \alpha_0) \cos(\mu_0 - \delta)}{\sin(\mu_0 - \alpha_0) \cos(\mu_0 + \delta)} \qquad (3.12)$$

After simplifying:

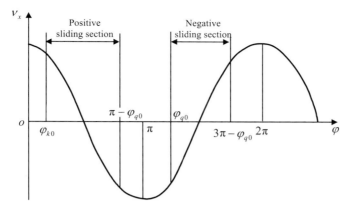

Fig. 3.2 Velocity curve of x direction of working surface and positive sliding angle φ_{k0} and negative sliding angle φ_{q0}

$$\delta = \arctan\frac{1-c}{(1+c)f_0} \quad (3.13)$$

in which $c = \frac{D_q}{D_k}\frac{\sin(\mu_0+\alpha_0)}{\sin(\mu_0-\alpha_0)}$.

When the positive and negative sliding indices D_k and D_q are selected and the vibrating direction angle obtained, the mechanical intensity needed (or mechanical index) can be determined from Eqs. (3.14a, 3.14b):

$$K = D_k\frac{\sin(\mu_0-\alpha_0)}{\cos(\mu_0-\delta)} \quad (3.14a)$$

or

$$K = D_q\frac{\sin(\mu_0+\alpha_0)}{\cos(\mu_0+\delta)} \quad (3.14b)$$

Since $K = \frac{\omega^2\lambda}{g}$, $\omega = \frac{2\pi n}{60}$ (n is the revolution number per minute). After the amplitude is selected, the vibration number needed from Eqs. (3.14a, 3.14b) is:

$$n = 30\sqrt{\frac{D_k g \sin(\mu_0-\alpha_0)}{\pi^2\lambda\cos(\mu_0-\delta)}} \quad (3.15a)$$

or

$$n = 30\sqrt{\frac{D_q g \sin(\mu_0+\alpha_0)}{\pi^2\lambda\cos(\mu_0+\delta)}} \quad (3.15b)$$

If the vibration number n is selected first, then one-sided vibration amplitude needed is:

$$\lambda = 900 \frac{D_k g \sin(\mu_0 - \alpha_0)}{\pi^2 n^2 \cos(\mu_0 - \delta)} \quad (3.16a)$$

or

$$\lambda = 900 \frac{D_q g \sin(\mu_0 + \alpha_0)}{\pi^2 n^2 \cos(\mu_0 + \delta)} \quad (3.16b)$$

(3) Positive (negative) Sliding Angle and Positive (negative) Sliding Coefficient

The time elapsed from the positive sliding beginning to its end is called positive sliding time, denoted as t'_{mk}, $t'_{mk} = t'_m - t'_k$ (t'_m and t'_k are the time of positive sliding end and positive sliding start respectively). Within the time period t'_{mk} the phase angle that the working surface vibration experienced is called the positive sliding angle, denoted as θ_k, $\theta_k = \varphi'_m - \varphi'_k = \omega(t'_m - t'_k)$. The positive sliding time t'_{mk} is directly proportional to the vibration period $2\pi/\omega$, and is called the positive sliding coefficient, denoted as i_k, then

$$i_k = \frac{t'_{mk}}{\frac{2\pi}{\omega}} = \frac{\varphi'_m - \varphi'_k}{2\pi} = \frac{\theta_k}{2\pi} \quad (3.17)$$

where φ'_m is the real positive sliding ending angle and φ'_k is the real positive sliding starting angle.

After the material starts positively sliding, its equation of motion is:

$$m(a_x + \Delta \ddot{x}) = G \sin \alpha_0 - f F_n \quad (3.18)$$

where f is the dynamic friction coefficient of the material to the working surface.

Substituting the normal pressure F_n in Eq. (3.6) into Eq. (3.18) and expressing the dynamic friction coefficient f by the dynamic friction angle μ, i.e. $f = \tan\mu$, then we have:

$$m\Delta \ddot{x} = -m a_x + G \sin \alpha_0 - \tan \mu (m a_y + G \cos \alpha_0)$$

Substituting a_y, a_x in Eq. (3.4) into equation above, the relative velocity of the positive sliding can be computed as follows:

$$\Delta \dot{x}_k = \int_{t'_k}^{t} \Delta \ddot{x} dt = g(\sin \alpha_0 - \tan \mu \cos \alpha_0) \frac{\varphi - \varphi'_k}{\omega} - \omega \lambda \sin \delta \tan \mu$$
$$\times (\cos \varphi - \cos \varphi'_k) - \omega \lambda \cos \delta (\cos \varphi - \cos \varphi'_k)$$

After simplifying the equation above is:

$$\Delta \dot{x}_k = \frac{\omega \lambda \cos(\mu - \delta)}{\cos \mu} [\cos \varphi'_k - \cos \varphi - \sin \varphi_k (\varphi - \varphi'_k)] \quad (3.19)$$

in which

$$\sin \varphi_k = \frac{\sin(\mu - \alpha_0)}{K \cos(\mu - \delta)} \quad (3.20)$$

in which φ_k is the assumed positive sliding starting angle.

When the static friction coefficient is equal to the dynamic friction coefficient, $\mu_0 = \mu$, $\varphi_k = \varphi_{k0}$. In general, $\mu < \mu_0$, thus φ_k is smaller than φ_{k0}.

For most of the vibrating machines, the working surface incline angle $\alpha_0 < \mu$, when $\varphi = \varphi'_m$ after a while, the relative velocity $\Delta \dot{x} = 0$, the positive sliding ends. According to Eq. (3.19), the condition for ending the positive sliding is:

$$\cos \varphi'_k - \cos \varphi'_m - \sin \varphi_k (\varphi'_m - \varphi'_k) = 0 \quad (3.21)$$

Since $\varphi'_m = \varphi'_k + \theta_k$, Eq. (3.21) can be simplified and the relation between the real positive sliding starting angle φ'_k and the positive sliding angle θ_k is:

$$\tan \varphi'_k = \frac{1 - \cos \theta_k}{\frac{\sin \varphi_k}{\sin \varphi'_k} \theta_k - \sin \theta_k} = \frac{1 - \cos 2\pi i_k}{\frac{\sin \varphi_k}{\sin \varphi'_k} 2\pi i_k - \sin 2\pi i_k} \quad (3.22)$$

Figure 3.3 shows the relation curves among the real sliding starting angle $\varphi'_k(\varphi'_q)$, the real sliding ending angle $\varphi'_m(\varphi'_e)$, and the velocity coefficients $P_{km}(P_{qe})$. Assume that the positive sliding starting angle φ_k can be computed from (3.20), the real positive sliding starting angle $\varphi_k{'}$ can be determined by the material movement state, it's equal to the minimum positive sliding starting angle φ_{k0}, or in the range of $\varphi_{k0} \sim (180° - \varphi_{k0})$. After φ_k and φ'_k are determined, the positive sliding ending angle φ'_m can be determined directly from the Fig. 3.3. Since the positive sliding ending angle $\varphi'_m \leq 360°$, the positive sliding angle is definitely larger than 360°.

For example, given $\varphi'_k = 23°30'$, $\varphi_k = 15°30'$, from Fig.3.3, $\varphi'_m = 258°30'$, $P_{km} = 3$. In Fig. 3.3, P is P_{km} or P_{qe}. By the same method, the negative sliding angle θ_q and negative sliding ending angle φ'_e can be obtained.

The negative sliding equation can be written as:

$$m(a_x + \Delta \ddot{x}) = G \sin \alpha_0 + f F_n \quad (3.23)$$

Substituting Eqs. (3.4) and (3.6) into Eq. (3.23) the relative velocity of the negative sliding can be obtained:

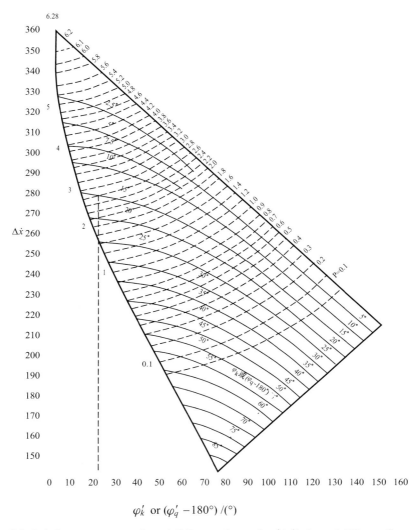

Fig. 3.3 Relation curves among the real sliding starting angle $\varphi'_k(\varphi'_q)$, the real sliding ending angle $\varphi'_m(\varphi'_e)$, and the velocity coefficients $P_{km}(P_{qe})$

$$\Delta \dot{x} = \frac{\omega \lambda \cos(\mu + \delta)}{\cos \mu} [\cos \varphi'_q - \cos \varphi - \sin \varphi_q (\varphi - \varphi'_q)] \quad (3.24)$$

in which

$$\sin \varphi_q = -\frac{\sin(\mu + \alpha_0)}{K \cos(\mu + \delta)} \quad (3.25)$$

3.1 Theory and Technological Parameter Computation ...

where φ'_q is the real negative sliding starting angle; φ_q is the assumed negative sliding starting angle.

When φ is equal to the negative sliding ending angle φ_e', the negative sliding relative velocity $\Delta \dot{x}$ is equal to zero and the negative sliding is ended. Thus the relation between the real negative sliding starting angel φ'_q and the negative sliding angle θ_q can be obtained:

$$\tan \varphi'_q = \frac{1 - \cos \theta_q}{\frac{\sin \varphi_q}{\sin \varphi'_q} \theta_q - \sin \theta_q} = \frac{1 - \cos 2\pi i_q}{\frac{\sin \varphi_q}{\sin \varphi'_q} 2\pi i_q - \sin 2\pi i_q}$$

in which

$$\varphi'_e = \varphi'_q + \theta_q \qquad i_q = \frac{\theta_q}{2\pi} \qquad (3.26)$$

where φ'_e is the real negative sliding ending angle; θ_q is the negative sliding angle; i_q is the negative sliding coefficient.

The assumed negative sliding starting angle φ_q can be obtained from Eq. (3.25), the real negative sliding starting angle φ'_q is related to the material movement state, it may be equal to the minimum negative sliding starting angle φ_{q0}, or in the range of $\varphi_{q0} \sim (540° - \varphi_{q0})$. After φ_q and φ_q' are determined, the negative sliding ending angle φ'_e can be obtained from Fig. 3.3. However, when using Fig. 3.3, φ_k must be replaced by $\varphi_q - 180°$, and φ'_k by $\varphi'_q - 180°$. The real φ'_e is obtained by summing φ'_m and $180°$.

(4) Averaged Velocity of the Positive and Negative Sliding

Integrating the relative velocity of the positive sliding with respect to time t results in the relative displacement Δx_k, from the positive sliding staring time t'_k to the sliding end time t'_m the displacement Δx_{km} of the material relative to the working surface divided by the vibrating period $2\pi/\omega$ gives the averaged relative velocity of the positive sliding of the material:

$$v_k = \frac{\omega}{2\pi} \int_{t'_k}^{t'_m} \Delta \dot{x}_k dt = \frac{\omega \lambda}{2\pi} \frac{\cos(\mu - \delta)}{\cos \mu} \int_{\varphi'_k}^{\varphi'_m} [-\cos \varphi + \cos \varphi'_k - \sin \varphi_k (\varphi - \varphi'_k)] d\varphi$$

$$= \frac{\omega \lambda}{2\pi} \frac{\cos(\mu - \delta)}{\cos \mu} [-(\sin \varphi'_m - \sin \varphi'_k) + \cos \varphi'_k (\varphi'_m - \varphi'_k) - \sin \varphi_k \frac{(\varphi'_m - \varphi'_k)^2}{2}]$$

According to conditions of Eq. (3.21), the term in the above equation can be written as

$$\cos \varphi'_k (\varphi'_m - \varphi'_k) - \sin \varphi_k \frac{(\varphi'_m - \varphi'_k)^2}{2} = \frac{\sin^2 \varphi'_m - \sin^2 \varphi'_k}{2 \sin \varphi_k}$$

Assuming

$$\sin \varphi'_k = b'_k, \quad \sin \varphi'_m = b'_m, \quad \sin \varphi_k = b_k$$

then the averaged velocity will be:

$$\begin{aligned} v_k &= \omega\lambda \frac{\cos(\mu-\delta)}{2\pi \cos\mu} \left[\frac{b'^2_m - b'^2_k}{2b_k} - (b'_m - b'_k) \right] \\ &= \omega\lambda \cos\delta (1 + \tan\mu \tan\delta) \frac{P_{km}}{2\pi} \end{aligned} \qquad (3.27)$$

in which

$$P_{km} = \frac{b'^2_m - b'^2_k}{2b_k} - (b'_m - b'_k)$$

According to φ_k and $\varphi_{k'}$, P_{km} can be found directly from Fig. 3.3. When $\varphi'_k = \varphi_k$, the velocity coefficient is:

$$P_{km} = \frac{(b_m - b_k)^2}{2b_k}$$

Using the same method, the averaged negative sliding velocity can be obtained as:

$$\begin{aligned} v_q &= \frac{\omega}{2\pi} \int_{t'_q}^{t'_e} \Delta \dot{x}_q dt = \frac{\omega\lambda}{2\pi} \frac{\cos(\mu+\delta)}{\cos\mu} \left[\frac{b'^2_e - b'^2_q}{2b_q} - (b'_e - b'_q) \right] \\ &= -\omega\lambda \cos\delta (1 - \tan\mu \tan\delta) \frac{P_{qe}}{2\pi} \end{aligned} \qquad (3.28)$$

where

$$-P_{qe} = \frac{b'^2_e - b'^2_q}{2b_q} - (b'_e - b'_q), \quad \sin\varphi'_e = b'_e, \quad \sin\varphi'_q = b'_q, \quad \sin\varphi_q = b_q \qquad (3.29)$$

When $\varphi'_q = \varphi_{q0} = \varphi_q$, the velocity coefficient P_{qe} is

$$-P_{qe} = \frac{(b_e - b_q)^2}{2b_q} \qquad (3.30)$$

From the φ'_q and φ_q, the P_{qe} can be found directly from Fig. 3.3.

3.1 Theory and Technological Parameter Computation … 67

(5) Relationship Between the Sliding Motion State and Kinematics of Material Sliding

On the variety of line vibrating machines on which the materials are sliding, the different sets of the kinematics parameters α_0, δ, λ and ω will determine the four following states of the motions as shown in Fig. 3.4.

The differences among the operation states are that the combination patterns are different of three basic operation states: relative still, positive and negative sliding. The operation states are used in different vibrating machines.

i. The operation states of the positive and negative sliding in one direction (Fig. 3.4a). Trough type of vibrating coolers and the vibrating centrifugal hydroextractor are all operating in this state. The operating state is the combination of two basic operation states: the relative still and positive sliding. $D_k > 1$, $D_q < 1$, $D < 1$ (D is the throwing index); the real p; positive sliding starting angle φ'_k is equal to the minimum positive sliding starting angle φ_{k0} and its theoretical average velocity can be computed from Eq. (3.27).

ii. The operation state in which there are two discontinuities in the positive and negative sliding (Fig. 3.4b). The state is used in the trough type of vibration

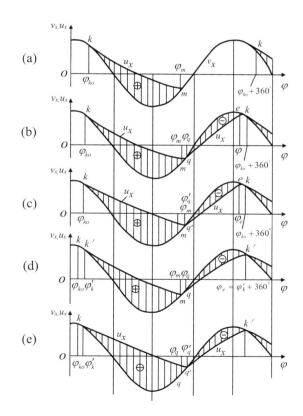

Fig. 3.4 Material sliding movement states. **a** motion state for positive motion; **b** there are two intermittent between positive and negative sliding; **c** and **d** there is an intermittent between positive and negative sliding

coolers and low speed vibrating screens and a few of the vibrating conveyors. The negative sliding is not significant, even harmful sometimes, it will still occur because the kinematics parameters are constrained in some conditions. The characteristics of the state are $D_k > 1$, $D_q > 1$, $D < 1$. Its basic pattern combinations of the movements are relative still—positive sliding—relative still—negative sliding; the minimum negative sliding starting angle φ_{q0} is larger than the positive sliding end angle φ'_m, the minimum positive sliding starting angle φ_{k0} is larger than the negative sliding end angle $\varphi'_e - 360°$, i.e. $\varphi_{q0} > \varphi'_m$, $\varphi_{k0} > \varphi'_e - 360°$. So $\varphi'_k = \varphi_{k0}$, $\varphi'_q = \varphi_{q0}$, its theoretical average speed is equal to sum of the averaged velocity of the positive sliding [Eq. (3.27)] and the average velocity of the negative sliding [Eq. (3.28)].

iii. The operation state in which there is only one discontinuity in the positive and negative sliding (Fig. 3.4c, d). The necessary conditions for this state to occur are: $D_k > 1$, $D_q > 1$, and $D < 1$. The first combination of the operation states is: relative still—positive sliding—negative sliding—relative still, i.e. the real positive sliding end angle φ'_m, is equal to the real negative sliding starting angle φ'_q, while the negative sliding end angle $\varphi'_e < \varphi_{k0} + 360°$, so that $\varphi'_k = \varphi_{k0}$ and $\varphi'_q > \varphi_{q0}$. The second combination of the operation states is: positive sliding—relative still—negative sliding—positive sliding, i.e. $\varphi'_m < \varphi_{q0}$, so $\varphi'_q = \varphi_{q0}$, while $\varphi'_e > \varphi_{k0} + 360°$ so that $\varphi'_k = \varphi'_e - 360°$. Its theoretical average speed is equal to sum of the averaged velocity of the positive sliding [Eq. (3.27)] and the averaged velocity of the negative sliding [Eq. (3.28)].

iv. The operation state in which there is no discontinuity in the positive and negative sliding (Fig. 3.4e). The necessary conditions for this state to occur are: $D_k > 1$, $D_q > 1$, and $D < 1$, at the same time, $\varphi'_k > \varphi_{k0}$, $\varphi'_q > \varphi_{q0}$, since $\varphi'_k > \varphi_{k0}$, $\varphi'_e > \varphi_{k0} + 360°$, so $\varphi'_q = \varphi'_m$, $\varphi'_k = \varphi'_e - 360°$. In order to find out the φ_k' and φ_q in the stable conditions, one must first find out the φ_k' and φ_q with the given φ_k and φ_{k0} from Fig. 3.3, then find out the positive sliding end angle φ_m' in the first cycle, replacing the φ_q' with the φ_m' (since $\varphi_m' > \varphi_{q0}$) just found to compute the φ_q. Then find out the φ_e' in the first cycle, replacing φ_e' with $\varphi_e' - 360°$ (since $\varphi_e' - 360° > \varphi_{k0}$) and using φ_k to find out the φ_m' in the second cycle. Replacing φ_q' with φ_m' and making use of φ_q' to find out φ_e' in the second cycle. In the sequence and after several cycles, φ_k', φ_m' φ_q' and φ_e' after the stable operations can be found, then use them to compute the theoretical velocity by Eqs. (3.27) and (3.28). To distinguish the operation states of the vibrating machines on which the materials are sliding, one must find out the φ_{k0}, φ_k, φ_{q0} and φ_q, using the curves in Fig. 3.3 to look up the sliding end angles φ_k' and φ_e'. Then using the values of φ_m', φ_e', φ_{k0}' and φ_{q0}' one can determine the operation states of the material sliding.

When the dynamic friction coefficient is equal to the static friction coefficient, the four operation states mentioned above can be directly determined by the values of D_k and D_q (or φ_{k0} and φ_{q0}) and the variety of states shown in the regions of Fig. 3.5.

Fig. 3.5 State regions of the sliding operation states

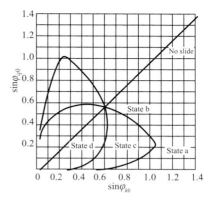

3.1.2 Theory of Material Throwing Movement

(1) Throwing Index D

The relative acceleration $\Delta y = 0$ along the y direction and the normal force $F_n = 0$ at the instance when the materials start to be thrown. The normal force F_n can be computed by Eqs. (3.6) and (3.4):

$$F_n = -m\omega^2 \lambda \sin\delta \sin\varphi_d + G\cos\alpha_0 = 0$$

in which φ_d is the vibration phase angle when the material starts to be thrown.

The vibration phase angle φ_d can be computed from the equation below:

$$\varphi_d = \arcsin \frac{1}{D} \tag{3.31}$$

in which

$$D = K \frac{\sin\delta}{\cos\alpha_0}$$

When the throwing index $D > 1$, φ_d in Eq. (3.31) is solvable and material may be thrown and the throw starting angle φ_d is in the range of 0°–180°.

When $D < 1$, φ_d in Eq. (3.31) has no solutions and the materials can not be thrown.

The vibration strength $K = \omega^2\lambda/g$, $\omega = 2\pi n/60$, given the vibration amplitude λ, the vibration number can be found:

$$n = 30\sqrt{\frac{Dg\cos\alpha_0}{\pi^2\lambda\sin\delta}} \tag{3.32}$$

If the vibration number n is selected first, the vibration amplitude is:

$$\lambda = \frac{900 D g \cos \alpha_0}{\pi^2 n^2 \sin \delta} \tag{3.33}$$

(2) The Throwing Angle θ_d and Throwing Index i_D

After the material leave the working surface, the normal force F_n must be zero. Substituting Eq. (3.4) into Eq. (3.6) yields the relative movement equation of the material in the vertical direction to the working surface:

$$m \Delta \ddot{y} = -G \cos \alpha_0 + m \omega^2 \lambda \sin \delta \sin \varphi \tag{3.34}$$

The displacements of the material relative to the working surface can be obtained by integrating the relative acceleration with respect to the time twice:

$$\Delta y = \lambda \sin \delta (\sin \varphi_d - \sin \varphi) + \lambda \sin \delta \cos \varphi_d \times (\varphi - \varphi_d) - \frac{1}{2} g \cos \alpha_0 \frac{(\varphi - \varphi_d)^2}{\omega^2} \tag{3.35}$$

When the relative displacement in the y direction is equal to zero again, the throwing movement ends. The vibration phase angle $\varphi = \varphi_z$ (φ_z is called throwing end angle). The difference between throwing end angle φ_z and the throw starting angle φ_d is called the throw-leaving angle $\theta_d = \varphi_z - \varphi_d$ or $\varphi_z = \varphi_d + \theta_d$.

The ending condition of the material throwing movement ($\varphi \to \varphi_z, \Delta y = 0$) can be found from Eq. (3.35):

$$\sin(\varphi_d + \theta_d) = \sin \varphi_d + \theta_d \cos \varphi_d - \frac{\theta_d^2 \sin \varphi_d}{2}$$

or

$$\cot \varphi_d = \frac{\frac{1}{2} \theta_d^2 - (1 - \cos \theta_d)}{\theta_d - \sin \theta_d} \tag{3.36}$$

Equation (3.36) describes the relation of the throw-starting angle φ_d and the throw-leaving angle θ_d.

If φ_d has been found out from Eq. (3.31), θ_d can be found out from Fig. 3.6.

The relation between throwing index D and the throw-starting angle φ_d exists as follows:

$$\cot \varphi_d = \frac{\cos \varphi_d}{\sin \varphi_d} = \sqrt{\frac{1}{\sin^2 \varphi_d} - 1} = \sqrt{D^2 - 1}$$

or

$$D = \sqrt{\cot^2 \varphi_d + 1} \tag{3.37}$$

3.1 Theory and Technological Parameter Computation ...

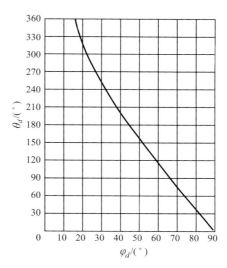

Fig. 3.6 Relation of the throw-starting angle φ_d and the throw-leaving angle θ_d

The relation between throw-leaving coefficient i_D (i.e. the ratio of the time it takes for one throwing to the one vibration period) and the throw-leaving angle θ_d exists as follows:

$$i_D = \frac{\theta_d}{2\pi} \qquad (3.38)$$

Substituting Eqs. (3.37) and (3.38) into Eq. (3.36) yields the relation between the throwing index D and the throw-leaving coefficient i_D:

$$D = \sqrt{\left(\frac{2\pi^2 i_D^2 + \cos 2\pi i_D - 1}{2\pi i_D - \sin 2\pi i_D}\right)^2 + 1} \qquad (3.39)$$

A curve can be drown to show the relation between D and i_D by Eq. (3.39) as shown in Fig. 3.7. According to D value, i_D can be computed, or given value i_D to compute D. When $i_D = 0, D = 1$; when $i_D = 1, D = 3.3$; when $i_D = 2$ or 3 $D = 6.36$ or 9.48. For most of the vibrating machines based on the throwing principles, it usually takes $D < 3.3$ i.e. the material will be thrown once for every vibration cycle of the working surface. This operation state will be beneficial for reducing the unnecessary energy consumption and increasing the working efficiency of the machines.

(3) The Theoretical Velocity of the Material Throwing Movements

After the materials are thrown (leaving the working surface), the equation of the motion along x direction is:

$$m(a_x + \Delta \ddot{x}) = mg \sin \alpha_0$$

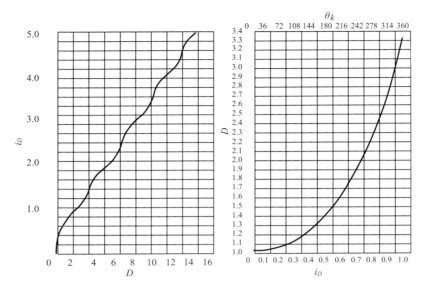

Fig. 3.7 Relation between throwing index D and throw-leaving coefficient i_D

i.e.

$$\Delta \ddot{x} = g \sin \alpha_0 + \omega^2 \lambda \cos \delta \sin \varphi \tag{3.40}$$

Integrating Δx twice with respect to time and replacing φ with φ_d yield the relative displacement of each throw movement:

$$\Delta x_z = \lambda \cos \delta (-\sin \varphi_z + \sin \varphi_d + \theta_d \cos \varphi_d + \frac{g \sin \alpha_0}{2\omega^2 \lambda \cos \delta} \theta_d^2)$$

According to Eq. (3.36):

$$-\sin \varphi_z + \sin \varphi_d + \theta_d \cos \varphi_d = \frac{1}{2} \theta_d^2 \sin \varphi_d$$

So that the relative displacement of each throw movement is:

$$\Delta x_z = \frac{1}{2} (\lambda \cos \delta) \theta_d^2 \sin \varphi_d (1 + \tan \alpha_0 \tan \delta)$$
$$= (\lambda \cos \delta) 2\pi^2 i_D^2 \frac{1}{D} (1 + \tan \alpha_0 \tan \delta) \tag{3.41}$$

The theoretical average velocity of the material throwing movement is equal to the relative displacement of each throwing movement divided by the vibration period $2\pi/\omega$:

3.1 Theory and Technological Parameter Computation ...

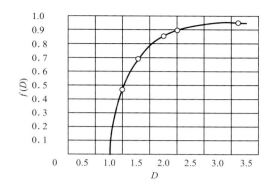

Fig. 3.8 Relation between the function $f(D)$ and D

$$v_d = \frac{\Delta x_z}{\frac{2\pi}{\omega}} = \omega \lambda \cos \delta \frac{\pi i_D^2}{D}(1 + \tan \alpha_0 \tan \delta) \quad (3.42)$$

When the working incline angle $\alpha_0 = 0$, the theoretical relative velocity is:

$$v_d = \omega \lambda \cos \delta \frac{\pi i_D^2}{D} \quad (3.43)$$

Take the dimension of 1 in the coefficient $\frac{\pi i_D^2}{D}$, when $D \leq 3.3$, i.e. $i_D \leq 1$, its maximum is:

$$\frac{\pi i_D^2}{D} = \frac{\pi \times 1^2}{3.3} = 0.95$$

It is known that when $D \leq 3.3$, the maximum theoretical average velocity of the material throwing movement on a horizontal vibrating conveyor does not exceed 0.95 times of the maximum velocity $\omega \lambda \cos \delta$ of the working surface in x direction. The $\omega \lambda \cos \delta$ is called the speed limit of the throwing movement:

$$v_l = \omega \lambda \cos \delta \quad (3.44)$$

The theoretical average velocity can be expressed by a product of a function $f(D)$ and the speed limit $\omega \lambda \cos \delta$:

$$v_{dl} = f(D) \omega \lambda \cos \delta \quad (3.45)$$

in which

$$f(D) = \frac{\pi i_D^2}{D}$$

$f(D)$ can be drown as a curve with a variable of D as shown in Fig. 3.8. It can be seen that when $D = 2–3.3$, $f(D)$ is in the range of 0.86–0.95. For the horizontal vibrating machine with $D = 2$–3.3, its average velocity of the material can be approximated as:

$$v_d = 0.9\omega\lambda \cos\delta \qquad (3.46)$$

The computing error using Eq. (3.46) is, in general, less than 5%. This is permissible for the general-purpose machines. When accurate results are required for the theoretical velocity designs, Eq. (3.45) must be used.

(4) Categories of the Material Throwing Movements

i. Classify with the values of throwing index D. The light throwing movement ($D = 1$–1.75) and the quick throwing movement ($D > 1.75$) (Fig. 3.9). In the former case it accompanies with a large positive sliding; for the latter case, the sliding movement is very light and even can be ignored.

Figure 3.10 is drown by Eq. (3.25). The figure shows the displacement in y direction and velocity in x direction of the material and the working surface. Some sliding movements occur before and after the throwing. It can be seen from Fig. 3.10 that the material sliding movements have some effects on the conveying velocity. In addition, when the material falls on the working surface, impacts on it and generates a instance friction, the sliding velocity of the material after the ending of sliding is smaller than that of the material before the impact.

In the quick throwing states, the states can be classified into two states: mid-speed throwing states ($D = 1.75$–3.3) and high-speed throwing state ($D > 3.3$).

Most of the vibrating machines are working in the mid-speed throwing state (Fig. 3.9b), for example the vibrating feeder, vibrating conveyers, and vibrating screen etc. When the throwing index $D = 1.75$–3.3, the throwing leave angle

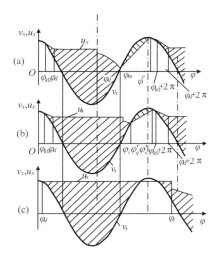

Fig. 3.9 The throwing movement states. **a** light throwing movement, **b** mid-speed throwing movement, **c** high-speed throwing movement

3.1 Theory and Technological Parameter Computation …

Fig. 3.10 Displacement in y direction and velocity in x direction of the material and working surface. **a** x direction velocity curve, **b** y direction displacement curve

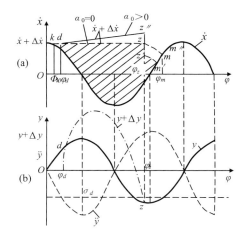

$\theta_d = 220°–360°$, the throwing leave coefficient $i_D = 0.67–1$, its theoretical average velocity can be computed by Eq. (3.42).

For the high-speed throwing state, the throw index $D > 3.3$ and the throwing leave coefficient $i_D > 1$. For the materials containing soils and some material, which is difficult to treat, the state is often used. The theoretical average velocity can be computed by Eq. (3.42). When $i_D = 1–2$, and $i_D = 2–3$, the theoretical average velocity must be divided by 2 and 3 respectively.

ii. The periodical and non-periodical throwing movement. When $D = 1–3.3$, 4.6–6.36 and 7.78–9.48, the periodical throwing movements occur. When $D = 3.3–4.6$, 6.36–7.78 and 9.48–10.94, the non-periodical throwing movements occur.

The so called periodical throwing movements are the movements in which every throwing movements of the materials have the same time and the same distance. When the material falling it is in the non-throwing range $(\pi - \varphi_d \sim 2\pi + \varphi_d)$, i.e. the acceleration of the working surface $-a_y < g\cos\alpha_o$, that is: $\omega^2\lambda\sin\delta \sin\varphi < g\cos\alpha_o$. Assume that at the time when the material falls, its impact with the working surface is a non-elastic impact (like most of the vibrating machines), it will be in contact with the working surface and slides a small distance. When φ reaches $2\pi + \theta_d$, the material is thrown again. The time and distance for the first and the second throwing should be the same at least theoretically. This is the periodical throwing movements.

The non-periodical throwing movements are the movements in which every throwing movements of the materials have the different time distance. When the material falling it is in the throwing range $(\varphi_d \sim \pi - \varphi_d)$, i.e. the acceleration of the working surface $-a_y > g\cos\alpha_o$, that is: $\omega^2\lambda\sin\delta \sin\varphi > g\cos\alpha_o$. So that after the material falls on the surface, it is thrown immediately for the second time. However the initial velocity of the throwing movement is different from the initial velocity of the previous throwing. Thus the period of the second throwing is different from the first period and every time the period is different. This is the non-periodical throwing movement.

Nowadays most of the vibrating machines used in industry are designed to work in the periodical throwing movement state.

3.1.3 Selections of Material Movement State and Kinematics Parameters

(1) The Selection of the Material Movement States

Since the materials keep contact with the working surface, and there is no mutual impact, the advantage of the material sliding state is that the noise is low, the material not vulnerable to be crushed. It's suitable for the materials that make noise when conveyed and those fragile materials. In addition, the materials always keep contact with the working surface and there is no gap change, unlike the throwing movement, the air cushion generated between the material layers and the working surface will affect the material movements when the aeration of the material layers is not so good. So this conveying means has a good adaptability to the powder materials. The disadvantage of the sliding movements is that the working surface is easy to be worn. Adding the wear-resisting metal pad or increasing the thickness of the working surface plate can overcome this shortcoming. The second means is to increase the vibrating amplitude to obtain higher conveyance speed.

When the materials are in the sliding state, it should avoid or reduce the negative sliding as much as possible and increase the positive sliding as much as possible. It is the positive sliding movement that conveys the material efficiently, the negative sliding cannot accomplish the convey material, it reduces the conveyance efficient and increases the surface wear-out.

When the material is in the throwing state, the contact time of the material with the working surface is short, this conveyance advantage is the working surface has less wear-out and conveyance speed is high. When selecting the throwing states for material screening, the materials on the upper and lower layers are subject to churning, thus the materials have big chances to be screened and the screen machine efficiency is increased. However the throwing state requires a large vibration strength (vibration acceleration), the dynamic stresses in the machine parts increase and thus the requirements for the part strengths are increased accordingly.

The selection of the material movement states is mainly based on the properties of the materials (such as fragile, viscosity, moisture, particle size, relative intensity, friction coefficient, etc.), the purposes of the machines and characteristics of the working surfaces. At the same time, the durability, output and working quality (such screening efficiency, and feeding precision etc.) of the machines must be considered as well. For example, for the trough type of vibrating coolers and vibrating centrifugal hydroextractor the positive sliding only state is used; for some vibrating conveyers, vibrating feeders and vibration screens the sliding states or the light throwing states are used; for most of the vibrating screens, vibrating feeders and vibrating conveyers, the mid-speed throwing states ($D = 1.75$–3.3) are used. In these states, the vibrating machines have high outputs, high working quality, less-energy consumption and lax requirements on the part stiffness and strength.

For those fragile materials that should not be crushed during the processes, the sliding state or light throwing movement state can be used. For those materials that contain dirt or those which are hard to be treated, or some special purpose machines, the high speed throwing movement states can be used ($D = 3.3$–5).

In the sliding states, in order for the materials to have a good sliding movements and obtain large conveying speed, the sliding index chosen is generally around $D_k = 2$–3.

In the case of selecting the throwing states, in order for the materials to have a good throwing movements and obtain large conveying speed, the throwing index chosen is generally around $D = 1.4$–5.

For a variety of the vibrating machines, the selecting range of the throwing index is different. For most of the long distance and large output vibrating conveyors, the throwing index is generally chosen as $D = 1.4$–2.5; for the electric vibrating feeders (distance is short), in order to obtain a large conveying speed, the throwing index is chosen as $D = 2.5$–3.3; the throwing index of the vibrating screens is determined on the basis of the properties of the materials to be treated. For the easy-to-screened materials, the index is chosen as $D = 2$–2.8; for the general materials, $D = 2.5$–3.3; and for the hard-to-screened materials, $D = 3$–5.

(2) Selections of Vibration Strength K, Vibration Number n and Vibration Amplitude λ

The selection of the vibration strength K (mechanical index) is mainly subject to the limit of the strength of material and the stiffness of parts. For most of the vibrating conveyance machines, the vibration strength K is chosen as $K = 4$–6 on the considerations of the long distance conveyance, large output and not-overly strengthening structure and the stiffness of the machine parts and the machine working durability. For the vibrating feeders, only minority of them have $K = 10$.

The working frequencies (vibration numbers) n and vibration amplitude λ of the vibrating conveyers, vibrating feeders, vibrating screens and the resonance screens vary in a large range. They are associated not only with the structure architectures, but also with the specific technological requirements. The selection should be dependent on the specific situations.

The relative vibration amplitudes of the electric–magnetic vibrating machines are limited to the working gap between the electric–magnetic airgaps. The increase of the gap will bring a lot of the bad effects (such as increasing the magnetic-excitation current), so it is generally used the high frequency and small amplitude. For $n = 3{,}000$ rpm, the single amplitude λ is 0.5–1 mm; for $n = 1{,}500$ rpm, one-sided amplitude λ is 1.5–3 mm. For some few of electric–magnetic vibrating machines, $n = 6{,}000$ rpm, or lower than $1{,}500$ rpm.

For the inertial type of vibrating machines, the mid-frequency and mid-amplitude are chosen, minority of them are chosen high-frequency and small amplitude. The vibration number n is generally taken 700–1,800 rpm, the single amplitude taken 1–10 mm. The over large amplitude would require large eccentric mass and the over high frequency would increase the pressure on the bearings and increase the dynamic stresses in parts.

For the elastic connecting rod type of the vibrating machines the low-frequency large amplitude are selected, only minority of them with mid-frequency and mid-amplitude. The vibration number n is taken 400–1,000 rpm, amplitude $\lambda = 3$–30 mm.

For the vibrating screens, the fine screens take small amplitude; rough screens take a large amplitude.

When selecting vibration number n and amplitude λ, the requirement for the allowed vibration strength $[K]$ must be satisfied. $[K]$ is 5–10. The vibration strength can be verified:

$$K = \frac{\omega^2 \lambda}{g} = \frac{\pi^2 n^2 \lambda}{900g} < [K] \tag{3.47}$$

(3) The Selection of the Vibration Directional Angle δ

The selection of the vibration direction angle δ is mainly dependent on the purpose of the vibration machines. For example, as the vibrating conveyors or feeders, the high conveying speed should be guaranteed; as the screening, the high screening efficiency and large output should be guaranteed. The next thing to be considered would be the properties and requirements of the materials to be treated, such as fragile, viscosity, moisture, particle size, relative intensity, etc. If the density is large or the powders with fine size of particles, the small vibration directional angle δ should be selected; for a material with large moisture and sticky, large vibration direction angle δ should be selected; for fragile material, small vibration directional angle δ should be selected to avoid the breaking of the material during the working process; for the wear-resistant materials, large vibration directional angle δ should be chosen to prevent the wear-out of the working surfaces.

When the sliding movement states are selected, to ensure the negative sliding movement as little as possible and obtain a high output, the vibration directional angle δ can be computed by Eq. (3.13) after the positive and negative sliding indices D_k and D_q are determined.

When the throwing movement states are selected, from the viewpoint of increasing conveying speed there is an optimal vibration directional angle δ corresponding to every vibration strength K for a different incline angle α_0. Figure 3.11 is the relation curves between the optimal vibration directional angle δ and the vibration strength K for different incline angle α_0.

Table 3.1 is the conveying speeds for different vibration directional angle δ and different vibration amplitude λ. The data are measured from a horizontally installed vibrating conveyer (installation incline angle $\alpha_0 = 0°$), with the vibration number n = 600 rpm, the material layer thickness is relative thin. The material used for the experiments are the mix of 30% fine iron ores and the tail core powder, their particle sizes are less than 200 mm.

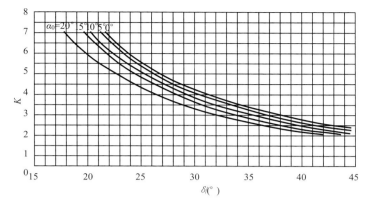

Fig. 3.11 The relation between the optimal vibration direction angle δ and the vibration strength K

Table 3.1 The conveying speeds versus the amplitude and vibration direction angle

Vibration direction angle/(°)	Single amplitude/mm						
	6.00	6.25	6.50	6.75	7.00	7.25	7.50
20	0.11–0.15	0.12–0.16	0.15–0.19	0.18–0.22	0.21–0.25	0.24–0.28	0.26–0.30
25	0.15–0.19	0.16–0.20	0.18–0.22	0.20–0.24	0.22–0.26	0.25–0.29	0.27–0.31
30	0.17–0.21	0.18–0.22	0.21–0.25	0.24–0.28	0.25–0.29	0.27–0.31	0.29–0.33
35	0.18–0.22	0.19–0.23	0.22–0.26	0.25–0.29	0.26–0.30	0.27–0.31	0.28–0.32
40	0.19–0.23	0.20–0.24	0.22–0.26	0.24–0.28	0.25–0.29	0.26–0.30	0.27–0.31

It can be seen from Table 3.1 that when the amplitudes are large (>7.5 mm) even the vibration directional angle vary in the ranges of 30°–40°, the variation of the conveying speeds of the materials is less than 10%. However when the amplitudes are small (6–6.5 mm) the vibration directional angle has obvious effects on the conveying speeds. The conveying speed at the vibration direction angle 40° is twice of that at the 20°. It is obvious that selection of the vibration direction angle between 20° and 30° is not appropriate and the angle should remain in the range of 30°–40°. Table 3.2 lists the relation between vibration strength and good vibration direction angle for the installation incline angle $\alpha_0 = 0°$.

For the vibration screen and resonance screens, under the premise of ensuring the screening quality, the high conveying speed and output should be considered.

Table 3.2 The relation between vibration strength and good vibration direction angle

Vibration strength K	2	3	4	5	6	7
Good vibration direction angle/(°)	40–50	30–40	26–36	22–32	20–30	18–28

(4) The Selection of the Installation Incline Angle α_0

For long distance vibration conveyors, they are installed horizontally if there are no special requirements. When it is required to convey the materials upwards, the maximum incline angle upwards can not exceed 15°–17° on the basis of the material properties (shapes, particle sizes). For large size or sphere-shape material which may have tendency of rolling downward, the maximum incline angle upwards can not exceed 12°. When it is required to convey the materials downwards, the downward incline angle normally is required not to exceed 15°–20° to avoid the trough body or pipes are subject to severe wearing.

For the vibration feeder, it is usually installed downward inclined about 10° unless some technological operation requires it installed horizontally. For sending the materials with large moistures, or stick characteristics, the downward incline angle can be increased to 15°–20°.

For the vibration feeder and the vibration separator screen making use of the different friction coefficients of the materials to be separated, to satisfy the requirements of the material moving upwards, the upward incline angles are selected between 15° and 20°.

3.1.4 Calculation of Real Conveying Speed and Productivity

(1) The Comprehensive Theory of the Conveying Speeds

For the vibration conveyors and feeders with designs for the sliding states ($D \leq 1$), their theoretical conveying speeds are the simple sum of the averaged positive and negative sliding speeds. For those with designs for the throwing states ($D > 1$), since the materials are subject to more or less throwing movements, the theoretical conveying speeds are the sum of the averaged positive, negative sliding and throwing speeds. The same computation cited by some domestic and international references ignored the effect of the material sliding movements on the conveying speeds, thus the results on the situation with a light throwing movement are far away from the real conveying speeds. Especially when the throwing index is near 1, the theoretical conveying speeds calculated by some references are near zero. In reality, due to the existence of the sliding movements, the averaged conveying speeds for the materials could still be around 0.05–0.2 m/s, even larger, and the machines can still work ideally. Therefore, when computing the average velocity of the materials by the vibration conveyors or feeders, the formulas used to compute the average velocity for the sliding only movements can be not used, or one can not simply apply the theoretical formulas for the average velocity of the pure throwing movements. When using the sliding movement formulas one must consider the throwing movement effects and when using the throwing movement formulas one must consider the sliding movement effects.

Table 3.3 lists the approximate values of the throwing index D, the effect coefficients C_D and C_w. It can be seen from the Table 3.3 that when $D = 1$–1.75, the sliding

3.1 Theory and Technological Parameter Computation … 81

Table 3.3 The effect coefficients C_D and C_w

Throwing index D	1	1.25	1.5	1.75	2	2.5	3
Throwing effect coefficient C_D	1	1.1–1.3	1.2–1.4	1.3–1.5	–	–	–
Sliding effect coefficient C_w	–	–	–	1.1–1.15	1.05–1.1	1–1.05	1

movement is dominant, using the sliding movement formulas will obtain the approximate results. However the results must be corrected by multiplying the coefficient C_D that the effect of the throwing movements has on the conveying velocity. When $D = 1.75$–2.5, the throwing movement is dominant, using the throwing movement formulas will obtain the approximate results. However the results must be corrected by multiplying the coefficient C_w that the effect of the sliding movements has on the conveying velocity. When $D > 2.5$ the sliding movement effect can be ignored and the average velocity formulas for the throwing movement can be used.

(2) Effect of Installation Incline Angle (Working Surface Incline) on Conveying Velocity

In using Eqs. (3.27) and (3.28) to compute the conveying velocity of the materials in the sliding states, the formulas is derived under the condition of the incline angles. However it is noticed that using Eq. (3.42) to compute the conveying velocity of a large incline angle for the throwing movement state will result in a big error compared to the measured data. Figure 3.12 shows the relations between the measured conveying velocity and incline angles for different materials.

It can be seen from the Fig. 3.12 that the conveying speeds for the concrete for a 15° incline angle is 70% higher than those for a horizontally installed. In the same condition it is 62% and 115% higher for the quartz sands and iron ores respectively. The measured velocity increases are much larger than the theoretically computed one. This is because the end velocity of the throwing movements will increase when the incline angles are large, the next throwing initial velocity will be larger than that

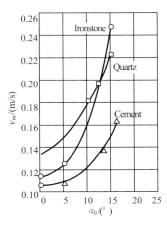

Fig. 3.12 Relations between the working incline angle and measured averaged velocity v_m

Table 3.4 The incline angel correction coefficients and effect coefficients

Incline angle $\alpha_0/(°)$	−15 to +5			10		15	
Incline angle correction coefficients to average velocity γ_α	1			1.2–1.3		1.25–1.6	
Incline angle $\alpha_0/(°)$	−15	−10	−5	0	5	10	15
Incline angle correction coefficients to average velocity C_α	0.6–0.8	0.8–0.9	0.9–0.95	1	1.05–1.1	1.3–1.4	1.5–2.0

of the working surface, the average velocity of the material movement will increase as a result. This actual situation has not been considered yet in the formulas derivation. According the measurement results, Table 3.4 lists the incline angle α_0 correction coefficients γ_α to the average velocity computed by Eq. (3.42) and the incline angle α_0 effect coefficients C_α to Eq. (3.45).

(3) Effect of the Material Properties on the Conveying Velocity

For the vibration conveyors, feeders and cooler in the sliding movement states, the effects of the material properties on the average conveying velocity have been somehow considered in the theoretical formulas. However when working in the throwing movement states, the material property effects on the conveying velocity have not been considered in the derivation. The tests show that the real throw starting angle of the material throwing movement is lag behind some angles to the theoretical material throw starting angle due to the existence of the friction and other resistances. The real average velocity is usually smaller than the theoretical average velocity. The amount of the lag angles is dependent on the amount of velocity decreases and is associated with the material properties (particle size, intensity, moisture, friction coefficients) and other resistances. Figure 3.13 is the comparison of the theoretical average conveying velocity to the real average conveying velocity. Due to lack of the experimental data, the effect coefficients C_m of the properties of the materials on the conveying velocities are approximated values. For a bar-shape material $C_m = 0.8$–0.9; for a pellet-shape material $C_m = 0.9$–1.0 and for a powder material $C_m = 0.6$–0.7.

(4) The Effect of the Material Layer Thickness on the Conveying Velocity

For the vibration conveyors, feeders and coolers in the sliding and throwing movement states, the effects of the material layer thickness on the average conveying velocity have not been considered in the theoretical formulas. The experiments show that the material layer thickness has obvious effects on the average conveying velocities. When

Fig. 3.13 Theoretical average velocity and measure average velocity

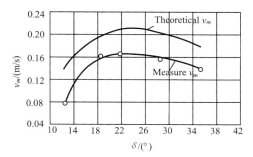

Fig. 3.14 The effect of material thickness on absolute and relative displacements. 1-Extra-thin layer; 2-Thin layer; 3-Mid-thin layer; 4-Thick layer

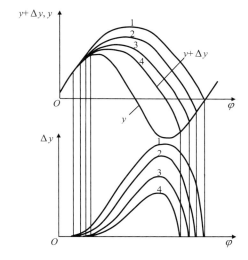

the layer is thicker, the material velocities at different material layers are different. The further it is from the work surface, the larger the lag angle of the material's real throw starting angle than the theoretical throw starting angle. Figure 3.14 shows the effect of the material layer thickness on the material's absolute and relative displacements. It can be seen from the figure that the average velocities of the materials vary in a large range along the material layer thickness. The thicker the material layer the much lower the real velocities than the theoretical average velocities. Figure 3.15 shows the relation curves of the measured average velocities versus the material layer thickness for three materials. It can be seen from Fig. 3.15 that for a bar-shaped and pellet-shaped materials (pebbles, quartz) the material conveying velocity changes slowly with the material layer thickness; for powder materials (concrete, talcum powder), the material conveying velocities have large changes with the material layer thickness changes. Here is the explanation: when thickness of the powder materials is large the ventilation capability in the layers becomes worse and the air cushions are formed between the material and the working surface, which affects the normal throwing of the materials. However for some powder materials when the thickness

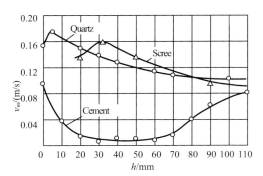

Fig. 3.15 Relation between the average velocity and material layer thickness

increases the conveying velocity is not decreased but increased instead. According to the observations and analysis on the real situations the possible explanation for this special phenomena is that when the thickness is large to some extent, due to the vibrations the material layers become porous and the materials are flowing in the layers.

For the need of computations, Table 3.5 gives the effect coefficient C_h of the material layer thickness on the conveying velocity. The values in the Table 3.5 are valid only to the bar-shaped, pellet-shaped materials and the powder materials in the sliding movement state. For powder materials the rule of the thumb is to take the low limit values. For the powder materials in the throwing movement states, since the changes of the thickness of the materials have large effect on the conveying velocity and the changes are very complex, it'd better be determined by tests. If the tests can not be carried out it can be referenced by the similar powder materials data. If no other data available, it can be taken as $C_h = 0.5$–0.6.

(5) Computation of the Real Average Velocity

For the vibration machine working in the sliding movement states, its average velocity can be computed by the following formulas considering the effects of the material layer thickness:

$$v_m = C_h(v_k + v_q) \tag{3.48}$$

in which v_k and v_q can be computed by Eqs. (3.27) and (3.28) respectively.

For the vibration machine working in the light-throwing movement states ($D \leq 1.75$), its real average velocity is:

Table 3.5 The effect coefficient of the material layer thickness

Thickness of the material layer	Thin layer	Mid-thin layer	Thick layer
Effect coefficient of the material layer thickness C_h	0.9–1	0.8–0.9	0.7–0.8

3.1 Theory and Technological Parameter Computation ...

$$v_m = C_h C_D (v_k + v_q) \tag{3.49}$$

For the vibration machine working in the violate-throwing movement states ($D >$ 1.75), its real average velocity is:

$$v_m = C_\alpha C_h C_m C_w v_{dl} \tag{3.50}$$

in which v_{dl} is the theoretical average velocity in the horizontal conveyance and can be computed by Eq. (3.45).

The real average velocity is:

$$v_m = \gamma_\alpha C_h C_m C_w v_d \tag{3.51}$$

in which v_d can be computed by Eq. (3.42).

The friction coefficients f_0, f and friction angle μ_0, μ for different materials are listed in Table 3.6.

(6) Productivity Calculation

The production rate Q (t/h) of the vibration machines, the vibration conveyors, the vibration feeders, the vibration coolers and the vibration screens is:

$$Q = 3600\, hBv_m\gamma \tag{3.52}$$

in which B is the working surface width (m); γ is the material porous density (t/m^3); h is the thickness of the material (m).

For the trough type of vibration feeders and conveyors, $h = 0.7$–$0.8H$ (H is the height of the trough), it can be a little bit larger for the one with rear fender plate. For pipe type of the vibration conveyors, $h \leq (1-2)a$ (a is the size of the screen hole). For general screens, $h \leq (3-5)a$, for thick layer screens $h \leq (10-20)a$.

Table 3.6 Friction coefficients f_0, f and friction angle μ_0, μ for different materials

Material names	Relax density/(t/m³)	Nature angle of rest/(°)		Friction coefficients of materials					
				Steel		Wood		Cement	
		Move	Still	f	f_0	f	f_0	f	F_0
Anthracite	0.8–0.95	27	45	0.29	0.84	0.47	0.84	0.51	0.90
Coke	0.36–0.53	35	50	0.47	1.00	0.80	1.00	0.84	1.00
Clay, sandy	1.4–1.9	30	45	0.58	1.00	–	–	–	–
Mine	1.3–3	30	50	0.59	1.19	–	–	–	–
Granulated sugar	–	50	70	1.00	2.14	–	–	–	–
Cement	0.9–1.7	35	40–50	0.50–0.60	1.00	–	–	–	–
Limestone	1.2–1.5	30–35	40–50	0.50–0.60	1.00	–	–	–	–

If the product output is given, the material thickness needed is:

$$h = \frac{Q}{3600Bv_m\gamma} \quad (m) \qquad (3.53)$$

3.1.5 Examples

Example 3.1 A vibration conveyor is used to convey a material which is not easy-to-be-crushed, its conveyance density is 1.6 t/m³ and its conveyance distance is 15 m. The dynamic and static friction coefficient of the material to the trough body is 0.6 and 0.9 respectively. Select and compute the kinematics and technological parameters of the machine.

Solution

(1) Movement State Selection. The sliding state is chosen by the requirement of not-crushed in conveyance. The throwing index $D < 1$. The positive sliding index $D_k \approx$ 3–4, the negative sliding index $D_q \approx 1$.
(2) Trough Body Incline Angle α_0 and Vibration Direction Angle δ. For the long distance vibration conveyor $\alpha_0 = 0$. The vibration direction angle δ is

$$\delta = \arctan\frac{1-c}{(1+c)f_0} = \arctan\frac{1-0.33}{(1+0.33)\times 0.9} = 29°14'$$

Taking $\delta = 30°$, then

$$c = \frac{D_{q0}}{D_{k0}}\frac{\sin(\mu_0+\alpha_0)}{\sin(\mu_0-\alpha_0)} = \frac{1}{3.0}\frac{\sin(42°+0°)}{\sin(42°-0°)} = 0.33$$

in which

$$\mu_0 = \arcsin 0.9 = 42°$$

(3) Amplitude and Vibration Time. According to the structure, take $\lambda_1 = 5$ mm.

The vibration time:

$$n = 30\sqrt{\frac{D_{k0}g\sin(\mu_0-\alpha_0)}{\pi^2\lambda_1\cos(\mu_0-\delta)}} = 30\sqrt{\frac{3.0\times 9.80\sin(42°-0°)}{\pi^2\times 0.005\cos(42°-30°)}} = 606 \text{ r/min}$$

Verify the vibration strength:

3.1 Theory and Technological Parameter Computation ...

$$K = \frac{\omega^2 \lambda_1}{g} = \frac{\pi^2 n^2 \lambda_1}{900g} = \frac{3.1416^2 \times 606^2 \times 0.005}{900 \times 9.80} = 2.05 < 7{-}10$$

(4) Material Average Velocity. The positive and negative sliding indices are:

$$D_{k0} = \frac{\omega^2 \lambda_1 \cos(\mu_0 - \delta)}{g \sin(\mu_0 - \alpha_0)} = \frac{63.46^2 \times 0.005 \cos(42° - 30°)}{9.8 \sin(42° - 0°)} = 3.0$$

$$D_{q0} = \frac{\omega^2 \lambda_1 \cos(\mu_0 + \delta)}{g \sin(\mu_0 + \alpha_0)} = \frac{63.46^2 \times 0.005 \cos(42° + 30°)}{9.8 \sin(42° + 0°)} = 0.949$$

Throwing index:

$$D = \frac{\omega^2 \lambda_1 \sin \delta}{g \cos \alpha_0} = \frac{63.46^2 \times 0.005 \sin 30°}{9.8 \cos 0°} = 1.0$$

There is no negative sliding and nor throwing movements.
The real positive sliding start angle:

$$\varphi'_k = \varphi_{k0} = \arcsin \frac{1}{D_{k0}} = \arcsin 0.333 = 19°28'$$

Assume the positive sliding start angle

$$\varphi_k = \arcsin \frac{1}{D_k} = \arcsin \frac{g \sin(\mu - \alpha_0)}{\omega^2 \lambda_1 \cos(\mu - \delta)}$$
$$= \arcsin \frac{9.80 \sin(30°58' - 0')}{63.46^2 \times 0.005 \cos(30°58' - 30°)} = \arcsin 0.25 = 14°30'$$

in which

$$\mu = \arctan 0.6 = 30°58'$$

From Fig. 3.3, φ'_k and φ_k, $\varphi'_m = 269°30'$ and

$$b'_k = \sin \varphi'_k = 0.333, \quad b_k = \sin \varphi_k = 0.25, \quad b'_m = -1$$

Compute P_{km}:

$$P_{km} = \frac{b'^2_m - b'^2_k}{2b_k} - (b'_m - b'_k) = \frac{(-1)^2 - 0.333^2}{2 \times 0.25} - (-1 - 0.333) = 3.11$$

Or directly from Fig. 3.3, $P_{km} = 3.11$.
The theoretical average velocity is:

$$v_k = \omega\lambda_1 \cos\delta(1 + \tan\mu\tan\delta)\frac{P_{km}}{2\pi}$$

$$= 63.46 \times 0.005 \cos 30°(1 + 0.6 \times \tan 30°)\frac{3.11}{2\pi}$$

$$= 0.275 \times (1 + 0.346) \times 0.5 = 0.19 \text{ m/s}$$

The real average velocity is

$$v = C_h v_k = 0.8 \times 0.19 = 0.15 \text{ m/s}$$

(5) Trough Body Section Dimension

$$Q = 3600\, Bhv\gamma = 3600\, Bh \times 0.15 \times 1.6 \text{ t/h}$$

When taking $h = 0.6H$ (H is the trough body height)

$$BH = \frac{Q}{3600 \times 0.6 v\gamma} = \frac{30}{3600 \times 0.6 \times 0.15 \times 0.16} = 0.0579 \text{ m}^2$$

When taking $h = 0.75B$, $B = \sqrt{\frac{0.0579}{0.75}} = 0.278$ m, take $B = 0.28$ m; $H = 0.75 \times 0.28 = 0.21$ m, so take $H = 0.22$ m.

Example 3.2 Given a double pipe vibration conveyor and it is used to convey a material at 1.6 t/m³. Required output is 100 t/h and distance is 32 m, select the throwing movement state. Determine the kinematics and technological parameters.

Solution

(1) Select Throwing Index D and Vibration Strength K.

$$D = 1.5\text{--}2.5, K = 3\text{--}5$$

(2) Trough Body Incline Angle α_0 and Vibration Directional Angle δ. For long distance conveyance, $\alpha_0 = 0$; for the throwing state, when vibration strength $K = 4$, the optimal vibration direction angle $\delta \approx 30°$. A good choice would be in the range of $25°\text{--}35°$.

(3) Amplitude λ and Vibration Time n. According to the structure, if taking $\lambda = 7\text{--}8$ mm, the vibration number can be computed:

$$n = 30\sqrt{\frac{Dg \cos\alpha_0}{\pi^2 \lambda \sin\delta}} = 30\sqrt{\frac{2 \times 9.8 \cos 0°}{\pi^2 \times 0.008 \sin 30°}} = 670 \text{ r/min}$$

Now taking $n = 680$ r/min, K and D are respectively:

$$K = \frac{\omega^2 \lambda}{g} = \frac{\pi^2 n^2 \lambda}{900g} = \frac{3.14^2 \times 680^2 \times 0.008}{900 \times 9.8} = 4.14$$

$$D = K \sin \delta = 4.14 \sin 30° = 2.07$$

(4) Material Average Velocity. The theoretical average velocity is:

$$v_d = \omega\lambda \cos\delta \frac{\pi i_D^2}{D}(1 + \tan\alpha_0 \tan\delta)$$
$$= \frac{\pi \times 680}{30} \times 0.008 \cos 30° \frac{\pi \times 0.77^2}{2.07}(1 + \tan 0° \times \tan 30°)$$
$$= 71.2 \times 0.008 \times 0.866 \times 0.9 = 0.444 \,\text{m/s}$$

The real average velocity is:

$$v = C_m C_h C_\alpha C_w v_d = 0.8 \times 0.8 \times 1.0 \times 1.0 \times 0.444 = 0.284 \,\text{m/s}$$

(5) Trough Body Section Area and Width. Take the material layer thickness $h = 0.1$ m. When using double pipes, the trough width is:

$$B = \frac{Q}{2 \times 3600 h v \gamma} = \frac{100}{2 \times 3600 \times 0.1 \times 0.284 \times 1.6} = 0.306 \,\text{m}$$

Now take $B = 350$ mm.

(6) Material Mass on the Trough Body

$$G_m = \frac{QL}{2 \times 3600 v} = \frac{100 \times 22}{2 \times 3600 \times 0.284} = 1.076 \,\text{t}$$

L is the trough length and taking $L = 22$ m.

3.2 Theory and Technological Parameter Computation of Circular and Ellipse Vibration Machine

3.2.1 *Displacement, Velocity and Acceleration of Vibrating Bed*

The circular and ellipse vibration machines are used in addition to the line vibration machines. The vibration machines with the circular movements include the single-axle inertial screen, the single-axle vibration feeder, and the vibration conveyor etc. The vibration machines with the ellipse movements include the inertial type of the near resonance feeder, conveyors and ellipse screens. When the bodies of the line and circular vibration machines swing vibrate about their center of mass, the ellipse trajectory will occur.

The equations of movements of the two aforementioned machines will be given later in its dynamics analysis, its general expressions are:

$$S_y = a \sin \omega t + b \cos \omega t = \lambda_y \sin(\omega t + \beta_y)$$
$$S_x = c \sin \omega t + d \cos \omega t = \lambda_x \sin(\omega t + \beta_x) \quad (3.54)$$

in which

$$\lambda_y = \sqrt{a^2 + b^2} \quad \lambda_x = \sqrt{c^2 + d^2}$$
$$\beta_y = \arctan \frac{b}{a} \quad \beta_x = \arctan \frac{d}{c}$$

where a, b, c, d are constants; λ_y, λ_x are the amplitude in y and x directions; β_y, β_x are the initial phase angle in y and x direction; ω is the angular velocity and t is time.

The velocity and acceleration of the working bed can be derived from Eq. (3.54):

$$\begin{cases} v_y = \omega a \cos \omega t - \omega b \sin \omega t = \omega \lambda_y \cos(\omega t + \beta_y) \\ v_x = \omega c \cos \omega t - \omega d \sin \omega t = \omega \lambda_x \cos(\omega t + \beta_x) \\ a_y = -\omega^2 a \sin \omega t - \omega^2 b \cos \omega t = -\omega^2 \lambda_y \sin(\omega t + \beta_y) \\ a_x = -\omega^2 c \sin \omega t - \omega^2 d \cos \omega t = -\omega^2 \lambda_x \sin(\omega t + \beta_x) \end{cases} \quad (3.55)$$

Figure 3.16 shows the displacement curve of the working bed of the ellipse vibration machine.

It can be seen from Eqs. (3.54) and (3.55) that the line and circular machines are a special case of the ellipse vibration machines.

For the circular vibration machines, let $\beta_y = \beta_x - \pi/2$, i.e. $b/a = -c/d$ and satisfy the condition: $\sqrt{a^2 + b^2} = \sqrt{c^2 + d^2} = \lambda$ then Eq. (3.54) will be converted into a circular equation:

$$S_y = \lambda \sin(\omega t + \beta_y) = \lambda \sin \omega t'$$

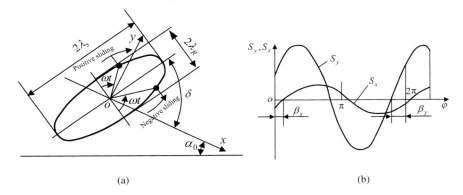

Fig. 3.16 The displacement curve of the ellipse vibration working bed. **a** Working bed movement trajectory; **b** y and x direction displacement curves

3.2 Theory and Technological Parameter Computation ...

$$S_x = -\lambda \cos(\omega t + \beta_y) = -\lambda \cos \omega t' \qquad (3.56)$$

in which $\omega t' = \omega t + \beta_y$. For the line vibration machines, $\beta_y = \beta_x$, i.e. $\frac{b}{a} = \frac{d}{c}$, Eq. (3.54) becomes the line vibration equation:

$$\begin{aligned} S_y &= \lambda \sin \delta \sin(\omega t + \beta_y) = \lambda \sin \delta \sin \omega t' \\ S_x &= \lambda \cos \delta \sin(\omega t + \beta_x) = \lambda \cos \delta \sin \omega t' \end{aligned} \qquad (3.57)$$

in which

$$\omega t' = \omega t + \beta_y = \omega t + \beta_x$$

$$\lambda = \sqrt{\lambda_y^2 + \lambda_x^2} = \sqrt{a^2 + b^2 + c^2 + d^2}$$

$$\delta = \arctan \frac{\lambda_y}{\lambda_x} = \arctan \sqrt{\frac{a^2 + b^2}{c^2 + d^2}}$$

Therefore the conclusions drawn from the ellipse vibration machines can be applied directly to the line and circular vibration machines.

In the ellipse movement of the vibration machines, based on the dynamic kinematics parameters the materials would have 4 types of basic movement forms just as the line vibration machines: relative still, positive sliding, negative sliding and throwing movements and the above movements will form the movements of the materials. The states can be divided into two categories: sliding movement and throwing movement.

3.2.2 Theory of Material Sliding Movements

(1) The Positive and Negative Sliding Index D_k and D_q for Ellipse Vibration Machines

In the ellipse vibration machines, its acceleration of the working surface is different from those of the line vibration machines, its positive sliding index D_k and negative sliding index D_q are not the same as those of the line vibration machines. However, the sum of the components in x direction of the inertial force and gravity of the materials are the same as that of Eq. (3.3), the normal pressure of the materials on the working surface is the same as that in Eq. (3.6), and the limit friction force can also be expressed as the same form of Eq. (3.7).

Substituting the acceleration a_y, a_x of Eq. (3.55) into Eqs. (3.5) and (3.6), taking Δy and Δx equal to zero, and letting the resultant F of the inertial force and gravity components in Eq. (3.55) be equal to the limit friction force F_0 in Eq. (3.7) one obtains:

$$m\omega^2(c\sin\omega t + d\cos\omega t) + G\sin\alpha_0$$
$$\mp f_0[-m\omega^2(a\sin\omega t + b\cos\omega t) + G\cos\alpha_0] = 0 \quad (3.58)$$

Simplifying Eq. (3.58) yields the nominal positive sliding start angle φ_{k0} and nominal negative sliding start angle φ_{q0}:

$$\varphi_{k0} = \arcsin\frac{1}{D_{k0}} \qquad \varphi_{q0} = \arcsin\frac{1}{D_{q0}} \quad (3.59)$$

in which

$$\begin{cases} D_{k0} = \frac{\omega^2 \lambda_{k0}}{g\sin(\mu_0 - \alpha_0)} & D_{q0} = \frac{\omega^2 \lambda_{q0}}{g\sin(\mu_0 + \alpha_0)} \\ \lambda_{k0} = \frac{a\sin\mu_0 + c\cos\mu_0}{\cos\rho_{k0}} & \lambda_{q0} = \frac{-a\sin\mu_0 + c\cos\mu_0}{\cos\rho_{q0}} \\ \rho_{k0} = \arctan\frac{b\sin\mu_0 + d\cos\mu_0}{a\sin\mu_0 + c\cos\mu_0} & \rho_{q0} = \arctan\frac{-b\sin\mu_0 + d\cos\mu_0}{-a\sin\mu_0 + c\cos\mu_0} \end{cases} \quad (3.60)$$

where D_{k0}, D_{q0} are the positive and negative sliding index respectively, $\lambda_{k0}, \lambda_{q0}$ are the calculated positive and negative sliding amplitude respectively; and ρ_{k0}, ρ_{q0} are the calculated positive and negative sliding phase angle respectively.

The positive sliding start angle ωt_{k0} and negative sliding start angle ωt_{q0} are:

$$\omega t_{k0} = \varphi_{k0} - \rho_{k0} \qquad \omega t_{q0} = \varphi_{q0} - \rho_{q0} \quad (3.61)$$

in which t_{k0} and t_{q0} are the positive and negative sliding start time.

If the positive sliding index $D_{k0} \leq 1$, the positive sliding for the material could not occur. When $D_{k0} > 1$, φ_{k0} has a solution and the positive sliding for the material occurs. By the same token, when the negative sliding index $D_{q0} \leq 1$, the negative sliding for the material could not occur; when $D_{q0} > 1$, the negative sliding for the material could occur. Similar to the line vibration machines, the possible positive sliding range is $\varphi_{k0} \sim (180° - \varphi_{k0})$, while the possible negative sliding range is $\varphi_{q0} \sim (540° - \varphi_{q0})$.

The aforementioned positive and negative sliding indices D_{k0}, D_{q0} for the ellipse vibration machines are also valid to the circular and line vibration machines.

For the circular vibration machines, when $b = c, a = -d = \lambda, \rho_{k0} = \mu_0 - \pi/2$ in Eq. (3.60), $\lambda_{k0} = \lambda, \rho_{q0} = -(\mu_0 + \pi/2), \lambda_{q0} = \lambda$, thus the positive and negative sliding indices are respectively:

$$D_{k0} = -\frac{\omega^2 \lambda}{g\sin(\mu_0 - \alpha_0)} \qquad D_{q0} = -\frac{\omega^2 \lambda}{g\sin(\mu_0 + \alpha_0)} \quad (3.62)$$

For the line vibration machines, when $b = d = 0$, $\sqrt{a^2 + c^2} = \lambda$, $\frac{a}{c} = \tan\delta$ then $\rho_{k0} = 0$ in Eq.(3.60), $\lambda_{k0} = \lambda\cos(\mu_0 - \delta)$, $\rho_{q0} = 0$ and $\lambda_{q0} = \lambda\cos(\mu_0 + \delta)$ in Eq. (3.60), thus the positive and negative sliding indices are respectively:

3.2 Theory and Technological Parameter Computation ...

$$D_{k0} = \frac{\omega^2 \lambda \cos(\mu_0 - \delta)}{g \sin(\mu_0 - \alpha_0)} \quad D_{q0} = \frac{\omega^2 \lambda \cos(\mu_0 + \delta)}{g \sin(\mu_0 + \alpha_0)} \quad (3.63)$$

The results in Eq. (3.63) are the same as those in Eq. (3.11).

For the ellipse vibration machines working in the sliding movement states, the positive sliding indices D_{k0} can be chosen in the range of 2–3 and negative sliding index can be taken $D_{q0} \approx 1$.

The vibration number n can be calculated using the selected D_{k0} and D_{q0}:

$$n = 30\sqrt{\frac{gD_{k0} \sin(\mu_0 - \alpha_0)}{\pi^2 \lambda_{k0}}}$$

or

$$n = 30\sqrt{\frac{gD_{q0} \sin(\mu_0 + \alpha_0)}{\pi^2 \lambda_{q0}}} \quad (3.64)$$

If the vibration number n is determined first then the nominal vibration amplitude is:

$$\begin{cases} \lambda_{k0} = \dfrac{900gD_{k0} \sin(\mu_0 - \alpha_0)}{\pi^2 n^2} \\ \lambda_{q0} = \dfrac{900gD_{q0} \sin(\mu_0 + \alpha_0)}{\pi^2 n^2} \end{cases} \quad (3.65)$$

(2) The Positive (Negative) Sliding Angle and Positive (Negative) Sliding Coefficients

Substituting the accelerations a_y, a_x of the ellipse vibration machines into Eq. (3.18) one obtains the positive and negative sliding relative velocities of the materials:

$$\Delta \dot{x} = g(\sin \alpha_0 - f \cos \alpha_0)\frac{\varphi - \varphi_k}{\omega} - f\lambda_y \omega[\cos(\omega t + \beta_y) - \cos(\omega t_k + \beta_y)] + \lambda_x \omega \cos(\omega t_k + \beta_x) - \lambda_x \omega \cos(\omega t + \beta_x)$$

and

$$\Delta \dot{x} = g(\sin \alpha_0 + f \cos \alpha_0)\frac{\varphi - \varphi_q}{\omega} - f\lambda_y \omega[\cos(\omega t + \beta_y) - \cos(\omega t_q + \beta_y)] + \lambda_x \omega \cos(\omega t_q + \beta_x) - \lambda_x \omega \cos(\omega t + \beta_x) \quad (3.66)$$

in which

$$\omega t_k = \varphi_k - \rho_k \quad \omega t_q = \varphi_q - \rho_q$$

$$\varphi_k = \arcsin \frac{1}{D_k} \qquad \varphi_q = \arcsin \frac{-1}{D_q}$$

$$D_k = \frac{\omega^2 \lambda_k}{g \sin(\mu - \alpha_0)} \qquad D_q = \frac{\omega^2 \lambda_q}{g \sin(\mu + \alpha_0)}$$

$$\lambda_k = \frac{a \sin \mu + c \cos \mu}{\cos \rho_k} \qquad \lambda_q = \frac{-a \sin \mu + c \cos \mu}{\cos \rho_q}$$

$$\rho_k = \arctan \frac{b \sin \mu + d \cos \mu}{a \sin \mu + c \cos \mu} \qquad \rho_q = \arctan \frac{-b \sin \mu + d \cos \mu}{-a \sin \mu + c \cos \mu} \qquad (3.67)$$

where ωt_k and ωt_q are the assumed positive and negative sliding start angles; φ_k and φ_q are the assumed nominal positive and negative sliding start angles; D_k and D_q are the assumed positive and negative sliding index; λ_k and λ_q are the positive and negative sliding vibration amplitude calculated by the dynamic friction angles;. ρ_k and ρ_q the positive and negative sliding angles calculated by the dynamic friction angles respectively.

Simplifying Eq. (3.66) one obtains:

$$\Delta \dot{x} = \frac{\omega \lambda_k}{\cos \mu} [\cos \varphi_k' - \cos \varphi - \sin \varphi_k (\varphi - \varphi_k')]$$

and

$$\Delta \dot{x} = \frac{\omega \lambda_q}{\cos \mu} [\cos \varphi_q' - \cos \varphi - \sin \varphi_q (\varphi - \varphi_q')] \qquad (3.68)$$

When the positive and negative sliding relative velocities reach to zero, the positive and negative sliding end. The conditions for ending the material sliding for the ellipse vibration machines are the same as the line vibration machines:

$$\tan \varphi_k' = \frac{1 - \cos \theta_k}{\frac{\sin \varphi_k}{\sin \varphi_k'} \theta_k - \sin \theta_k} = \frac{1 - \cos 2\pi i_k}{\frac{\sin \varphi_k}{\sin \varphi_k'} 2\pi i_k - \sin 2\pi i_k}$$

and

$$\tan \varphi_q' = \frac{1 - \cos \theta_q}{\frac{\sin \varphi_q}{\sin \varphi_q'} \theta_q - \sin \theta_q} = \frac{1 - \cos 2\pi i_q}{\frac{\sin \varphi_q}{\sin \varphi_q'} 2\pi i_q - \sin 2\pi i_q} \qquad (3.69)$$

in which

$$\theta_k = \varphi_m' - \varphi_k' \qquad \theta_q = \varphi_e' - \varphi_q' \qquad i_k = \frac{\theta_k}{2\pi} \qquad i_q = \frac{\theta_q}{2\pi}$$

the denotes are the same as those in Eqs. (3.22) and (3.26).

When φ_k', φ_k, φ_q' and φ_q are known, the positive or negative sliding angle θ_k and θ_q and the positive or negative sliding end angles φ_m' and φ_e' can be found in Fig. 3.3, then the positive and negative sliding coefficients i_k and i_q can be calculated.

(3) Positive and Negative Sliding Averaged Velocity

$$\Delta x_k = \frac{\lambda_k}{\cos \mu} \int_{\varphi_k'}^{\varphi_m'} [-\cos \varphi + \cos \varphi_k' - \sin \varphi_k (\varphi - \varphi_k')] d\varphi$$

$$= \frac{\lambda_k}{\cos \mu} [-(\sin \varphi_m' - \sin \varphi_k') + \cos \varphi_k' (\varphi_m' - \varphi_k') + \sin \varphi_k \frac{(\varphi_m' - \varphi_k')^2}{2}]$$

$$= \frac{\lambda_k}{\cos \mu} [\frac{b_m'^2 - b_k'^2}{2b_k} - (b_m' - b_k')] \quad (3.70)$$

where $b_k' = \sin \varphi_k'$, $b_k = \sin \varphi_k$, $b_k' = \sin \varphi_k'$, $b_k = \sin \varphi_k$, $b_m' = \sin \varphi_m'$, $\varphi_m' = \varphi_k' + \theta_k$.

$$\Delta x_q = \frac{\lambda_q}{\cos \mu} \int_{\varphi_q'}^{\varphi_e'} [-\cos \varphi + \cos \varphi_q' - \sin \varphi_q (\varphi - \varphi_q')] d\varphi$$

$$= \frac{\lambda_q}{\cos \mu} \left[\frac{b_e'^2 - b_q'^2}{2b_q} - (b_e' - b_q') \right] \quad (3.71)$$

where $b_q' = \sin \varphi_q'$, $b_q = \sin \varphi_q$, $b_e' = \sin \varphi_e'$, $\varphi_e' = \varphi_q' + \theta_q$.

The positive and negative sliding relative velocities are:

$$v_k = \frac{\Delta x_k}{\frac{2\pi}{\omega}} = \omega \lambda_k \frac{P_{km}}{2\pi \cos \mu} \quad (3.72)$$

where

$$P_{km} = \frac{b_m'^2 - b_k'^2}{2b_k} - (b_m' - b_k')$$

$$-P_{qe} = \frac{b_m'^2 - b_q'^2}{2b_q} - (b_e' - b_q')$$

The velocity coefficients P_{km} and P_{qe} can be found in Fig. 3.3 directly given the φ_k', φ_k, φ_q', φ_q. The real average velocity is equal to the product of the theoretical average velocity and the layer thickness effect coefficients C_h:

$$v_m = C_h(v_k + v_q) \tag{3.73}$$

where C_h can be found in Fig. 3.3.

3.2.3 Theory of Material Throwing Movements

(1) The Throwing Index of the Ellipse Vibration Machines

At the time of the material begins its throwing movements, the relative acceleration along y direction $\Delta \ddot{y} = 0$, the normal pressure $F_n = 0$. Substituting a_y of Eq. (3.55) into Eq. (3.6) yields:

$$F_n = -\frac{G}{g}\omega^2 \lambda_y \sin \varphi_{dy} + G \cos \alpha_0 = 0 \tag{3.74}$$

$$\varphi_{dy} = \omega t_{dy} + \beta_y$$

where φ_{dy} is the nominal throwing start angle; and ωt_{dy} is the throwing start angle.

The nominal throwing start angle can be derived in Eq. (3.74):

$$\varphi_{dy} = \arcsin \frac{1}{D} \qquad D = \frac{\omega^2 \lambda_y}{g \cos \alpha_0} \tag{3.75}$$

where D is the throwing index.

The throwing start angle is:

$$\omega t_{dy} = \varphi_{dy} - \beta_y \tag{3.76}$$

When the throwing index $D < 1$, no solution for φ_{dy}. When $D > 1$ the throwing movements occur. When the throwing index is chosen, the vibration number n or the vertical vibration amplitude λ_y can be found:

$$n = 30\sqrt{\frac{gD \cos \alpha_0}{\pi^2 \lambda_y}}$$

3.2 Theory and Technological Parameter Computation ...

$$\lambda_y = \frac{900gD\cos\alpha_0}{\pi^2 n^2} \qquad (3.77)$$

The formula for throwing index D of the aforementioned ellipse vibration machines are valid for the circular and line vibration machines. For the ellipse vibration machines, when $b = c = 0$, $\beta_y = 0$, $\lambda_y = \lambda$, the throwing index from Eq. (3.75) is:

$$D_{\text{circular}} = \frac{\omega^2 \lambda}{g\cos\alpha_0} \qquad (3.78)$$

For the line vibration machines, when $b = d = 0$, $\beta_y = 0$, $\lambda_y = \lambda\sin\delta$, the throwing index from Eq. (3.75) is:

$$D_{\text{line}} = \frac{\omega^2 \lambda \sin\delta}{g\cos\alpha_0} \qquad (3.79)$$

When the dynamic parameters are determined, Eqs. (3.75), (3.78) and (3.79) can be used to calculate the throwing indices, then the nominal throwing start angle φ_{dy} can be calculated by Eq. (3.75), its values normally fall into the range of 0°–180°.

(2) The Throwing Leave Angle θ_d and Throwing Coefficient i_D for Ellipse Vibration Machine

Referring to Eq. (3.6), due to the throwing movements, the normal pressure $F_n = 0$, and the equations for the relative movements in y direction can be written as:

$$m\Delta\ddot{y} = -G\cos\alpha_0 - ma_y = -G\cos\alpha_0 + m\omega^2\lambda_y \sin\varphi_y \qquad (3.80)$$

$$\varphi_y = \omega t + \beta_y$$

Integrating twice of Δy with respect to t yields the relative displacement:

$$\Delta y = \lambda_y \left[\sin\varphi_{dy} - \sin\varphi_y + \cos\varphi_{dy}(\varphi_y - \varphi_{dy}) - \frac{1}{2}\sin\varphi_{dy}(\varphi_y - \varphi_{dy})^2 \right] \qquad (3.81)$$

When $\varphi_y = \varphi_{zy}$ (φ_{zy} is the throwing stop angle), $\Delta y = 0$, the throwing movements end. Simplifying Eq. (3.81) yields:

$$\cot\varphi_{dy} = \frac{\frac{\theta_d}{2} - (1 - \cos\theta_d)}{\theta_d - \sin\theta_d} \qquad (3.82)$$

$$\theta_d = \varphi_{zy} - \varphi_{dy}$$

where θ_d is the throwing leave angle.

The relation between the throwing index D and the throwing leave coefficient $i_D = \theta_d/2\pi$ can be obtained by simplifying Eq. (3.82):

$$D = \sqrt{\left(\frac{2\pi^2 i_D^2 + \cos 2\pi i_D - 1}{2\pi i_D - \sin 2\pi i_D}\right)^2 + 1} \qquad (3.83)$$

From Eqs. (3.82) and (3.83) the relation curves between φ_{dy} and θ_d and D and i_D can be generated (Fig. 3.7). When the throwing start angle is known, the throwing leave angle θ_d can be found in Fig. 3.6 or when the throwing index D is known, the throwing index i_D can be found in Fig. 3.7.

When the times of material jumps are equal to the vibration period of the vibration machines, i.e. $i_D = 1$, $\theta_d = 360°$, $D = 3.3$, the first critical vibration time of the material throwing movement is:

$$n_{e1} = 54\sqrt{\frac{g\cos\alpha_0}{\pi^2\lambda_y}} \qquad (3.84)$$

When the times of material jumps are equal to the two or three vibration periods of the vibration machines, i.e. $i_D = 2, 3$; $\theta_d = 720°, 1080°$, $D = 6.36, 9.48$ from Fig. 3.7. The second and the third critical vibration times are:

$$n_{e2} = 75\sqrt{\frac{g\cos\alpha_0}{\pi^2\lambda_y}} \qquad n_{e3} = 95\sqrt{\frac{g\cos\alpha_0}{\pi^2\lambda_y}} \qquad (3.85)$$

The vibration time for the most of the vibration machines is a little bit less than the first critical vibration time or the throwing index D is a little bit less than 3.3.

(3) The Theoretical Average Velocity of the Material Throwing Movements

The relative acceleration of the material normal to the working surface of the ellipse vibration machines can be expressed as:

$$\Delta\ddot{x} = g\sin\alpha_0 - a_x = g\sin\alpha_0 + \omega^2\lambda_x\sin\varphi_x \qquad (3.86)$$

$$\varphi_x = \omega t + \beta x$$

Integrating twice of Δx with respect to t and replacing φ with φ_{dx} yield the relative displacement each of the throwing movements:

$$\Delta x_d = \lambda_x(-\sin\varphi_{zx} + \sin\varphi_{dx} + \theta_d\cos\varphi_{dx}) + \frac{g\sin\alpha_0}{2\omega^2}\theta_d^2$$

or

3.2 Theory and Technological Parameter Computation ...

$$\Delta x_d = \lambda_x [\sin \varphi_{dx}(1 - \cos \theta_d) + \cos \varphi_{dx}(\theta_d - \sin \theta_d)] + \frac{\theta_d^2}{2} \lambda_y \tan \alpha_0 \sin \varphi_{dy}$$
(3.87)

$$\varphi_{dx} = \omega t_d + \beta_x, \quad \varphi_{zx} = \omega t_z + \beta_x = \varphi_{dx} + \theta_d$$
$$\theta_d = \varphi_{zx} - \varphi_{dx} = \varphi_{zy} - \varphi_{dy} = 2\pi i_D$$

Replacing θ_d with $2\pi i_D$ one has:

$$\Delta x_d = \lambda_x [-\sin(\varphi_{dx} + 2\pi i_D) + \sin \varphi_{dx} + 2\pi i_D \cos \varphi_{dx}] + \frac{2\pi^2 i_D^2}{D} \lambda_y \tan \alpha_0$$

or

$$\Delta x_d = \lambda_x [\sin \varphi_{dx}(1 - \cos 2\pi i_D) + \cos \varphi_{dx}(2\pi i_D - \sin 2\pi i_D)] + \frac{2\pi^2 i_D^2}{D} \lambda_y \tan \alpha_0$$
(3.88)

$$\varphi_{dx} = \omega t_d + \beta_x - \beta_y + \beta_y = \varphi_{dy} + \beta_x - \beta_y = \arcsin \frac{1}{D} + \beta_x - \beta_Y$$

The theoretical average velocity of the materials:

$$v_d = \frac{\Delta x_d}{\frac{2\pi}{\omega}} = \frac{\omega \lambda_x}{2\pi}[-\sin(\varphi_{dx} + \theta_d) + \sin \varphi_{dx} + \theta_d \cos \varphi_{dx}] + \frac{\theta_d^2}{4\pi D} \omega \lambda_y \tan \alpha_0$$
$$= \frac{\omega \lambda_x}{2\pi}[-\sin(\varphi_{dx} + 2\pi i_D) + \sin \varphi_{dx} + 2\pi i_D \cos \varphi_{dx}] + \frac{\pi i_D^2}{D} \omega \lambda_y \tan \alpha_0$$
(3.89)

or

$$v_d = \frac{\omega \lambda_x}{\frac{2\pi}{\omega}}[\sin \varphi_{dx}(1 - \cos \theta_d) + \cos \varphi_{dx}(\theta_d - \sin \theta_d)] + \frac{\theta_d}{4\pi D} \omega \lambda_y \tan \alpha_0$$
$$= \frac{\omega \lambda_x}{2\pi}[\sin \varphi_{dx}(1 - \cos 2\pi i_D) + \cos \varphi_{dx}(2\pi i_D - \sin 2\pi i_D)] + \frac{\pi i_D^2}{D} \omega \lambda_y \tan \alpha_0$$

It can be simplified as:

$$v_d = \frac{\omega \lambda_x \cos(\beta_x - \beta_y)}{2\pi D} \left\{ (\theta_d - \sin \theta_d)\left[\sqrt{D^2 - 1} - \tan(\beta_x - \beta_y)\right] \right.$$
$$\left. + (1 - \cos \theta_d)[(1 + \sqrt{D^2 - 1} \tan(\beta_x - \beta_y)]\right\} + \frac{\theta_d^2}{4\pi D} \omega \lambda_y \tan \alpha_0 \quad (3.90)$$

The theoretical average velocity of the materials of the ellipse vibration machines is also valid to the line and circular vibration machines. For the circular vibration machines, $\lambda_x = \lambda_y = \lambda$, $\beta_x = \beta_y - \pi/2$, the theoretical average velocity is:

$$\begin{aligned} V_{\text{dcircular}} &= \frac{\omega\lambda}{2\pi D}[\theta_d - \sin\theta_d - \sqrt{D^2-1}(1-\cos\theta_d) + \frac{\theta_d^2}{2}\tan\alpha_0] \\ &= \frac{\omega\lambda}{2\pi D}[2\pi i_D - \sin 2\pi i_D - \sqrt{D^2-1}(1-\cos 2\pi i_D) + 2\pi^2 i_D^2 \tan\alpha_0] \end{aligned}$$
(3.91)

For the line vibration machines, $\lambda_x = \lambda\cos\delta$, $\lambda_y = \lambda\sin\delta$, $\beta_x = \beta_y$, the theoretical average velocity is:

$$V_{\text{dline}} = \omega\lambda\cos\delta\frac{\theta_d^2}{4\pi D}(1+\tan\delta\tan\alpha_0) = \omega\lambda\cos\delta\frac{\pi i_D^2}{D}(1+\tan\delta\tan\alpha_0) \quad (3.92)$$

The real average velocity of the material movements can be expresses as:

$$v_m = \gamma_a C_h C_m C_w v_d \quad (3.93)$$

The coefficients γ_a, C_h, C_m, C_w can be found by Tables 3.3, 3.4 and 3.5.

In the ellipse vibration machines, the movement states of the materials throwing are similar to those of line vibration machines.

(4) The Relative Impact Velocity when the Material is Falling

The relative impact velocity of the material falling is:

$$\Delta\dot{y} = \lambda_y\omega\left(\cos\varphi_{dy} - \cos\varphi_{zy} - g\cos\alpha_0\frac{\theta_d}{\omega^2\lambda_y}\right)$$

After simplifying, it can be written as:

$$\begin{aligned} \Delta\dot{y} &= \frac{g\cos\alpha_0}{\omega}[\sqrt{D^2-1}(1-\cos\theta_d) - (\theta_d - \sin\theta_d)] \\ &= \frac{g\cos\alpha_0}{\omega}[\sqrt{D^2-1}(1-\cos 2\pi i_D) - (2\pi i_D - \sin 2\pi i_D)] \end{aligned}$$
(3.94)

For those materials not required to be crushed, the throwing index D should be chosen with a small impact velocity to reduce the possibility to be crushed during the conveying process.

(5) Practical Example

Example 3.3 Given a double body inertial near resonance vibration conveyor, to be used to convey lime. The density of the lime is 1.1 t/m³, the conveying trough incline angle $\alpha_0 = 0$. The trough body has the ellipse movement. The long axis of the ellipse has 30° angle with the trough bottom surface. The vibration amplitudes in two directions are 4 mm and 0.8 mm respectively. The vibration frequency is 98.4 rad/s. Find the average velocity of the material for the positive and reverse rotations.

3.2 Theory and Technological Parameter Computation …

Solution

(1) From Eq. (3.54), compute the vibration amplitudes, λ_x and λ_y in x and y direction and its phase angles β_x, β_y:

$$\lambda_y = \sqrt{a^2 + b^2} = \sqrt{(4\sin 30°)^2 + (0.8\cos 30°)^2} = 2.12\,\text{mm}$$

$$\lambda_x = \sqrt{c^2 + d^2} = \sqrt{(4\cos 30°)^2 + (0.8\sin 30°)^2} = 3.49\,\text{mm}$$

$$\beta_y = \arctan\frac{b}{a} = \arctan\frac{0.8\cos 30°}{4\sin 30°} = 19°6'$$

$$\beta_x = \arctan\frac{d}{c} = \arctan\frac{-0.8\cos 30°}{4\sin 30°} = -6°35'$$

For the reverse rotation,

$$\beta_y = \arctan\frac{-0.8\cos 30°}{4\sin 30°} = -19°6'$$

$$\beta_x = \arctan\frac{0.8\cos 30°}{4\sin 30°} = 6°35'$$

(2) Throwing Index

$$D = \frac{\omega^2 \lambda_y}{g\cos\alpha_0} = \frac{98.4^2 \times 2.12}{9800 \times \cos 0°} = 2.09$$

(3) Nominal Throwing Start Angle and Throwing Angle

$$\varphi_{dy} = \arcsin\frac{1}{D} = \arcsin\frac{1}{2.09} = 28°35'$$

$$\omega t_{dy} = \varphi_{dy} - \beta_y = 28°35' - 19°6' = 9°29'$$

For the reverse rotation:

$$\omega t_{dy} = \varphi_{dy} - \beta_y = 28°35' + 19°6' = 47°41'$$

(4) Throwing Leave Coefficient i_D. Finding it with D give: $i_D = 0.76$.
(5) Theoretical Average Velocity of the Material Movement

$$v_d = \frac{\omega\lambda_x}{2\pi}[-\sin(\varphi_{dx} + 2\pi i_D) + \sin\varphi_{dx} + 2\pi i_D \cos\varphi_{dx}] + \frac{\pi i_D^2}{D}\omega\lambda_y \tan\alpha_0$$

$$= \frac{98.4 \times 3.49}{2\pi}[-\sin(2°54' + 360° \times 0.76) + \sin 2°54' + 2\pi \times 0.76\cos 2°54']$$

$$= 54.68 \times (0.99 + 0.05 + 4.77) = 317.7\,\text{mm}/\text{s} = 0.318\,\text{m}/\text{s}$$

in which

$$\varphi_{dx} = \arcsin \frac{1}{D} + \beta_x - \beta_y = 28°35' - 6°35' - 19°6' = 2°54'$$

When reverse rotation:

$$v_d = \frac{98.4 \times 3.49}{2\pi}[-\sin(54°16' + 360° \times 0.76) + \sin 54°16' + 2\pi \times 0.76 \cos 54°16'] = 225.8 \text{ mm/s} = 0.226 \text{ m/s}.$$

in which

$$\varphi_{dx} = \arcsin \frac{1}{D} + \beta_x - \beta_y = 28°35' + 6°35' + 19°16' = 54°16'$$

(6) Real Average Velocity

$$v_m = C_h C_m C_w v_d = 0.8 \times 0.8 \times 1.05 \times 0.318 = 0.214 (\text{m/s})$$

When reverse rotation:

$$v_m = C_h C_m C_w v_d = 0.8 \times 0.8 \times 1.05 \times 0.226 = 0.15 (\text{m/s})$$

3.3 Basic Characteristics of Material Movement in Non-harmonic Vibration Machines

3.3.1 Initial Conditions for Positive and Negative Sliding Movements

The displacement for this vibration machine body:

$$x = \lambda_1 \sin \omega t + \lambda_2 \sin(2\omega t + \theta_2) \quad (3.95)$$

in which

$$\lambda_1 = -\frac{m_{01} r_1}{m} \quad \lambda_2 = -\frac{m_{02} r_2}{m}$$

When $\theta_2 = -\pi/2$, its displacement, velocity and acceleration can be expressed as:

3.3 Basic Characteristics of Material Movement …

$$x = \lambda_1 \sin \omega t + \lambda_2 \sin(2\omega t - \frac{\pi}{2})$$
$$\dot{x} = \omega\lambda_1 \cos \omega t + 2\omega\lambda_2 \cos(2\omega t - \frac{\pi}{2})$$
$$\ddot{x} = -\omega^2\lambda_1 \sin \omega t - 4\omega^2\lambda_2 \sin(2\omega t - \frac{\pi}{2}) \qquad (3.96)$$

Figure 3.17 shows the displacement, velocity and acceleration curves calculated by the above equations.

When the materials move together with the body, its acceleration is equal to the body's acceleration. The inertial force of the materials is equal to $-m\ddot{x}$. At this time, the body generates a friction force which is the same in amplitude and opposite in direction as the $-m\ddot{x}$. The friction force F is less than or equal to the limit friction force $f_0 F_n$ (F_n is the normal force, $F_n = mg$ when the body is installed horizontally and vibrates in the horizontal direction. When the material has the positive sliding tendency, the friction is negative and when it has the negative sliding tendency, the friction is positive. Thus the condition for the material to start the positive and negative sliding is:

$$-m\ddot{x} \mp f_0 mg = 0$$

or

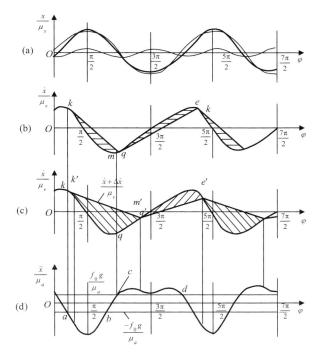

Fig. 3.17 The displacement, velocity and acceleration curves and material movement curves of the non-harmonic vibration machines. **a** displacement curve, **b** discontinuous sliding velocity curve, **c** continuous sliding velocity curve, **d** acceleration curves

$$-\ddot{x} \mp f_0 g = 0 \qquad (3.97)$$

The displacement is x/μ_s, velocity $\dot{x}/\mu_v$ and acceleration $\ddot{x}/\mu_a$ in Fig. 3.17. In Fig. 3.17d there are two straight lines: $\pm f_0 g/\mu_a$. There are four intersection points between the line $\pm f_0 g/\mu_a$ and the acceleration curve which satisfies Eq. (3.97). The interval $[a, b]$ is the positive sliding region; the interval $[c, d]$ is the negative sliding region. When the material is fed into these two intervals, the positive or negative sliding will occur. As mentioned in Sect. 3.1 for the line vibration machines, when the positive or negative sliding has discontinuity, its discontinuity starts at point a and b respectively. When there is a discontinuity for the positive and negative sliding, the velocity line, line km, is as shown in Fig. 3.17b. When the positive and negative sliding have no discontinuity, the material steady sliding velocity straight line can be drawn using the continuous drawing methods (line $k'm'$ and $q'e'$ in Fig. 3.17c).

3.3.2 Stopping Conditions for Positive and Negative Sliding Movements

When the material starts positive sliding, the friction force will decelerate the material movement. In each vibration period T, the velocity reduction is $f_0 gT$. The slope in the velocity straight line is negative, i.e., the velocity reduction is $f_0 gT/\mu_v$ in an interval of 2π. When the material starts negative sliding, the friction is positive, the velocity of the material will increase $f_0 gT$ in each vibration period. In the velocity figure, the velocity increase is $f_0 gT/\mu_v$ in an interval of 2π. Drawing a vertical line up from point a at the acceleration curve, the line intersects the velocity curve at point k. Starting from point k draw a velocity curve of the material starting to make a sliding movement which intersects itself at point m. If point m is falling in the non-negative sliding region, then the movement is discontinuous, and negative sliding starts at point q. From the point q draw a negative sliding velocity curve and it intersects itself at point e. If point e falls into a non-positive sliding region, then the movement is discontinuous. If point m falls into a negative sliding region, the movement is non-discontinuous, the negative sliding starts from point m, i.e. point q', and ends at point e. If point e falls into a positive sliding region, then the positive sliding starts from point e, i.e. k', and ends at point e. Repeating this processes several times one can obtain the positive and negative sliding movement stable velocity graphs and the average velocity of the materials can be calculated by the velocity graphs.

3.3.3 Calculations of Averaged Material Velocity

The Area Meter can be used to measure the relative displacement of the material positive sliding (the shadow area under the relative velocity in the velocity graphs).

The relative displacement of the real positive sliding Δx_k (in cm) is:

$$\Delta x_k = F_k \mu_v \mu_\varphi \frac{1}{\omega} \quad (3.98)$$

in which μ_v is the velocity scale, i.e. the real velocity ((cm/s)/mm) represented by a millimeter in the vertical axis in the velocity graph; μ_φ is the rotation angle scale, i.e. the radian represented by a millimeter (rad/mm) in a horizontal axis in the velocity graphs.

By the same token, the area meter can also be used to measure the negative sliding relative displacement and the real relative displacement Δx_q (in cm) is:

$$\Delta x_q = -F_k \mu_v \mu_\varphi \frac{1}{\omega} \quad (3.99)$$

The average velocity for the material sliding v_m (cm/s) is

$$v_m = \frac{\omega}{2\pi}(\Delta x_k + \Delta x_q) \quad (3.100)$$

Comparing among the average velocities for the different parameters can help determine the better dynamic parameters (such as frequency ω, vibration amplitude λ_1 and vibration amplitude ratio λ_1/λ_2) for the non-harmonic vibration machines.

For the discussions on installation incline angle, the similar analysis can be conducted as before.

3.4 Theory on Material Movement in Vibrating Centrifugal Hydroextractor

The hydroextractors are used for extracting water contents from the fine particle coals etc. The vibration centrifugal hydroextractors have two types: upright and horizontal. The main shafts of the upright and horizontal are installed upright and horizontally respectively. Their bodies are a cone-shaped with mesh holes. The water contents in the materials are thrown off from the mesh holes by the centrifugal force. The cone body is vibrating along the cone axis direction and the material left is moved from the small end to the large end and is discharged out.

The force diagram is shown in Fig. 3.18.

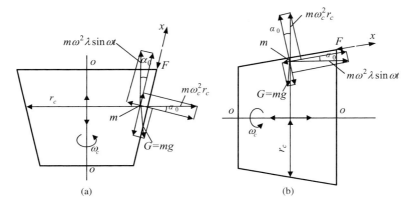

Fig. 3.18 Force analysis of materials in the vibration hydroextractors cone. **a** Upright vibration hydroextractor, **b** horizontal vibration hydroextractor

3.4.1 Basic Characteristics of Material Movement on Upright Vibration Hydroextractor

When the materials in the vibration hydroextractor move with the cone body without relative movement, the forces acting on the materials are: the inertial force of the materials due to its rotation with the cone body $Q = m\omega_c^2 r_c$ (m is the mass of the material, ω_c is the angular velocity of the cone body rotation, r_c is the radius of the cone body where the material resides); the gravity of the material itself $G = mg$ (it can be ignored compared to the centrifugal force); the inertial force of the material along the cone vertical direction $P_u = m\omega^2 \lambda \sin \omega t$ (ω is the vibration angular frequency, λ is the vibration amplitude) and the friction of the material to the cone body.

When the material slides along the cone generatrix, the normal forces of the material acting on the cone surface are:

$$F_n = m\omega_c^2 r_c \cos \alpha_0 + m(g - \omega^2 \lambda \sin \omega t) \sin \alpha_0 \qquad (3.101)$$

The sum of the force components of gravity and inertial force along the generatrix is:

$$P = m\omega_c^2 r_c \sin \alpha_0 + m[(\omega^2 \lambda \sin \omega t - g) \cos \alpha_0 - \Delta \ddot{x}] \qquad (3.102)$$

where α_0 is the angle between the cone surface and its axis line; $\Delta \ddot{x}$ is the relative acceleration of the material to the cone body along the generatrix.

The cone body friction to the material is:

$$F \leq f_0[m\omega_c^2 r_c \cos \alpha_0 + m(g - \omega^2 \lambda \sin \omega t) \sin \alpha_0, \quad \text{when } \Delta \ddot{x} = 0$$
$$F = f[m\omega_c^2 r_c \cos \alpha_0 + m(g - \omega^2 \lambda \sin \omega t) \sin \alpha_0, \quad \text{when } \Delta \ddot{x} \neq 0$$

3.4 Theory on Material Movement in Vibrating …

where f_0 is the static friction of the material to the cone body; f is the dynamic friction of the material to the cone body.

(1) Positive Sliding Index D_k, Negative Sliding Index D_q and Throwing Index D

In order to make the vibration centrifugal hydroextractor work effectively, the materials must keep contact often with the cone surface. That is, the normal force F_n of the material to the cone surface F_n should be larger than zero, from Eq. (3.101):

$$F_n = m\omega_c^2 r_c \cos\alpha_0 + m(g - \omega^2 \lambda \sin\omega t)\sin\alpha_0 > 0$$

When $\sin\omega t = 1$, the left hand term on the equation above is the minimum, after simplifying, one obtains:

$$m\sin\alpha_0(\omega_c^2 r_c \cot\alpha_0 + g)(1 - D) > 0 \qquad (3.103)$$

where D is the throwing index.

For the material to keep touch with the cone surface, the throwing index D should be less than 1: $D < 1$.

When the components in x direction of the inertial and the gravity force and friction force are equal, the material will begin the positive and negative sliding:

$$m\omega_c^2 r_c \sin\alpha_0 + m\left[(\omega^2 \lambda \sin\omega t - g)\cos\alpha_0 - \Delta\ddot{x}\right] \mp f_0 F_n = 0 \qquad (3.104)$$

where the sign of the last term: "−" denotes positive sliding and "+" the negative sliding.

Substituting F_n in Eq. (3.101) into Eq. (3.104), letting $\Delta x = 0$ and simplifying the results one obtains the positive and negative sliding start angle φ_{k0} and φ_{q0}:

$$\begin{aligned}
\varphi_{k0} &= \arcsin \frac{1}{D_{k0}} \\
\varphi_{q0} &= \arcsin \frac{1}{D_{q0}} \\
D_{k0} &= \frac{\omega^2 \lambda}{\omega_c^2 r_c \tan(\mu_0 - \alpha_0) + g} \\
D_{q0} &= \frac{\omega^2 \lambda}{\omega_c^2 r_c \tan(\mu_0 + \alpha_0) - g}
\end{aligned} \qquad (3.105)$$

It can be seen from Eq. (3.105) that the positive sliding start angle φ_{k0} is in the range of 0°–180° while the negative sliding start angle φ_{q0} in the range of 180°–360°.

The condition for the hydroextractor working normally is that the positive sliding index $D_{k0} > 1$, while the negative sliding is no practical meaning the negative sliding index $D_{q0} < 1$ or $D_{q0} \approx 1$.

(2) The Approximate Calculation of the Theoretical Average Velocity for the Positive and Negative Sliding

The approximate calculation is based on two assumptions: the relative velocity of the material along the circumference direction are too small to be ignored and r_c can be considered as a constant because it doesn't change much for each of the sliding process.

Under the assumptions, referring to Eq. (3.102), the equation of the motion of the materials along the cone generatrix direction can be written as:

$$\Delta \ddot{x} = \omega_c^2 r_c \sin \alpha_0 + \left(\omega^2 \lambda \sin \omega t - g\right) \cos \alpha_0$$
$$\mp f \left[\omega_c^2 r_c \cos \alpha_0 + (-\omega^2 \lambda \sin \omega t + g) \sin \alpha_0\right] \quad (3.106)$$

The relative velocities of the positive and negative sliding are:

$$\Delta \dot{x}_k = \frac{\omega \lambda \cos(\mu - \alpha_0)}{\cos \mu} \left[\cos \varphi_{k0} - \cos \varphi - \sin \varphi_k (\varphi - \varphi_{k0})\right] \quad (3.107)$$

in which $\sin \varphi_k = \frac{1}{D_k}$, $D_k = \frac{\omega^2 \lambda}{\omega_c^2 r_c \tan(\mu - \alpha_0) + g}$.

$$\Delta \dot{x}_q = \frac{\omega \lambda \cos(\mu + \alpha_0)}{\cos \mu} \left[\cos \varphi_{q0} - \cos \varphi - \sin \varphi_q (\varphi - \varphi_{q0})\right] \quad (3.108)$$

in which $\sin \varphi_q = \frac{1}{D_q}$, $D_q = \frac{\omega^2 \lambda}{\omega_c^2 r_c \tan(\mu - \alpha_0) - g}$.

When $\varphi = \varphi_{m0}$ and $\varphi = \varphi_{e0}$, the ending conditions for the positive and negative sliding are:

$$\cos \varphi_{k0} - \cos \varphi_{m0} - \sin \varphi_k (\varphi_{m0} - \varphi_{k0}) = 0$$
$$\cos \varphi_{q0} - \cos \varphi_{e0} - \sin \varphi_q (\varphi_{e0} - \varphi_{q0}) = 0 \quad (3.109)$$

where φ_{m0} is the positive sliding end angle and φ_{e0} is the negative sliding end angle.

After the positive sliding start angle φ_{k0} and the assumed positive sliding start angle φ_k is found by Eqs. (3.105) and (3.10) one can find the positive sliding end angle φ_{m0} and similarly find the negative sliding end angle φ_{e0}.

The relative displacement of each material sliding is the same as that for the line vibration machines:

$$\Delta x_k = \int_{\varphi_{k0}}^{\varphi_{m0}} \Delta \dot{x}_k \mathrm{d}t = \int_{\varphi_{k0}}^{\varphi_{m0}} \frac{\omega \lambda \cos(\mu - \alpha_0)}{\cos \mu} \left[\cos \varphi_{k0} - \cos \varphi - \sin \varphi_k (\varphi - \varphi_{k0})\right] \mathrm{d}\varphi$$
$$= \frac{\lambda \cos(\mu - \alpha_0)}{\cos \mu} P_{km}$$

3.4 Theory on Material Movement in Vibrating ...

$$\Delta x_q = \int_{\varphi_{q0}}^{\varphi_{e0}} \Delta \dot{x}_q dt = \int_{\varphi_{q0}}^{\varphi_{e0}} \frac{\omega\lambda \cos(\mu + \alpha_0)}{\cos\mu}[\cos\varphi_{q0} - \cos\varphi - \sin\varphi_q(\varphi - \varphi_{q0})]d\varphi$$

$$= -\frac{\lambda\cos(\mu+\alpha_0)}{\cos\mu}P_{qe} \qquad (3.110)$$

in which
$$P_{km} = \frac{\sin^2\varphi_{m0} - \sin^2\varphi_{k0}}{2\sin\varphi_k} - (\sin\varphi_{m0} - \sin\varphi_{k0})$$

$$-P_{qe} = \frac{\sin^2\varphi_{e0} - \sin^2\varphi_{q0}}{2\sin\varphi_q} - (\sin\varphi_{e0} - \sin\varphi_{q0})$$

The velocity P_{km} and P_{qe} can be found via Fig. 3.3.
The average velocities of the material positive and negative sliding are:

$$v_k = \omega\lambda\cos\alpha_0 \frac{1}{2\pi}(1+f\tan\alpha_0)P_{km} \quad (m/s)$$
$$v_q = -\omega\lambda\cos\alpha_0 \frac{1}{2\pi}(1-f\tan\alpha_0)P_{qe} \quad (m/s) \qquad (3.111)$$

For the vibration hydroextractors used in the fine particle coal and dirt separation the static friction coefficient f_0 is 0.4–0.6 and the dynamic friction coefficient f is 0.15–0.4.

(3) The Accurate Calculation of the Theoretical Average Velocity for the Positive and Negative Sliding

The accurate calculation is removed one of the assumption in the approximate calculation, i.e. the radius is not considered as a constant any more:

$$r_c = r_0 + \Delta x \sin\alpha_0 \qquad (3.112)$$

where r_0 is the radius at the sliding start point; α_0 is the angle between the cone surface and the vertical line; Δx is the distance between the position after the sliding and initial position.

Substituting r_c in Eq. (3.112) into Eq. (3.106) yields the accurate equation of motion of the material along the cone generatrix:

$$\Delta \ddot{x} = \omega_c^2(r_0 + \Delta x \sin\alpha_0)\sin\alpha_0 + (\omega^2\lambda\sin\omega t - g)\cos\alpha_0$$
$$\mp f[\omega_c^2(r_0 + \Delta x \sin\alpha_0)\cos\alpha_0 + (-\omega^2\lambda\sin\omega t + g)\sin\alpha_0] \qquad (3.113)$$

in which "−" denotes the positive sliding and "+" the negative.
Simplifying Eq. (3.113) one gets:

$$\Delta \ddot{x} + \omega_c^2 \frac{\sin(\mu - \alpha_0) \sin \alpha_0}{\cos \mu} \Delta x =$$

$$-\frac{1}{\cos \mu}\left[\omega_c^2 r_0 \sin(\mu - \alpha_0) + g \cos(\mu - \alpha_0)\right] + \frac{\omega^2 \lambda \cos(\mu - \alpha_0)}{\cos \mu} \sin \omega t \quad (3.114)$$

Assume:

$$\omega_0^2 = \omega_c^2 \frac{\sin(\mu - \alpha_0) \sin \alpha_0}{\cos \mu}$$

$$q_0 = -\frac{\cos(\mu - \alpha_0)}{\cos \mu}\left[\omega_c^2 r_0 \tan(\mu - \alpha_0) + g\right]$$

$$q = \omega^2 \lambda \frac{\cos(\mu - \alpha_0)}{\cos \mu}$$

and substituting them into Eq. (3.114):

$$\Delta \ddot{x} + \omega_0^2 \Delta x = q_0 + q \sin \omega t \quad (3.115)$$

Equation (3.115) is non-homogeneous second order ordinary differential equation. It can be seen from Eq. (3.114) that for the vibration hydroextractor μ is generally larger than α_0 and ω_0^2 is positive. Thus Eq. (3.115) has the following solution:

$$\Delta x = C_1 \sin \omega_0 t + C_2 \cos \omega_0 t + \frac{q_0}{\omega_0^2} + \frac{q}{\omega_0^2 - \omega^2} \sin \omega t \quad (3.116)$$

in which C_1, C_2 are integral constants.

C_1 and C_2 are determined by initial conditions. When $\omega t_0 = \varphi_{k0}, \Delta x = 0, \Delta \ddot{x} = 0$, i.e. at the instance when the material starts to slide the relative displacement and velocity are equal to zero. Substituting $t_0 = \varphi_{k0}/\omega$ Eq. (3.116) yields:

$$C_1 \sin \omega_0 t_0 + C_2 \cos \omega_0 t_0 + \frac{q_0}{\omega_0^2} + \frac{q}{\omega_0^2 - \omega^2} \sin \omega t_0 = 0$$

$$C_1 \omega_0 \cos \omega_0 t_0 - C_2 \omega_0 \sin \omega_0 t_0 + \frac{q}{\omega_0^2 - \omega^2} \omega \cos \omega t_0 = 0 \quad (3.117)$$

Solving C_1 and C_2 yields:

$$C_1 = -\frac{1}{\omega_0}\left[\left(\frac{q_0}{\omega_0^2} + \frac{q}{\omega_0^2 - \omega^2} \sin \omega t_0\right) \omega_0 \sin \omega_0 t_0 + \frac{q\omega}{\omega_0^2 - \omega^2} \cos \omega_0 t_0 \cos \omega t_0\right]$$

$$C_2 = -\frac{1}{\omega_0}\left[\left(\frac{q_0}{\omega_0^2} + \frac{q}{\omega_0^2 - \omega^2} \sin \omega t_0\right) \omega_0 \cos \omega_0 t_0 - \frac{q\omega}{\omega_0^2 - \omega^2} \sin \omega_0 t_0 \cos \omega t_0\right]$$

$$(3.118)$$

3.4 Theory on Material Movement in Vibrating ... 111

Substituting C_1 and C_2 into Eq. (3.116) yields:

$$\Delta x = \frac{q_0}{\omega_0^2}[1 - \cos \omega_0(t - t_0)] - \frac{q}{\omega_0^2 - \omega^2}$$
$$\times \left[\sin \omega t_0 \cos \omega_0(t - t_0) + \frac{\omega}{\omega_0} \cos \omega t_0 \sin \omega_0(t - t_0) - \sin \omega t\right] \quad (3.119)$$

The relative velocity is:

$$\Delta \dot{x} = \frac{q_0}{\omega_0} \sin \omega_0(t - t_0) + \frac{q}{\omega_0^2 - \omega^2}\{\omega_0 \sin \omega t_0 \sin \omega_0(t - t_0)$$
$$+ \omega[\cos \omega t - \cos \omega t_0 \cos \omega_0(t - t_0)]\} \quad (3.120)$$

When $t = t_m = \frac{\varphi_m}{\omega}$, the positive sliding ends. Replacing t in Eq. (3.120) with t_m and letting $\Delta x = 0$, the condition of the positive sliding ending is:

$$\cos \varphi_m - \cos \varphi_{k0} \cos \frac{\omega_0}{\omega}(\varphi_m - \varphi_{k0}) + \frac{\omega_0}{\omega} \sin \varphi_{k0} \sin \frac{\omega_0}{\omega}(\varphi_m - \varphi_{k0})$$
$$+ \frac{q_0}{q} \frac{\omega_0^2 - \omega^2}{\omega_0^2} \frac{\omega_0}{\omega} \sin \frac{\omega_0}{\omega}(\varphi_m - \varphi_{k0}) = 0$$

or

$$\cos(\varphi_{k0} + \theta_k) = \cos \varphi_{k0} \cos \frac{\omega_0}{\omega}\theta_k - \frac{\omega_0}{\omega}\left(\sin \varphi_{k0} + \frac{q_0}{q}\frac{\omega_0^2 - \omega^2}{\omega_0^2}\right) \sin \frac{\omega_0}{\omega}\theta_k$$
$$(3.121)$$

In which $\theta_k = \varphi_m - \varphi_{k0}$; φ_m is the positive sliding stop angle; θ_k is the positive sliding angle.

The positive sliding start angle φ_{k0} can be found by Eq. (3.105):

$$\varphi_{k0} = \arcsin \frac{1}{D_{k0}}, \quad D_{k0} = \frac{\omega^2 \lambda}{\omega_c^2 r_0 \tan(\mu_0 - \alpha_0) + g} = -\frac{q}{q_0}$$

When ω_0/ω is small, Eq. (3.121) can be simplified as:

$$\cos(\varphi_{k0} + \theta_k) = \cos \varphi_{k0} - \theta_k \sin \varphi_k \quad (3.122)$$

$$\varphi_k = \arcsin \frac{1}{D_k}, \quad D_k = \frac{\omega^2 \lambda}{\omega_c^2 r_0 \tan(\mu - \alpha_0) + g} = -\frac{q}{q_0} \quad (3.123)$$

After φ_{k0} and φ_k were found via Eqs. (3.105) and (3.123), the sliding start angle θ_k and sliding end angle φ_m can be found in Fig. 3.3.

The relative displacement of the positive sliding is obtained by Eq. (3.119):

$$\Delta x_k = \frac{\lambda \omega^2 \cos(\mu - \alpha_0)}{(\omega^2 - \omega_0^2) \cos \mu} \left[\frac{\omega}{\omega_0} \cos \varphi_{k0} \sin \frac{\omega_0}{\omega} \theta_k + \left(\sin \varphi_{k0} + \frac{\omega^2 - \omega_0^2}{\omega_0^2 D_k} \right) \right.$$
$$\left. \times \cos \frac{\omega_0}{\omega} \theta_k - \sin(\varphi_{k0} + \theta_k) - \frac{\omega^2 - \omega_0^2}{\omega_0^2 D_k} \right] \tag{3.124}$$

When $\omega^2 \gg \omega_0^2$ the above equation can be simplified as:

$$\Delta x_k = \frac{\lambda \cos(\mu - \alpha_0)}{\cos \mu} \left[\theta_k \cos \varphi_{k0} + \sin \varphi_{k0} - \sin(\varphi_{k0} + \theta_k) - \frac{\theta_k^2}{2} \sin \varphi_k \right]$$
$$= \frac{\lambda \cos(\mu - \alpha_0)}{\cos \mu} P_{km} \tag{3.125}$$

This result is similar to the one obtained by the theoretical velocity calculation formula for the line vibration machine's positive sliding. Thus as long as φ_{k0} and φ_k can be computed, the relative displacement Δx_k can be found.

The average velocity of the material positive sliding can be accurately calculated as:

$$v_k = \frac{\Delta x_k}{\frac{2\pi}{\omega}} = \frac{\lambda \omega^2 \cos(\mu - \alpha_0)}{2\pi (\omega^2 - \omega_0^2) \cos \mu}$$
$$\times \left[\frac{\omega}{\omega_0} \cos \varphi_{k0} \sin \frac{\omega_0}{\omega} \theta_k + \left(\sin \varphi_{k0} + \frac{\omega^2 - \omega_0^2}{\omega_0^2 D_k} \right) \right.$$
$$\left. \times \cos \frac{\omega_0}{\omega} \theta_k - \sin(\varphi_{k0} + \theta_k) - \frac{\omega^2 - \omega_0^2}{\omega_0^2 D_k} \right] \tag{3.126}$$

If $\tan \mu$ is replaced by f, then Eq. (3.126) can be simplified as:

$$v_k = \lambda \omega \cos \alpha_0 \frac{\omega^2}{\omega^2 - \omega_0^2} \frac{1 + f \tan \alpha_0}{2\pi}$$
$$\times \left[\frac{\omega}{\omega_0} \cos \varphi_{k0} \sin \frac{\omega_0}{\omega} \theta_k + \left(\sin \varphi_{k0} + \frac{\omega^2 - \omega_0^2}{\omega_0^2} \frac{1}{D_k} \right) \right.$$
$$\left. \times \cos \frac{\omega_0}{\omega} \theta_k - \sin(\varphi_{k0} + \theta_k) - \frac{\omega^2 - \omega_0^2}{\omega_0^2} \frac{1}{D_k} \right] \tag{3.127}$$

When ω_0/ω is given, and when $D_k = D_{k0}$, i.e. $\varphi_k = \varphi_{k0}$ (when $\mu = \mu_0$) then θ_k can be found in Fig. 3.19 (curves of Eq. (3.121)).

When the material is in the negative sliding, its average velocity can be derived by the similar method:

$$v_q = \lambda \omega \cos \alpha_0 \frac{\omega^2}{\omega^2 - \omega_0'^2} \frac{1 - f \tan \alpha_0}{2\pi} \left[\frac{\omega}{\omega_0'} \cos \varphi_{q0} \sin \frac{\omega_0'}{\omega} \theta_q \right.$$

3.4 Theory on Material Movement in Vibrating …

Fig. 3.19 Curve φ_{k0} when $\omega_0/\omega = 0$, and 0.2

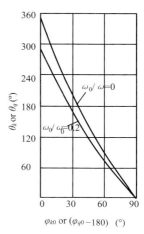

$$+ \left(\sin \varphi_{q0} + \frac{\omega^2 - \omega_0'^2}{\omega_0'^2} \frac{1}{D_q} \right) \cos \frac{\omega_0'}{\omega} \theta_q - \sin(\varphi_{q0} - \theta_q) - \frac{\omega^2 - \omega_0'^2}{\omega_0'^2} \frac{1}{D_q} \Bigg] \tag{3.128}$$

in which

$$\omega_0' = \omega_c^2 \frac{\sin(\mu + \alpha_0) \sin \alpha_0}{\cos \mu}$$

$$D_{q0} = \frac{\lambda \omega^2}{\omega_c^2 r_c \tan(\mu_0 + \alpha_0) + g}$$

$$\varphi_{q0} = \arcsin -\frac{1}{D_{q0}}$$

$$D_q = \frac{\lambda \omega^2}{\omega_c^2 r_c \tan(\mu + \alpha_0) + g}$$

$$\varphi_q = \arcsin -\frac{1}{D_q}$$

The real material average velocity is:

$$v_m = C_k(v_k + v_q) \tag{3.129}$$

In which C_k is the influence coefficient of the thickness of the material layer.

3.4.2 Characteristics of Material Movement on Horizontal Vibration Hydroextractor

The difference between the forces acted on the materials of the horizontal vibration hydroextractor and the upright vibration centrifugal hydroextractor is: the centrifugal force is in the same vertical plane as the gravity for the horizontal hydroextractor. Thus the positive sliding index D_{k0} and D_k and the negative sliding index D_{q0} and D_q should be written as:

$$D_{k0} = \frac{\omega^2 \lambda}{\omega_c^2 r_c + g \sin \omega_c t} \cot(\mu_0 - \alpha_0)$$

$$D_k = \frac{\omega^2 \lambda}{\omega_c^2 r_c + g \sin \omega_c t} \cot(\mu - \alpha_0)$$

$$D_{q0} = \frac{\omega^2 \lambda}{\omega_c^2 r_c + g \sin \omega_c t} \cot(\mu_0 + \alpha_0) \quad (3.130)$$

$$D_q = \frac{\omega^2 \lambda}{\omega_c^2 r_c + g \sin \omega_c t} \cot(\mu + \alpha_0)$$

In which $\omega_c t$ is the angle the cone body rotates after time t.

Since the centrifugal force $\omega_c^2 r_c$ and g are in the same vertical plane, and $\omega_c^2 r_c \gg g$, the ignorance of the last term in the denominator would cause a big calculation error. Equation (3.130) can be rewritten as:

$$D_{k0} = \frac{\omega^2 \lambda}{\omega_c^2 r_c} \cot(\mu_0 - \alpha_0)$$

$$D_k = \frac{\omega^2 \lambda}{\omega_c^2 r_c} \cot(\mu - \alpha_0)$$

$$D_{q0} = \frac{\omega^2 \lambda}{\omega_c^2 r_c} \cot(\mu_o + \alpha_0) \quad (3.131)$$

$$D_q = \frac{\omega^2 \lambda}{\omega_c^2 r_c} \cot(\mu + \alpha_0)$$

The minimum sliding start angles φ_{k0}, φ_{q0} for the positive and negative sliding and the assumed sliding start angle φ_k, φ_q can be found respectively:

$$\varphi_{k0} = \arcsin \frac{1}{D_{k0}}$$

$$\varphi_k = \arcsin \frac{1}{D_k}$$

3.4 Theory on Material Movement in Vibrating ...

$$\varphi_{q0} = \arcsin -\frac{1}{D_{q0}}$$

$$\varphi_q = \arcsin -\frac{1}{D_q} \tag{3.132}$$

Once φ_{k0}, φ_{q0}, φ_k, φ_q are found, θ_k and θ_q can be found from Fig. 3.3. Equations (3.126) and (3.128) can be used to find the material theoretical velocity and Eq. (3.126) can be used to find out the rear average velocity.

3.4.3 Computation of Kinematics and Technological Parameters of Vibration Centrifugal Hydroextractor

The kinematics parameters for the vibration centrifugal hydroextractor include the cone angle α_0, cone rotation speed n_c, cone body vibration time n and vibration amplitude λ, material movement average velocity v and staying time in the cone t etc. The technological parameters of it include the output and hydro-extracting efficiency.

(1) Cone Rotation Speed n_c and Diagram d_c

The centrifugal strength can be chosen to satisfy the technological requirements and material properties to be treated by the vibration centrifugal machines. The centrifugal strength is defined as the ratio of the centripetal force acted on the material $\omega_c^2 r_c$ to the gravity acceleration g:

$$K_c = \frac{\omega_c^2 r_c}{g} = \frac{\pi^2 n_c^2 r_c}{900g} = \frac{\pi^2 n_c^2 d_c}{1800g} \tag{3.133}$$

in which r_c and d_c is the centrifugal cone radius and diameter respectively, n_c is the cone rotation revolution.

After K_c is selected, the cone rotation revolution n_c or diameter d_c can be determined by Eq. (3.133). Once d_c is selected, the n_c is:

$$n_c = 30\sqrt{\frac{2K_c g}{\pi^2 d_c}} \tag{3.134}$$

If the rotation is selected first, the diameter is:

$$d_c = \frac{1800 K_c g}{\pi^2 n_c^2} \tag{3.135}$$

The diameter d_c of the vibration centrifugal machines is in general computed for the larger end the cone body. The centrifugal strength K_c is generally selected in the range of 40–120.

(2) The Incline Angle of the Cone Body

The incline angle of the cone body is ½ of the cone angle. The practice has shown that the centrifugal vibration hydroextractor has higher hydro-extracting efficiency and higher output per unit area than other types of the centrifugal hydroextractors. The discharge rates can be changed by adjusting some of the design parameters. Those advantages can only be valid if the cone surface incline angle α_0 is less than the friction angle μ. So the incline angle is taken as:

$$\alpha_0 < \mu \tag{3.136}$$

(3) The Material Operation States and Selection of the Sliding Indices

The negative sliding and jump are not welcomed in the vibration centrifugal hydroextractors because the negative sliding increases the wear-out of the screen net while the jumps decrease the material centrifugal force and reduce the hydro-extracting efficiency. The single sliding movement is selected for the vibration centrifugal hydroextractor. The selected positive and negative sliding indices and throwing index are respectively:

$$D_{k0} = 1.2\text{--}2, \quad D_{q0} < 1, \quad D < 1 \tag{3.137}$$

(4) The Vibration Time and Amplitude of the Cone Body Axial Vibration

Referring to Eq. (3.105), the vibration time n and the vibration amplitude λ can be computed as follows:

For the upright hydroextractor:

$$n = \frac{30}{\pi}\sqrt{\frac{D_{k0}}{\lambda}\left[\omega_c^2 r_c \tan(\mu_0 - \alpha_0) + g\right]}$$

or

$$\lambda = \frac{900\left[\omega_c^2 r_c \tan(\mu_0 - \alpha_0) + g\right] D_{k0}}{\pi^2 n^2} \tag{3.138}$$

For the horizontal hydroextractor:

$$n = \frac{30}{\pi}\sqrt{\frac{D_{k0}}{\lambda}\omega_c^2 r_c \tan(\mu_0 - \alpha_0)}$$

or

$$\lambda = \frac{900\omega_c^2 r_c \tan(\mu_0 - \alpha_0) D_{k0}}{\pi^2 n^2} \tag{3.139}$$

3.4 Theory on Material Movement in Vibrating ...

(5) The Average Velocity v of Material Sliding on Cone Surface and Staying Time in The Cone

The total distance the material sliding in the cone body is $H/\cos\alpha_0$ (H is the cone height). It is the sum of the relative displacements for each sliding:

$$\frac{H}{\cos\alpha} = \sum_{i=1}^{j} S_i = S_1 + S_2 + \cdots + S_j \tag{3.140}$$

in which S_i is the relative displacement for the ith sliding; j is the total number of material sliding in the cone body.

When the total sliding number j is determined, the total staying time for the material in the cone body is:

$$t = j\frac{60}{n} \tag{3.141}$$

The average velocity of the material sliding on the whole height of the cone body is:

$$v_m = \frac{H}{t\cos\alpha_0} = \frac{nH}{60j\cos\alpha_0} \tag{3.142}$$

If the feeding and the discharging end average velocities have no big difference, then the total average velocity of the material is:

$$v_m \approx \frac{v_a + v_b}{2} \tag{3.143}$$

After the average velocity is found, the total staying time for the material in the cone body is:

$$t = \frac{H}{v_m \cos\alpha_0} \tag{3.144}$$

It is obvious that the longer the materials stay in the cone body the less the water contents of the products, the lower the outputs. For an appropriate material thickness the proper average velocity should be selected.

(6) Relation of Material Thickness h to Relative Velocity v and Radius r

Assume that the solid material discharged from the screen holes is $b\%$ of the fed solid materials, the ratio of the material thickness h_b at the discharging end to thickness h_a at the feeding end is:

$$\frac{h_b}{h_a} = \frac{(100-b)}{100} \times \frac{2\pi r_a v_a}{2\pi r_b v_b} = \frac{(100-b)}{100} \times \frac{r_a v_a}{r_b v_b} \tag{3.145}$$

in which r_b, r_a are the radii at the feeding end and discharging end of the cone.

The material thickness at the feeding end and discharging end should have a big difference to avoid the water extracting effects.

(7) Productivity of the Centrifugal Hydroextractor

The productivity computed by the water-extracted solid materials is:

$$Q_B = 3600 \times 2\pi r_b h_b v_b \gamma \quad (t/h) \qquad (3.146)$$

in which v_b is the average velocity of the materials at the discharging end (m/s); r_b is the average radius at the discharging end (m); h_b is the material thickness at the discharging end (m); γ is the material porosity (t/m³).

When the water contents in the discharged material are considered ($a\%$ of the discharged), the productivity is:

$$Q_0 = Q_B \frac{100}{100-a} \quad (t/h) \qquad (3.147)$$

If the filtered liquid in the solid material is $b\%$ of the solid materials in the raw materials, the water contents in the raw material is $c\%$; the productivity computed by the raw materials is:

$$Q_1 = Q_B \frac{100}{100-b} \times \frac{100}{100-c} \quad (t/h) \qquad (3.148)$$

(8) The Material Mass in the Cone Body

The mass of the solid materials in the cone body is computed by the real productivity:

$$m_m = \frac{Q_B H \times 1000}{3600 \cos \alpha_0 v} = \frac{Q_B H}{3.6 \cos \alpha_0 v} \qquad (3.149)$$

When the water contents in the materials are considered, the mass of the materials is:

$$m'_m = \frac{Q_B H}{3.6 \cos \alpha_0 v} \times \frac{100}{100-b'} \qquad (3.150)$$

in which b' is the average water contents in the materials.

Fig. 3.20 The screening probability of a material falling on the screen surface vertically

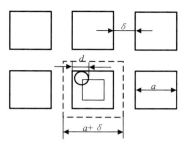

3.5 Probability Theory on Material Screening Process

The screening process of the materials on the vibration screens is based on the principles of the probability theory. There are many articles published to discuss the mathematics theories of the screening process. However, there are no significant positive results on how to increase the productivity and efficiency of the general screening machines by using the probability theory. 20 years ago, The Swedish Megginson proposed the probability screening method (Megginson Screening Method) based on the probability of the materials through the screen nets and created a quick screening machines (Megginson Screen) which can screen 5–10 times materials more than a general screening machines.

3.5.1 Probability of Screening for Material Particle Per Jump

There are differences between a general and a probability screen in structures, while the working principles are the same.

In order for the materials to be screened on the screening surfaces, the movement of the materials relative to the working surface must be ensured to allow the fine particles to go through the screen nets. In the screen machines, the materials jump on the working surfaces continuously. There is a part of the fine particles going through the screen nets in each jumping process. The percentage of the material going through the nets is called "screening probability" for a material. For example, the relative particle size is d/a (d is the size of a material particle, a is the screen hole size) and its mass is unit (that is 100%). If the probability of material going through the screen is 20% each jump, then screening probability for the material of this grade each jump is 0.2.

As shown in Fig. 3.20, when a material falls vertically on the screen surface whose screen hole diameter is a and the screen thread diameter is δ, the material particle would in general go through the screen net hole if it doesn't touch the thread. A few of them would go through the holes when they impact the thread and refract into the net holes. Thus for a particle of size d its screening probability each jump is equal to the ratio of the area of the passable screen holes $n(a - d + \Psi\delta)^2$ to the total screen

Fig. 3.21 Efficiency of particles falling from incline direction

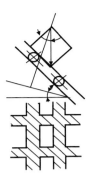

surface $n(a+\Psi\delta)^2$:

$$C_{x0} = \frac{n(a-d+\psi\delta)^2}{n(a+\delta)^2} = \frac{(1-\frac{d}{a}+\psi\frac{\delta}{a})^2}{(1+\frac{\delta}{a})^2} \tag{3.151}$$

in which n is the number of the net holes; a is the diameter of the net hole; d is the diameter of the material; δ is the diameter of the screen thread and Ψ is the coefficient for the material impacts the thread but still can fall into the net holes.

Coefficient Ψ is less than 1. When $x = d/a = 0.3, 0.4, 0.6$ and 0.8, Ψ is $0.2, 0.15, 0.10$, and 0.05 respectively.

It can be seen from Eq. (3.151) that the less the relative size d/a, i.e. the less the ratio of the size of the particle to the size of the net holes, the higher the screening probability. This is why that the finer material is easier to pass the net hole while the particle whose size is just a little bit less than the size of the net holes is hard to pass and it is intuitive. When d/a approaches to 1, the screening probability approaches to zero. The material with $d/a = 0.7$–1 is called "hard screened particle". It can also be seen from Eq. (3.151) that the larger the ratio of the screen hole area to the screen surface area (the ratio is called the effective area coefficient) $a^2/(a+\delta)^2$ the higher the screen probability. When the screen machines are designed, the screen surfaces with large effective area coefficient should be selected. This has significant importance to increase the screen quality and speed up the screen processes.

The aforementioned formulas are valid for the particles, which fall on the screen surface vertically. However in most of the screen machines, the particles fall on the surface along an incline angle. Assume that the angle between the direction of the falling particle relative to the working surface and the vertical line to the working surface is β and the incline angle of the screen surface is α_0 (Fig. 3.21), the screen probability of the material particle of the relative size $x = d/a$ can be written as:

$$C_{x0} = \frac{(a-d+\psi\delta)[(a+\delta)\cos(\alpha_0+\beta)-\delta-d+\psi\delta]}{(a+\delta)^2\cos(\alpha_0+\beta)}$$

3.5 Probability Theory on Material Screening Process

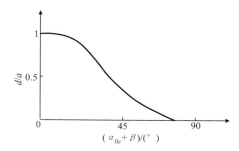

Fig. 3.22 The relationship of the critical angle α_{0e} versus relative size d/a

$$= \frac{\left(1 - \frac{d}{a} + \psi\frac{\delta}{a}\right)\left[\left(1 + \frac{\delta}{a}\right)\cos(\alpha_0 + \beta) - \frac{\delta}{a} - \frac{d}{a} + \psi\frac{\delta}{a}\right]}{\left(1 + \frac{\delta}{a}\right)^2 \cos(\alpha_0 + \beta)} \quad (3.152)$$

It can be seen from Eq. (3.153) that when the screen surface is inclined the theoretical screen probability is equal to zero if the numerator is equal to zero. The critical incline angle by which the material can go through the screen is:

$$\alpha_{0e} = \arccos\left[\frac{\frac{d}{a} + (1-\psi)\frac{\delta}{a}}{1 + \frac{\delta}{a}}\right] - \beta \quad (3.153)$$

It can be seen from Eq. (3.153) that the directional angle β of the particle falling has direct effect on the critical incline angle: the larger the β value, the smaller the screen probability α_{0e}. Some of the references ignored the effect of the β value and do not fit the real situation. Figure 3.22 shows the relationship between the sum of the critical angle and the falling directional angle with the relative size of the particle d/a. When the particle falls vertically, $\beta = 0$, the curve shows the relation of α_{0e} to d/a. It can be seen from the Figure that when $\alpha_{0e} + \beta = 0$, $d/a = 1$, when $\alpha_{0e} + \beta$ increases d/a decreases. In the incline screen machines it is intuitive that the large screen hole size is selected to screen the particles whose sizes are much smaller than the holes. In the probability screens their screen hole sizes are about 2–10 times that of the particle to be screened. Therefore the screen processes of the probability screens are conducted with much less particles, which are hard to be screened. In this way the screen speed is increased and the hole blocked phenomena is alleviated. The screen process of the general screens are conducted on the horizontal or the small incline surfaces, the quantity of the hard-to-be-screened particle is large, screen speed is low and the screen holes are easily to be blocked.

We have discussed the theoretical screen probability formulas on the sphere-shaped particles of relative size d/a. The matter of a fact is that the material particles are not all the sphere-shaped, besides, the thickness of the materials has some effect on the screen probability, the water and dirt contents in the materials have somehow effects on the screen probability. Therefore the theoretical screen probability should be modified. There exists the relationship between the real screen probability C_x and theoretical screen probability C_{x0}:

$$C_x = \gamma_p \gamma_h \gamma_w C_{x0} \qquad (3.154)$$

in which γ_p is the effect coefficient of the particle shape on the screen probability, γ_h is the effect coefficient of the material thickness on the screen probability and γ_w is the effect coefficient of the water and dirt contents on the screen probability. The three coefficients are all less than 1 and can be determined by tests.

Example 3.4 Given a probability screen to be used for screening mixed materials. The size of the screen hole is 10 mm, the diameter of the screen net thread is 2.5 mm, and the angle between the falling directional angle and vertical line of the screen surface is $\alpha_0 + \beta = 45°$. Find the screen probability of the relative size particle $d/a = 0.2$ and 0.4 each jump.

Solution

(1) The computation of the theoretical screen probability.

From Eq. (3.152), the screen probability for a relative size particle of 0.2:

$$\begin{aligned}C_{x0} &= \frac{\left(1 - \frac{d}{a} + \psi\frac{\delta}{a}\right)\left[\left(1 + \frac{\delta}{a}\right)\cos(\alpha_0 + \beta) - \frac{\delta}{a} - \frac{d}{a} + \psi\frac{\delta}{a}\right]}{\left(1 + \frac{\delta}{a}\right)^2 \cos(\alpha_0 + \beta)} \\ &= \frac{(1 - 0.2 + 0.25 \times 0.25)[(1 + 0.25)\cos 45° - 0.25 - 0.2 + 0.25^2]}{(1 + 0.25)^2 \cos 45°} = 0.387\end{aligned}$$

By the same token, the screen probability for $d/a = 0.4$ is 0.1776.

(2) The real screen probability computation.

If the given material is dry, $\gamma_w = 1$. The material layer is thin, $\gamma_h = 1$. If the relative size is 0.2 and 0.4 and its effect coefficient γ_p is 0.85 then the real probability is computed by Eq. (3.154):

For relative size $d/a = 0.2$:

$$C_x = \gamma_p C_{x0} = 0.85 \times 0.387 = 0.33$$

For relative size $d/a = 0.4$:

$$C_x = \gamma_p C_{x0} = 0.85 \times 0.1776 = 0.15$$

(3) The maxim size screened. Letting the numerator in Eq. (3.152) be zero:

$$\frac{d}{a} = \left(1 + \frac{\delta}{a}\right)\cos(\alpha_0 + \beta) - \frac{\delta}{a} + \psi\frac{\delta}{a} = (1 + 0.25)\cos(45°) - 0.25 + 0.25^2 = 0.696$$

3.5 Probability Theory on Material Screening Process

The theoretical maximum size

$$d = 0.696a = 0.696 \times 10 = 6.96 \text{(mm)}$$

The real maximum size is generally taken as 0.7—0.9 of the theoretical maximum size, if we take it as 0.8 then we have:

$$d = 0.8 \times 6.96 = 5.57 \text{(mm)}.$$

3.5.2 Falling Incline Angle and Number of Jumps of Materials on Screen Length

(1) The Falling Incline Angle β

Using the methods in Sects. 3.1 and 3.2 the relative velocities in the x and y direction of the materials falling on to the screen surfaces can be computed as:

$$\begin{aligned}\Delta v_y &= \omega \lambda_y (\cos \varphi_z - \cos \varphi_d) + g \cos \alpha_0 \frac{\varphi_z - \varphi_d}{\omega} \\ &= \omega \lambda_y (\cos \varphi_z - \cos \varphi_d + \theta_d \sin \varphi_d) \\ \Delta v_x &= -\omega \lambda_x (\cos \varphi_z - \cos \varphi_d) + g \cos \alpha_0 \frac{\varphi_z - \varphi_d}{\omega} \\ &= -\omega \lambda_x (\cos \varphi_z - \cos \varphi_d + \theta_d \sin \varphi_d \tan \alpha_0 \frac{\lambda_y}{\lambda_x}) \end{aligned} \quad (3.155)$$

in which $\theta_d = \varphi_z - \varphi_d$.

For a line vibration machines, the vibration amplitudes vertical and parallel to the working surface are respectively:

$$\lambda_y = \lambda \sin \delta, \quad \lambda_x = \lambda \cos \delta \quad (3.156)$$

When Δv_x and Δv_y are known, the angle between the falling material movement direction and the vertical line of the screen surface can be found by the following equation:

$$\alpha_0 + \beta = \arctan \frac{\Delta v_x}{\Delta v_y} \quad (3.157)$$

The falling incline angle (angle between the falling direction and the vertical line) is:

$$\beta = \arctan \frac{\Delta v_x}{\Delta v_y} - \alpha_0 \qquad (3.158)$$

For a general screen machine or the probability screen the falling incline angle β is not zero.

(2) Total Jumping Number

The total jumping number of the material on the whole length of the screen surface has some direct effect on the screening quality. For the general and the probability screen, the throwing index D is often selected to be less than 3.3, a few of them 3.3–5. When $D < 3.3$, the material jumps once when the screen surface jumps one time. Therefore the total number m of the material jumps on the screen surface is equal to the total time T of material spent on the screen surface divided by the vibration period $T_0 = \frac{2\pi}{\omega}$:

$$m = \frac{T}{T_0} = \frac{T\omega}{2\pi} \qquad (3.159)$$

The total time a material particle spent on the whole screen length is equal to the length of the screen surface divided by the material real average velocity v_m, that is:

$$T = \frac{L}{v_m} \qquad (3.160)$$

The total jumping number is:

$$m = \frac{L\omega}{2\pi v_m} \qquad (3.161)$$

in which L is the screen surface length (m); v_m is the material real average velocity (m/s).

3.5.3 Calculation of Probability of Material Going Through Screens for a General Vibration Screen

In a general or a probability screen machine, the materials are screened on a single screen surface or multiple screen surfaces and a particle may experience tens or even hundreds of jumps in its screen process. The screen total probability is the accumulation of the variety grades of screen process. First, let's discuss how to calculate the total screen probability for a single material layer in a general screen machine. When the probability is different for each jump, the amount of the material of the relative size d/a of a particle still left on the screen surface after 1, 2, 3, …, m jumps: $(1 - C_{x1})$, $(1 - C_{x1})(1 - C_{x2})$, $(1 - C_{x1})(1 - C_{x2})(1 - C_{x3})$,

3.5 Probability Theory on Material Screening Process

..., $\prod_{m=1}^{m}(1-C_{xm})$ ($C_{x1}, C_{x2}, ..., C_{xm}$ are the screen probability of material 1st, 2nd, 3rd, ..., mth jump). The material amounts that are screened through the screen surface are: $C_{x1}, C_{x2}(1-C_{x1}), C_{x3}(1-C_{x1})(1-C_{x2})$ and $C_{xm}\prod_{m=1}^{m}(1-C_{xm})$. The amount of material discharged from the screen surface P_x and screened material through the screen surface Q_x are respectively:

$$P_x = \prod_{m=1}^{m}(1-C_{xm})$$

$$Q_x = 1 - \prod_{m=1}^{m}(1-C_{xm}) = \sum_{m=1}^{m} C_{xm} \prod_{m=1}^{m-1}(1-C_{xm}) \tag{3.162}$$

When screening the mixed materials, the total amount of the material left on the screen surface and the total amount of material screened through are respectively:

$$P = \sum_{x=0}^{D/a} a_x \prod_{m=1}^{m}(1-C_{xm})$$

$$Q = \sum_{X=0}^{D/a} a_x \left[1 - \prod_{m=1}^{m}(1-C_{xm})\right] \tag{3.163}$$

in which D is the maximum size of particles.

For material whose size is less than the separation size, its screen probability is:

$$\eta = \frac{\sum_{x=0}^{d_g/a} a_x \left[1 - \prod_{m=1}^{m}(1-C_{xm})\right]}{\sum_{x=0}^{d_g/a} a_x} \tag{3.164}$$

If the screen probability is different for each jump, to increase the screen efficiency, it is tried to design that the fine materials are close to the screen surface while big size particles are on the top of the materials. Nowadays, the iso-layer screen methods or large thick screen methods are used to separate the fine and large material particle by making use of the vibration and fine particles get closer to the screen surface through the gaps of the large particles and thus screen efficiency is increased.

3.5.4 Calculation of Probability of Material Going Through Screens for a Multi-screen Vibrating Screen

A probability screen distinguishes in structures from a general screen in that its screen surfaces have multiple layers (3–6), large incline angle (30°–60°), large screen holes (2–10 time of the relative size of a particle). It also distinguishes itself from a general screen in the working principle. The probability theory has been used to the probability as the working principle to complete the screen process very quickly. The screen time is only 1/3–1/20 that of a general screen and its productivity is 5–10 that of a general screen. The screen process of the probability screen is a quick screen. To illustrate the basic principle of the probability screen's quick screening, let's analyze the total material screen probability of the two typical working states:

(1) The Screening for Different Screen Probability for Each Screen Surface

The multiple thin layer screenings have different screen hole sizes or different incline angles. In this situation, the materials with relative size $x = d/a$ has the same screen probability on the same layer and different screen probability on different surfaces. Figure 3.23 shows the material amounts left on the screen and screened through the surfaces after the material particles make 1st, 2nd, 3rd and gth jumps. It can be seen from the Figure that the material amounts discharged from the top of the 1st, 2nd, 3rd and gth layer screen surfaces are:

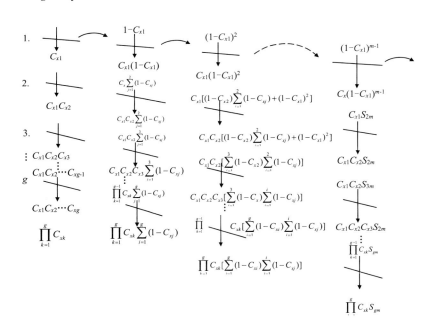

Fig. 3.23 The screening for different screen probability for different screen surface

3.5 Probability Theory on Material Screening Process

$$\begin{cases} P_{1x} = (1 - C_{x1})^m \\ P_{2x} = C_{x1}(1 - C_{x2})S_{2m} \\ P_{3x} = C_{x1}C_{x2}(1 - C_{x3})S_{3m} \\ \vdots \\ P_{gx} = \prod_{k=1}^{g-1} C_{xk}(1 - C_{xg})S_{gm} \end{cases} \quad (3.165)$$

In which

$$\begin{cases} S_{2m} = \sum_{i=0}^{m-1} (1 - C_{x1})^{m-1-i}(1 - C_{x2})^i \\ S_{3m} = \sum_{i=0}^{m-1} (1 - C_{x1})^{m-1-i} \sum_{j=0}^{i} (1 - C_{x2})^{i-j}(1 - C_{x3})^j \\ \vdots \\ S_{gm} = \sum_{i=0}^{m-1} (1 - C_{x1})^{m-1-i} \sum_{j=0}^{i} (1 - C_{x2})^{i-j} \cdots \sum_{v=0}^{u} (1 - C_{xg-1})^{u-v}(1 - C_{xg})^v \\ \qquad\qquad i \le m-1, \ j \le i, \ldots, \ v \le u \end{cases} \quad (3.166)$$

The amount discharged from the bottom end of the gth screen layer surface is:

$$Q_{gx} = \eta_{gx} = \sum_{m=1}^{m} \prod_{k=1}^{g} C_{gk} S_{gm} \quad (3.167)$$

When the particle of relative size x is a_x percent of the total materials, its total screen probability is:

$$\eta = \frac{\sum_{x=0}^{d_g/a} a_x \sum_{m=1}^{m} \prod_{k=1}^{g} C_{xk} S_{gm}}{\sum_{x=0}^{d_g/a} a_x} \quad (3.168)$$

(2) Different Probability for Each Jump Situation

When the material thickness is large, the screen probability of the material for each jump is different (Fig. 3.24). Assume that the screen probability of a material particle with relative size x on the gth screen surface and on the mth jump is C_{xgm}, and at this position, the materials on the screen surface and being screened through the surface are p_{xgm} and q_{xgm}, it is obvious that p_{xgm} and q_{xgm} have the following relationship:

$$\begin{aligned} p_{xgm} &= p_{xgm-1}(1 - C_{xgm-1}) + q_{xg-1m} \\ q_{xgm} &= p_{xgm}C_{xgm} \end{aligned} \quad (3.169)$$

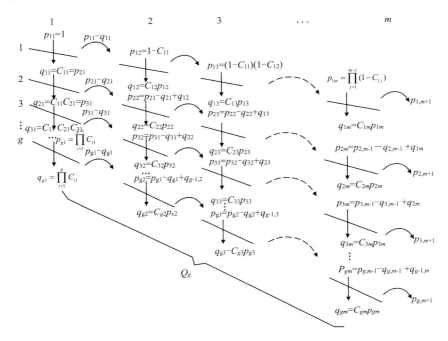

Fig. 3.24 Different probability for each jump

The material amounts, with relative size x, on the screen surface and screened through the surface after the mth jump are respectively:

$$\begin{cases} P_{xgm} = (1 - C_{xgm})p_{xgm} \\ Q_{xgm} = \sum_{m=1}^{m} q_{xgm} = \sum_{m=1}^{m} p_{xgm}C_{xgm} \end{cases} \quad (3.170)$$

When the grade contents are a_x, the screen efficiency for the mixed material is:

$$\eta = \frac{\sum_{x=0}^{d_g/a} a_x \sum_{m=1}^{m} q_{xgm}}{\sum_{x=0}^{d_g/a} a_x} \times 100\% \quad (3.171)$$

Once the material contents of different grades and the screen probability of the material for each jump are determined the material screen efficiency can be calculated.

3.5 Probability Theory on Material Screening Process   129

(3) The Theoretical Basis for the Quick Screening of a Probability Screen

i. The significance of the multi-layer screen surface for increasing the screen productivity and screening efficiency. To realize the quick screening, the condition of the thin layer must be created. Only under such conditions can the materials be screened by the principle stated in Sect. 3.5.1. Using the multiple screen layers can make the material to be screened distribute uniformly on each layer of the screen surfaces, decrease the thickness of the materials on each screen surfaces. Thus it creates a necessary condition for thinning material layers, and at the same time increases the productivity. However too many screen layers could make the machine more complex and decrease the screening speed. Therefore the basic measure for the screen probability method is to use 3~6 screen layer surfaces.

ii. Advantages of using large incline angle and large screen hole Surface. To accelerate the screening process speed and increase productivity per unit area of the probability screen, the movement velocity of the material on the screen surface must be increased. The basic measure for it is to increase the incline angle. The tests show that the selection of 30°–60° incline angle can increase the material movement velocity 3–4 time compared to that of the horizontal working surface. It is very obvious that under the given outputs, the increase of the material movement velocity could reduce the material thickness. This creates the necessary condition for the thin layer screen. Under the given material thickness the increase of the material movement velocity could increase the machine outputs.

Due to the increase of the incline angle, according to the screen probability principle, the screen hole size must be increased given the separation size. The increase of the real screen holes will decrease the effect on the hard-to-be-screened particles in the screen process and alleviate the blocking of the screen holes, and at the same time the screen process velocity increases dramatically and thus increases of the incline angle is the condition to realize the necessary condition for quick screening.

iii. The small vibration amplitude and high frequency can be used in the probability screen. In the screening machines, the selection of the vibration amplitudes is often based on the condition of no-screen net hole-blocking. The probability screen is used to the screen surface with large incline angles and thus alleviate greatly the net-hole blocking phenomena, and thus the vibration amplitudes can be reduced correspondingly. The amplitudes have direct effects on the net hole-blocking, increasing the amplitude can reduce the blocking phenomena. Under the condition of no-hole-blocking situation, the smaller vibration amplitude and large frequency should be used to increase the jumping numbers of the particle on the screen surface. According to the screen probability theory, increasing the jump number m has significant importance of quickening the screen speed and increasing the screen efficiency.

Table 3.7 The test results on a 5-layer surface probability screen

Grades	>8	>12	>16	>20	>24	>30	>36	>46	<46	m (kg)
Original materials	0.04	0.04	0.055	0.095	0.067	0.08	0.12			20
Materials on screen layer 1	0.30	0.52	0.14	0.04	0.002					2.2
Materials on screen layer 2		0.13	0.56	0.205	0.01		0.003	0.0025		1.7
Materials on screen layer 3			0.09	0.53	0.29	0.09	0.01	0.004		2.1
Materials on screen layer 4			0.004	0.03	0.41	0.42	0.115	0.025		2.5
Materials on screen layer 5					0.01	0.39	0.42	0.18		2.2
Materials under screen layer 5 (1)						0.025	0.205	0.18	0.60	0.6
Materials under screen layer 5 (2)							0.08	0.095	0.85	2.5
Materials under screen layer 5 (3)						0.005	0.07	0.05	0.88	4.2
Materials under screen layer 5 (4)						0.008	0.03	0.049	0.91	3.6

Note The materials screened through 1, 2, 3, 4, are arranged in the order discharged to the feed ends

Table 3.7 lists the test results for a 5-layer screen. It can be seen from the data that the probability screen has a good screening effect.

3.6 Classification of Screening Method and Probability Thick-Layer Screening Methods

3.6.1 Screening Methods

There are a variety of the screening methods. The common methods are: a general screen method, thin-layer screen method, probability screen method, thick-layer screen method etc. There are some advantages and shortcomings:

(1) The General Screening Method

The general screening method is the one used in industry for a long time. Its material thickness is moderate, and has the following characteristics:

i. Material thickness is about 3–6 times the sizes of the screen holes;
ii. The number of screen layers is 1–2;
iii. The material particles screened out are dependent on their sizes relative to that of the screen net holes. The particles whose size is less than that of the holes move along the length of the screen surface and are screened out through the net holes, while the particles whose sizes are larger than that of the net holes will move along the surface and are discharged out from the screen top.

In general, the net hole size and the maximum size of the screened out material particle should have the following relation: for an around hole, $a = (1.3–1.4)d_{max}$; for a square hole, $a = (1.1–1.15)d_{max}$; for a rectangular hole, $a = (0.7–0.8)d_{max}$ (a is the net hole size, for an around hole it is diameter, for a square or a rectangular hole, it's the shorter edge; d_{max} is the maximum size of the screened out particles).

The screen process by general screen methods is slow and at the same length of the screen surfaces the screen efficiency by this method is low. However the general screen methods are applicable: it can be used to screen either the granular particles, or the chunk type of material. It can be used to either dry screen, or wet screen. It is still used widely in the industries.

(2) Thin Layer Screen Method

The thin layer screen method is a type of screen for very thin layer material. The method is used for the vibration screens whose working surfaces are excited directly, such as the electric–magnetic vibration screens with screen-net vibration; a cam type of vibration screen, a line vibration screen etc. The characteristics of the method are:

i. The thickness of the materials is only about 1–2 times of the net holes: $h/a = 1–2$ (h is the material thickness, a is the size of the net hole).
ii. The screen surface is a single or a double layer.
iii. The high velocity of the materials are selected: the common velocity: 0.5–1 m/s.
iv. The high frequency and small vibration amplitude are used. The commonly used vibration frequencies are 1,500–6,000 rpm and the corresponding amplitudes are 0.3–0.12 mm.
v. The ratio the feeder width to the screen surface width is about 1:1.
vi. The applicable incline angle is about 25°–54°.
vii. The screen is suitable for screening the fine particles. Their size is about 0.05–5 mm, the most common is 0.1–3 mm.

(3) Probability Screen

The probability screen method is created by a Sweden Megginson. The screen has the following characteristics:

i. Multiple screen surfaces: 3–6 layers.
ii. Large incline angle: 25°–60°.

iii. Large net holes: the ratio of the size of net holes to the size of particle is 2–10.

The screen based on the method is called the probability screen, or Megginson Screen in abroad. The method makes it possible for a screen to finish a screen process actively based on the probability theory and the screen process becomes very quick. The materials stay on the screen about 3–6 s while it's 10–30 s for a general screen. The advantages of the probability screen are that the materials are screened through the net holes quickly due to its structures (multiple layers, large incline angle, and large net holes), at the same time overcome the shortcomings of the net hole blocks, eliminate the effects of critical particle, reduce the screen surface wear-out and increase the productivity per unit area (5 times larger than that of a general screen). The probability screen is applicable only to the approximate screening and the materials screened often contain certain rough particles. Sometimes it can not obtain a high screen efficiency.

(4) Thick-layer Screen Method

The thick-layer screen method is also called large thickness screen method. Its feature is that no matter what the percentage of the fed material whose sizes are less than that of the net holes is the thickness of the materials on the screen surfaces remains the same and never increases while for a general screen the thickness of the material on the surfaces is thin and decreases.

When a general screen is used to screen the materials, the acceleration of the screen surface to the materials is relatively small and the small particles have small velocity of subside and take long time to reach the screen surface. The materials do not layer well. Thus the screening capability per unit area for the general screen is about a quarter of the real screening capacity.

The acceleration of the materials at the feeding end for the thick-layer screen method is larger than that of the general screen, thus the acceleration of the materials is large, the material layers become thin and layered quickly. For the layered material groups, the same acceleration as that of the general screen will accelerate the screening process and make more materials go through the screen net. The method makes the probability of the small particles touching the screen surface increase dramatically and the screening capability per unit area is about 80% of the real screening capacity. Compared to the general method, the thick-layer method triples the capability.

In order to make the materials distribute as required the following methods will be used:

i. The thick layer screen method with a large throwing index. In the screen operations, the surface is a little bit of incline or horizontal, and divide into several intervals in which each interval has its own transmission mechanism. The amplitude (or frequency) in the first interval is large, and its vibration acceleration is about $10g$, and the amplitude (or frequency) is small and its vibration acceleration is about $(4-5)g$.

ii. The thick-layer screen with large incline angle installation. The line and circular vibration are connected in series and the incline angle of the circular vibration

3.6 Classification of Screening Method ...

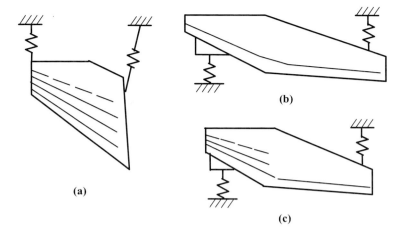

Fig. 3.25 Three screen methods. **a** Probability screen method, **b** thick layer screen method, **c** probability thick layer screen methods

screen is about 30°–40°, the materials have high movement velocities and the material layers are thin; the incline angle of the line vibration screen is about 0°–10°, and screening at the normal acceleration. The thick-layer screen method has been used in some of the large mineral separation factories in China.

3.6.2 Screening Methods for Probability Thick Layer Screens

We rolled the advantages of the probability screen and the thick-layer screen into one, overcame the shortcomings of the two and proposed the new screen method—the probability thick layer screen method.

The first stage of the probability thick-layer screen method is based on the probability screen principle, and the second stage is based on the thick layer screen principle to layer the materials quickly and increase the screen probability. This screen method could lead to large productivity and high screen efficiency and does not require a long screen surface.

According to this method, we designed a variety of the probability thick layer screen and they are used in industries and have achieved satisfactory results.

(1) The Background of the Probability Thick Layer Screen

In recent 40 years, there are big developments in the screen methods. Besides the general screen and thin layer screen methods, the probability screen and thick layer screen (also called iso-thick screen method) were created. We designed and manufactured the first self-synchronous probability screen of China (Fig. 3.25a) and successfully applied it to the screen process for the silicon-carbide in 1979 we designed and manufactured the first thick-layer screen of China (Fig. 3.25b).

In the early 1981, at the request from an industrial manufacture department, we designed a new type of screen machine with small size, large output and high screen efficiency. The length × width × height dimensions of the screen machine should be less than 3,500 mm × 2,200 mm × 1,400 mm, the product output is 675 m^3/h given the feeding material condition and its screen efficiency should be over 90%. Even though using either the probability screen or the thick-layer screen could satisfy the product output requirement, the probability screen hardly meet the requirement for the screen efficiency while the thick-layer screen is not applicable under the dimension constraints due to its length is too long (generally it's about 7–9 m). To resolve the key issues, we studied, designed and tested for different scenarios and proposed a new screen method which rolled the probability screen method and thick-layer method into one—the probability thick-layer screen method and manufactured the screen successfully, and then this new screen has been adopted by many industries. Recently we designed and manufactured another probability thick layer screen used for screening raw coals and gangues.

(2) The Characteristics of the Probability Thick Layer Screen Method

The working processes of the probability thick layer screen method include two phases: probability layer separation and thick layer screening.

i. The probability layer separation and screening phase

The difference of probability thick layer screen method distinguished from the thick layer screen is that the first phase of the probability screen is to separate the material layers quickly by the probability principles and at the same time to function screening process. Its length of the working surfaces is about 1.5–2 m. It can be said that this layer separation is a kind of forcing separation. While the first phase of the thick layer screen method is to separate the material layers naturally by the vibration of the working surface. Its working length needed is about 2.5–4 m. Its natural layer separation takes long time and is slow. Secondly the first phase of the probability thick layer screen method, besides the quick screen process, the fine particles pass through the screen holes of the different layers in different degree in fast speeds, partially realizing the screenings. While in the thick layer screen method, the material layers are separated naturally the speeds of the fine particles passing the screen holes will be far more slow than that of the probability thick layer screens, the fine materials screened out are less than that of the probability thick layer screens. Therefore, for the fine particle materials, the probability forcing layer separation process can obtain the better screening than the thick layer screen, while for the large size of materials this is only a layer separation process.

The working principles of the first phase of the probability thick layer screen have many similar points, compared to the probability screen, however there are many different points, the surface architectures have the following features:

3.6 Classification of Screening Method … 135

(a) Multiple Layers. There are 3–6 layers in the probability screen method and there are 3 layers for the probability thick layer screen method.
(b) Large Incline Angle. The incline angles are about 25°–60° for the probability screen method while the probability thick layer screen method incline angle for the first phase is about 20°–40°.
(c) Large Net Holes. The ratio of sizes of the net holes to the separation size is about 2–10, while the ratio for the probability thick layer screen method, the dimension for the lowest layer should be increased to a proper level. The separation size can be chosen as 1.3–1.7 times.

The characteristics of the screen surface above make the big differences for the first phase screen process for the probability thick layer screen method from that of the thick layer screen method. The major difference is that the first phase of the probability thick layer screen method accomplishes the material layer separation and partial fine particle screens obviously and effectively by the probability theory, while the time needed for the first phase and the surface length needed is far more shorter than that of the thick layer screen. One can say that the layer separation for the fine particle partial screening out is accomplished at an extremely fast speed. However for the thick layer screen method, the fine particle screening is not a major purpose but a secondary function. Of course, to increase the benefits of this function has significant importance. The screening effects for the fine materials, i.e. the screening efficiency for the screens, are guaranteed by the second phase. The layer separation of the screened materials and the distribution curves on different screen surfaces are shown in Fig. 3.26. It is obvious that the materials discharged from the top of the surfaces, from top to bottom layers, become finer and finer. The curve 1 in the Cartesian coordinate system is the composition of the screened materials, its horizontal coordinate is the relative size, i.e., the ratio of the material size to the size of the net holes, and the vertical coordinate represents the percentages of the different grades of the materials. The materials screened out from the first layer, the second layer and the third layer can be classified into two products, one on the screen surfaces and another is screened out, and the boundary curve for the two products

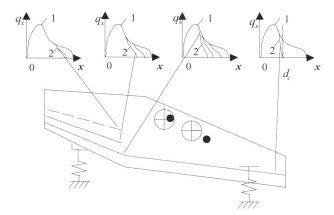

Fig. 3.26 The material distribution curves in different phases and on different layers

are represented by the curves 2: the difference between curve 2 and curve 3 represents the unscreened materials on the screen surface and the area between curve 2 and the horizontal axis is the materials screened out through the screen surfaces. It can be seen from Fig. 3.26 that after the layer separation and screening in the first phase, there are a lot of the fine particles which are not screened out and left on the surfaces. To screen them out more precisely, the second phase is necessary: the thick layer screening.

ii. The thick layer screen phase

In the second phase, the screening process of the materials is based on the thick layer screen method. The screen surface incline angle is in the range of 0°–10°, the material movement speed is low, the layer is thick (about 8–10 times of size of the net holes); the screen surface length is about 2.5–4 m, while the surface length for the second phase of the natural layer separation is in the range of 4–6 m. Since the screen process for this phase is under the condition of thick layers, the materials separate themselves into two layers: the large lump materials are on the top and the fine particle at the bottom. The fine particle is subject the pressures from the large size materials on the top and touch the surface screen nets. Therefore the screen probability for the fine particle is largely increased. Some reference mentioned that the screening capability per unit area for the thick layer screen is about 80% of the real screen out capability. While for the general screen its material layers are thin, the pressure subjected to the fine particles is less than the pressure for the thick layer screen, the screen probability is decreased. So that compared to the general screen, the thick layer screen could double increase the treatment capability. In the probability thick layer screen the probability quick layer separation is used (having the screening function as well), and thick layer screening and therefore the screen has big treatment capability and at the same time has large screening efficiency and its working surface length is much shorter than that of the thick layer screen.

The second phase distribution curves for the probability thick screen are shown on the top figures in Fig. 3.26. After the layer separation and screening in the first phase, the fine particles are not screened sufficiently and the screen precision is not high, the materials above the screen surface contain different grades of materials, especially the fine particles whose sizes are close that of the net holes. The area contained by curves 1 and 2 is composed of the sizes of the materials entered into the second phase. It can be seen from the Fig. 3.26 that the materials this portion materials are composed of much materials whose sizes are less than that of the separable particles and the second phase screen process is necessary. The area contained between the curve 3 and the horizontal axis represents the particle size of the materials screened out while the area between curve 3 and curve 1 is the particle size of the materials on the working surface. It can be seen from the Figure that even though the materials contains very little amount of the fine particle materials. It can be concluded that the probability thick layer screen method is very efficient with high precision.

(3) Measures of Increasing the Screen Effects of the Probability Thick Layer Screen

3.6 Classification of Screening Method … 137

As we analyzed that the screen processes are composed of two phases: the probability layer separation and screening and thick layer screening. The measures can be taken to increase the screening efficiency by the theory as we mentioned before:

i. Increasing the screen probability for every grade of material particles C_x, the big holes can be used in the first segment of the probability thick layer screen, while the thick layers can be used in the second segment t increase the pressures on the fine particles. This will favor the increase of the screen probability C_x and thus increase the screen efficiency of the screening process.
ii. Increasing the jump numbers of the materials on the screen surface is an important factor for the increase of the screening efficiency. The purpose for the second phase of the probability thick layer screen is to increase the jump numbers and thus increase the screening efficiency.
iii. Decreasing the contents of the hard-to-screened particles can increase the screening efficiency. The proper enlargement of the net holes in the probability layer separation and screening phases can increase the screening efficiency to some degree. However the screened out materials may contain more large size material particles. In the general screens, the efficiency increase can be achieved only by the increase of the jump numbers and probability for each jump.

The relationships between the screened amount with the jump numbers and the screen surface length for different grade of particles are shown in Fig. 3.27.

(4) Experimental Results on Probability Thick Layer Screen

The probability thick layer screen can be applied to screen the stones used for supporting the railway sleepers. The dirty and crushed stone whose size is less than

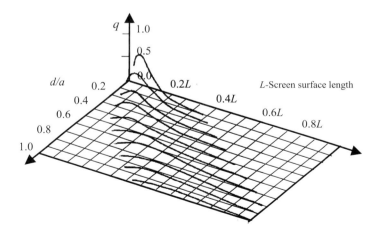

Fig. 3.27 The relation between the screened-out material with different relative size and the jump number and the screen surface length

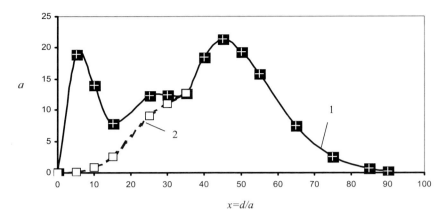

Fig. 3.28 The test results on the railway stone screen

20 mm should be removed. The experimental results are shown in Fig. 3.28. The separation size is 20 mm, the screen size for the lower screen surface is 20 mm × 30 mm. The size composition of the materials on the screen surface and the screened out materials after the screening process and the percentage contents of different grade materials on the screen surfaces over the materials on the surface are listed in Fig. 3.28.

The conclusions are drawn from the results:

i. The probability thick layer screen method is a new and high efficient screen method. It is developed on the basis of the probability screen method and the thick layer screen method. In other words, it replaces the natural layer separation in the thick layer methods with the forced probability layer separation (have the screening function as well). It has essential difference from the probability screen and thick layer screen methods.

ii. The probability thick layer screen method heritages the advantages of the probability screen method: large output per unit area and small dimensions and at the same time overcoming its shortcomings of the low screen efficiency, the rough size of particles for the screened materials, and the long length of the thick layer screen method. Therefore, the new method of the probability thick layer screen has become a perfect screen method (Table 3.8).

iii. The principles of the two phases for the probability thick layer screen method introduced in this chapter can be used to estimate qualitatively the screening results for the method.

iv. On the basis of studying the new screen method for the probability thick layer screen, authors have successfully developed a new type of a probability thick layer screen which has been used to the screening technological processes for the railway stones and gangue with excellent industry indices. This probability thick layer screen is a high efficiency screen method and has potential for more applications.

3.7 Dynamic Theory of Vibrating Machine Technological Processes

Table 3.8 The screen results of the probability thick layer screen

Sieve pore size/(mm)	Contents of each grade of original materials/(%)	Contents of each grade of screened out materials/(%)	Percentage contents of screened out materials/(%)	Percentage contents of each grade materials on screen surface/(%)	Percentage contents of each grade materials on screend out/(%)
0–10	19	18.4	96.84	0.6	3.16
10–20	7.8	7.5	96.15	0.3	3.85
20–30	12.3	5.3	43.09	7.0	56.91
30–40	12.7	0	0	12.7	100
40–50	21.4	0	0	21.4	100
50–60	15.8	0	0	15.8	100
60–70	9.1	0	0	9.1	100
+70	1.9	0	0	1.9	100
Total	100	31.2	–	68.8	–

Note The test was conducted under the condition of output 46.57 m³/h per unit area for a screen machine.

3.7 Dynamic Theory of Vibrating Machine Technological Processes

The vibration machines are usually used to accomplish a variety of technological processes, such as screening, hydro-extracting, cooling, drying, crushing etc. The technological effects of the processes are increased with the time elapsing. It is obvious that the technological effects are zero at the initial instance of the process and reaches to 100%, i.e. 1, when time goes to infinity.

Taking the material screening as an example, assume that the amount of the material whose size is less than the size of the net hole is $\sum_{x=0}^{a_0} q_{x0}$, x represents the average size of the various grades of the materials, a_0 should be the largest size of the materials screened out, q_{x0} is the percentage of the material contents of x grade. Assume also that the screen probability is c_x for size x of the material in a unit time, its elapsed time on the screen surface is y. In the interval Δt, the material amount screened out through the net hole for the x grade is Δq_x:

$$\Delta q_x = (q_{x0} - q_x)c_x \Delta t \quad \text{or} \quad dq_x = (q_{x0} - q_x)c_x dt \quad (3.172)$$

Integrating Eq. (3.172) one obtains:

$$\int \frac{dq_x}{q_{x0} - q_x} = \int c_x dt$$

i.e.

$$\ln|q_{x0} - q_x| + C = c_x t \tag{3.173}$$

The integral constant C can be determined by the following condition: when $t = 0$, $q_x = 0$, $C = \ln q_{x0}$, then

$$1 - \frac{q_x}{q_{x0}} = e^{-c_x t} \quad q_x = q_{x0}(1 - e^{-c_x t}) \tag{3.174}$$

If the materials stay on the screen surface, the material amount screened out through the net holes q_x and the material stayed on the surface p_x are respectively:

$$q_x = q_{x0}(1 - e^{-c_x t}), \; p_x = q_{x0} - q_x = q_{x0} e^{-c_x t} \tag{3.175}$$

Therefore, the screen efficiency for the size of z grade materials is:

$$\eta_x = 1 - e^{-c_x t} \tag{3.176}$$

If the screen probability c_x is a function of x, $c_x = c(x)$, then the theoretical screen probability for all grades should be:

$$\eta = \lim_{\Delta x \to 0} \sum_{x=0}^{a_0} \frac{q_{x0}}{q_0}(1 - e^{-c(x)t}) \Delta x = \frac{1}{q_0} \int_0^{a_0} q(x)(1 - e^{-c(x)t}) dx$$

$$= 1 - \frac{1}{q_0} \int_0^{a_0} q(x) e^{-c(x)t} dx \approx 1 - \frac{1}{q_0} \int_0^{a_0} q(x)(1 - c(x)t) dx \tag{3.177}$$

$$q_0 = \sum_{x=0}^{a_0} q_{x0} = \int_0^{a_0} q(x) dx \tag{3.178}$$

in which q_0 is the total amount of the materials screened out in theory, $q(x)$ is the granularity composition of the material screened out; $c(x)$ is the screen probability of materials whose size in x in a unit time. In general $c(x)$ is always less than 1.

In the material screening process, if the screen probability $c(x)$ is a constant, i.e. $c(x) = c$, then the screen probability is:

$$\eta = \frac{1}{q_0} \int_0^{a_0} q(x)(1 - e^{-c(x)t}) dx \approx 1 - e^{-ct} \tag{3.179}$$

For some technological process, the output rate in a unit time c can be approximated as a constant. In this case, the change of the discharging material amount in unit time is in exponential, as shown in Eq. (3.179).

3.7 Dynamic Theory of Vibrating Machine Technological Processes

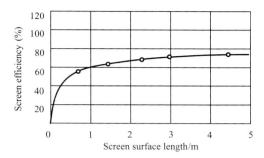

Fig. 3.29 The test curve of the relation between screen efficiency and time

As a matter of a fact there are a variety of the different grades of materials with different degrees of separations contained in the discharged materials in many technological processes. For example, for those materials whose size is close to that of the net holes, its screen probability is much less than that of the materials whose size is much less than that of the net holes. Therefore, the discharged amount of materials in a unit time from the net holes shows a big difference due to different grade contents. In this case $c(x)$ can not be treated as a constant to avoid big errors. When $c(x)$ is not a constant, Eq. (3.177) can be expressed as:

$$\eta = \frac{1}{q_0}\int_0^{a_0} q(x)(1 - e^{-c(x)t})dx \approx 1 - e^{-\sqrt[n]{c_x t}} \qquad (3.180)$$

where c_x is an equivalent weight screen probability; n is an index larger than 1.

For majority of the technological processes, the output rate in unit time in the initial times is large, while it becomes smaller and smaller after some of times. For example, for the material screen process, at the beginning the screen probability is high, because at that time, the materials whose size is much less than the net holes are on the surface, the screen probability rate is high. After a while, for the materials left on the surface their sizes are close to that of the net holes. The screen probability is low and screen efficiency is decreased.

Figure 3.29 shows the experimental curves for the relationship between the screen efficiency and time for a material. It can be seen from the experiment that at the beginning, the material screened out and material that goes through the net holes are increased quickly. As time goes, the screened material amount increases slowly and the tangent of the curve is small and this segment of the curves approaches to a horizontal straight line and takes long time. This indicates that the hard-screened granularity with size close to that of the net holes are very difficult to go through the net hole. This is a common characteristic for general technological processes.

Chapter 4
Linear and Pseudo Linear Vibration

4.1 Dynamics of Non-resonant Vibrating Machines of Planer Single-Axis Inertial Type

The inertial vibration machine can be classified by its number of bodies: single body type, double-body type and multi-body type. It can also be classified by its number of exciters: single axle type, double-axle type and multi-axle type; it can also be classified by its dynamic characteristics: linear-non-resonance type, linear near resonance, nonlinear type and impact type etc.

The plane movement single axle type of inertial vibration machines include the single axle type of inertial vibration screen, the vibration ball grinder, horizontal vibration feeder, etc. The mechanic model for mechanism figure of the single axle inertial vibration machines can be drawn (Fig. 4.1).

The vibration equation for the machine can be established using D'Alembert's Principle. It can be seen from Fig. 4.1 that the forces acting on the body, including, body inertial force, damping force, spring force and excitation force, the sum of the forces should be zero at a instance of vibration. In other words, the forces acting on the body m should be balanced each other, i.e.

$$\begin{aligned}&\text{In the } y \text{ direction: } (-m\ddot{y}) + (-f\dot{y}) + (-k_y y) + [-m_0\ddot{y} + F_y(t)] = 0\\&\text{In the } x \text{ direction: } (-m\ddot{x}) + (-f\dot{x}) + 0 + [-m_0\ddot{x} + F_x(t)] = 0\end{aligned} \quad (4.1)$$

in which, m is the calculated mass of the body, $m = m_0 + K_m m_m$ is the real mass of the body; K_m is the material united coefficient; m_m is the material mass on the surface; f is the equivalent resistant coefficient; k_y is the stiffness of the vibration isolation spring in y direction; m_0 is the eccentric mass; $F_x(t)$ and $F_y(t)$ are the relative inertial force (the inertial force of the rotation movement around the axle line) of the eccentric in x and y direction respectively; $\dot{x}$, $\ddot{x}$, $\dot{y}$, $\ddot{y}$ are the velocities and accelerations of the body in x and y directions respectively.

Fig. 4.1 The mechanics figure of the single axle inertial vibration machine

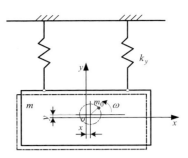

The inertial forces $F_y(t)$ and $F_x(t)$ of the eccentric relative to the rotation axle line can be expressed as follows respectively:

$$F_y(t) = m_0 \omega^2 r \sin \omega t$$
$$F_x(t) = m_0 \omega^2 r \cos \omega t \quad (4.2)$$

in which ω and r are the angular velocity and the eccentricity of the eccentric respectively.

Substituting Eq. (4.2) into Eq. (4.1) and simplifying one can obtain the vibration equations:

$$(m + m_0)\ddot{y} + f\dot{y} + k_y v = m_0 \omega^2 r \sin \omega t$$
$$(m + m_0)\ddot{x} + f\dot{x} = m_0 \omega^2 r \cos \omega t \quad (4.3)$$

i. The approximate solutions for Eq. (4.3). The damping force and the spring forces are ignored compared to the excitation force and inertial force of the body. The inertial force $m\lambda\omega^2$ generated by the body mass and the inertial force $m_0 r \omega^2$ generated by the eccentric are balanced each other. i.e. they are the same in amplitude but in the opposite direction:

$$m\omega^2 \lambda \approx m_0 \omega^2 r$$

or

$$\lambda \approx (m_0 r)/m \quad (4.4)$$

in which λ is the amplitude of the screen box; m_0, m is the mass of the eccentric and the computed mass (including the material united mass) of the vibration body (screen box).

Based on the aforementioned analysis (or proven by the tests), the body and eccentric are always in the two directions of the vibration center: when the body is on the top and the eccentric is on the bottom; or when the body is on the left and the eccentric is on the right, or vice versa. While the vibration center is the resultant mass center of the body and the eccentric.

4.1 Dynamics of Non-resonant Vibrating Machines ...

ii. Exact solutions for Eq. (4.3). Due to the existence of the damping, the free vibrations will decay exponentially to zero eventually, we study the forced vibration only of the vibration machines.

When the vibration machines work normally, the displacements of the body in y and x directions should have the following forms:

$$y = \lambda_y \sin(\omega t - a_y), \quad x = \lambda_x \cos(\omega t - a_x) \quad (4.5)$$

in which λ_y, λ_x are the vibration amplitudes in y and x directions respectively; a_y and a_x are the phase difference angles of the excitations relative to the displacements in x and y directions respectively.

The velocity and acceleration of the body can be found from Eq. (4.5):

$$\begin{cases} \dot{y} = \lambda_y \omega \cos(\omega t - a_y) \\ \ddot{y} = -\lambda_y \omega^2 \sin(\omega t - a_y) \\ \dot{x} = -\lambda_x \omega \sin(\omega t - a_x) \\ \ddot{x} = -\lambda_x \omega^2 \cos(\omega t - a_x) \end{cases} \quad (4.6)$$

Substituting $\dot{y}$, $\ddot{y}$ into Eq. (4.3) and expanding $\sin \omega t = \sin(\omega t - a_y + a_y) = \cos a_y \sin(\omega t - a_y) + \sin a_y \cos(\omega t - a_y)$, one can rewrite Eq. (4.3) as:

$$-(m+m_0)\omega^2 \lambda_y \sin(\omega t - a_y) + f\omega \lambda_y \cos(\omega t - a_y) + k_y \lambda_y \sin(\omega t - a_y) = m_0 \omega^2 r [\cos a_y \sin(\omega t - a_y) + \sin a_y \cos(\omega t - a_y)] \quad (4.7)$$

To keep the both sides equal, the coefficients of the triangle functions must satisfy the following condition:

$$\begin{array}{l} -(m+m_0)\omega^2 \lambda_y + k_y \lambda_y = m_0 \omega^2 r \cos a_y \\ f\omega \lambda_k = m_0 \omega^2 r \sin a_y \end{array} \quad (4.8)$$

The vibration amplitude of vibration body in y direction and phase difference angle can be derived from Eq. (4.8):

$$\lambda_y = \frac{m_0 \omega^2 r \cos a_y}{k_y - (m+m_0)\omega^2} = -\frac{m_0 r \cos a_y}{m'_y}$$

$$a_y = \arctan \frac{f\omega}{k_y - (m+m_0)\omega^2} = \arctan \frac{-f}{m'_y \omega} \quad (4.9)$$

in which m_y' is the computed mass of the body of the inertial vibration machines in y direction, $m'_y = m + m_0 - k_y/\omega^2$. By the same token, the vibration amplitude of vibration body in x direction and phase difference angle can be found:

$$\lambda_x = -\frac{m_0 r \cos a_x}{m+m_0} = -\frac{m_0 r \cos a_x}{m'_x}$$
$$a_y = \arctan \frac{-f}{(m+m_0)\omega} = \arctan \frac{-f}{m'_x \omega} \quad (4.10)$$

in which m'_x is the computed mass of the body of the inertial vibration machines in x direction, $m'_x = m + m_0$.

The damping forces are very small for the most of the machines, and $k \ll (m+m_0)\omega^2$, a_y and a_x are in the range of $170° - 180°$, so $\cos a_y \approx \cos a_x \approx 1$. Squaring Eqs. (4.9) and (4.10) and then adding them up, one can obtain the following ellipse equation:

$$\left(\frac{y}{\lambda_y}\right)^2 + \left(\frac{x}{\lambda_x}\right)^2 = 1 \quad (4.11)$$

When λ_y and λ_x approach each other, the locus of the movement is a circle.

Figure 4.2 shows the frequency amplitude response curves by Eq. (4.9) and Eq. (4.10). In y direction, when the working frequency is equal to the natural frequency $\omega_0 = \sqrt{k}/(m+m_0)$ the vibration amplitude is increase dramatically. In x direction, the spring stiffness constant is zero and the vibration amplitude λ_x is a constant. The non-resonant vibration machines usually work in the AB region far away from the resonance.

Figure 4.3 represents the displacement change curves of the body from the startup to normal operation, from normal operation to stop. It can be seen from the curve that amplitudes in y direction increase noticeably when it approaches a frequency; while the amplitudes in x direction keep a constant. This is due to the facts that the stiffness constant in x direction is zero and stiffness constant in y direction is not. It can also be seen from curve in y direction that in a period of time after startup there is a free vibration, and it decays to zero after a while, only does the forced vibration exist. Figure 4.3 shows the measured data on displacement of a vibration machineeccentric phase.

In some vibration machines the excitation force does not align to mass center of the body and the stiffness moment of the vibration isolation spring is non-zero. The vibration body will swing in some degree around its mass center.

Fig. 4.2 The frequency amplitude response curves of the inertial vibration machines

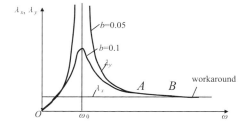

4.1 Dynamics of Non-resonant Vibrating Machines ...

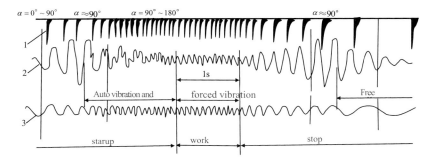

Fig. 4.3 The measured displacement curse of the inertial vibration machines. 1-Phase place of eccentric block; 2-y direction displacement; 3-x direction displacement

For most of the vibration machines, the elastic forces have little effect on the body vibration, in general not exceeding 2%–5%, it can be ignored in the approximate calculation (it should be considered for the exact solutions).

Referred to the Fig. 4.4, the equations for vibrations in x and y direction and swing vibration around the body mass center can be derived:

$$\begin{aligned} (m+m_0)\ddot{y} &= m_0\omega^2 r \sin \omega t \\ (m+m_0)\ddot{x} &= m_0\omega^2 r \cos \omega t \\ (J+J_0)\ddot{\phi} &= m_0\omega 2r(l_{0y}\cos \omega t - l_{0x}\sin \omega t) \end{aligned} \quad (4.12)$$

in which J, J_0 is the movement of inertial of the body and eccentric about the body mass center; l_{0y}, l_{0x} are the distances between the eccentric axle center and the mass center of the body in x and y direction respectively; $\ddot{\phi}$ is the angular acceleration of the swing vibration.

Figure 4.4 shows the swing vibration of a single axle inertial vibration machine. The special solutions for the above differential equations are:

Fig. 4.4 The rocking vibration of single axle inertial vibration machines

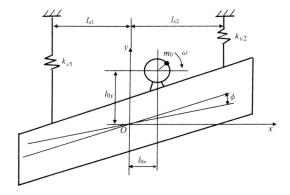

148   4 Linear and Pseudo Linear Vibration

$$y_0 = \lambda_y \sin \omega t$$
$$x_0 = \lambda_x \cos \omega t \qquad (4.13)$$
$$\phi = \lambda_{\phi x} \sin \omega t + \lambda_{\phi y} \cos \omega t$$

in which $\lambda_x, \lambda_y, \lambda_{\phi x}$ and $\lambda_{\phi y}$ are the vibration amplitudes and angles due to x, y excitation force and moments.

Differentiating Eq. (4.13) twice and substituting them into Eq. (4.12) one obtains:

$$\lambda_y = -\frac{m_0 r}{m+m_0}, \quad \lambda_x = -\frac{m_0 r}{m+m_0}$$
$$\lambda_{\phi x} = \frac{m_0 r l_{0x}}{J+J_0}, \quad \lambda_{\phi y} = -\frac{m_0 r l_{0y}}{J+J_0} \qquad (4.14)$$

Therefore the equations of movement of a point e on the body are:

$$y_e = y_0 - \phi l_{ex} = (\lambda_y - \lambda_{\phi x} l_{ex}) \sin \omega t - \lambda_{\phi y} l_{ex} \cos \omega t$$
$$x_e = x_0 + \phi l_{ey} = \lambda_{\phi x} l_{ey} \sin \omega t + (\lambda_x + \lambda_{\phi y} l_{ey}) \cos \omega t \qquad (4.15)$$

After obtaining the values of l_{ex}, l_{ey}, and λ_y, λ_x, $\lambda_{\phi x}$ and $\lambda_{\phi y}$, then dividing a period into 8, 12 or more equal parts, substituting them into the above equation, and finding the values of y_e and x_e when ωt is taken different values, one can determine the locus of any point on the body.

Example 4.1 Given a single axle inertial vibration screen, the total mass of the body and eccentric is 3,000 kg, the moment of inertial of the body and eccentric about the mass center of the body $J + J_0$ is 3,898 kg·m², the excitation force $F = m_0 \omega^2 r = 74,000$ N, the angular velocity $\omega = 78.51$ s-1, the coordinates for the eccentric axle to the mass center are $l_{0y} = 57$ cm, $l_{0x} = 0$ cm.

Find: Loci of the three points: A (0 cm, 132 cm), O (0 cm, 0 cm), B (100 cm, 132 cm).

Solutions: Substituting the given data into Eq. (4.14) one obtains:

$$\lambda_y = \lambda_x = \frac{-74000}{3000 \times 78.5^2} = -0.004(\text{m}) = -0.4(\text{cm})$$
$$\lambda_{\phi y} = \frac{-74000 \times 0.57}{3898 \times 78.5^2} = -0.00175(\text{rad}), \lambda_{\phi x} = 0$$

The equations of society at an arbitrary point e are:

$$y_e = -0.4 \sin \omega t + 0.00175 l_{ex} \cos \omega t$$
$$x_e = (-0.4 - 0.00175 l_{ey}) \cos \omega t$$

Substituting values of l_{ex}, l_{ey} and $\omega t = 0, \pi/4, \pi/2, 3\pi/4, ..., 2\pi$ into the above equations, one can calculate the values of y_e and x_e. The calculated values are listed in the following Table 4.1.

Using the data given in Table 4.1 the trajectory curves can be made as shown in Fig. 4.5. The loci at the two ends are ellipse.

4.1 Dynamics of Non-resonant Vibrating Machines … 149

Table 4.1 The computed values of y_e and x_e

Arbitrary point coordinate/cm		Displacement/cm	ωt							
			0	$\frac{\pi}{4}$	$\frac{\pi}{2}$	$\frac{3\pi}{4}$	π	$\frac{5\pi}{4}$	$\frac{3\pi}{2}$	$\frac{7\pi}{4}$
Approximate method	Point $A \begin{pmatrix} 0 \\ 132 \end{pmatrix}$	y_A	0.23	−0.12	−0.4	−0.44	−0.23	0.12	0.4	0.44
		x_A	−0.4	−0.28	0	0.28	0.4	0.28	0	−0.28
	Point $O \begin{pmatrix} 0 \\ 0 \end{pmatrix}$	y_0	0	−0.28	−0.4	−0.28	0	0.28	0.4	0.28
		x_0	−0.4	−0.28	0	0.28	0.4	0.28	0	−0.28
	Point $B \begin{pmatrix} 100 \\ 132 \end{pmatrix}$	y_B	−0.23	−0.44	−0.4	−0.12	0.23	0.44	0.4	0.12
		x_B	−0.23	−0.16	0	0.16	−0.23	0.16	0	−0.16

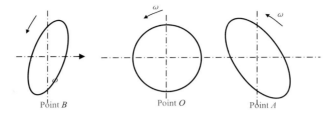

Fig. 4.5 The trajectories of points on the screen box

4.2 Dynamics of Non-resonant Vibrating Machines of Spatial Single-Axis Inertial Type

The spatial movements of the single inertial vibration machines include the vertical vibration polisher and insertion type of vibration rammer etc.

The vibration time of vertical vibration polisher is equal to the rotation time of the main axle. The vibration time of the insertion type of vibration rammer is related to its structure formations. For the vibration rammer shown in Fig. 4.6a, its vibration time n_1 can be computed as follows:

$$n_1 = \frac{n}{\frac{D}{d} - 1} \tag{4.16}$$

Since

$$n_1 \frac{D-d}{2} = n \frac{d}{2}$$

Fig. 4.6 Planet type insertion vibration rammers

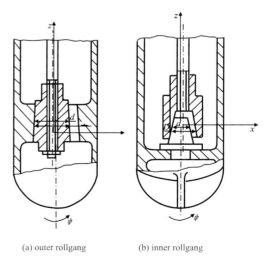

(a) outer rollgang (b) inner rollgang

in which n is the revolution of the drive shaft; D, d are the diameters of the polisher.

For the polisher insertion vibration rammer (Fig. 4.6b), the vibration time n_2 is:

$$n_2 = \frac{n}{1 - \frac{d}{D}} \quad (4.17)$$

since $n_2 \frac{(D-d)}{2} = n \frac{D}{2}$.

Please note that n_1 (or n_2) has opposite rotation directions with n.

The difference between the vertical polisher and insertion type of vibration rammer is that the former is supported on the springs and the latter has no spring supports. From the viewpoint of dynamics, their movements all include the mass center vibration along the vertical plane and swing vibrations of the body about the Ox and Oy axles. We take the polisher to illustrate its dynamic characteristics.

As shown in Fig. 4.7 this machine is driven by a single axle inertial exciter. The drive shaft is installed vertically. The eccentrics at the ends of the drive shaft have an angle γ. The exciting force $F(t)$ by the exciter in the horizontal plane and the exciting moment $M(t)$ about the horizontal axis are:

$$F(t) = \Sigma m_0 \omega^2 r \cos\frac{\gamma}{2}(\cos \omega t + i \sin \omega t) = \Sigma m_0 \omega^2 r \cos\frac{\gamma}{2} e^{i\omega t}$$

$$M(t) = \Sigma m_0 \omega^2 r[(\frac{1}{2}l_0 + l_1)\cos\frac{\gamma}{2}e^{i\omega t} + \frac{1}{2}l_0 \sin\frac{\gamma}{2}e^{i(\omega t - 90°)}] = \Sigma m_0 \omega^2 r L e^{i(\omega t - \beta)}$$

(4.18)

where

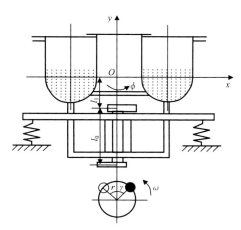

Fig. 4.7 The mechanics model of the vertical polisher

$$L = \sqrt{(\frac{1}{2}l_0 + l_1)^2 \cos^2 \frac{\gamma}{2} + \frac{1}{4}l_0^2 \sin^2 \frac{\gamma}{2}}$$

$$\beta = \arctan(\frac{1}{1 + \frac{2h}{l_0}} \tan \gamma)$$

in which l_0 is the vertical distance between the two eccentrics, l_1 is the vertical distance between the top eccentric to the mass center of the body.

The vertical polisher can be simplified as a 4-degree of freedom vibration system, its equation of movement can be expressed in a complex form:

$$\begin{aligned} m\ddot{\overline{x}} + f\dot{\overline{x}} + k\overline{x} &= \Sigma m_0 \omega^2 r \cos \frac{\gamma}{2} e^{i\omega t} \\ J\ddot{\overline{\phi}} + f_\phi \dot{\overline{\phi}} + k_\phi \overline{\phi} &= \Sigma m_0 \omega^2 r L e^{i(\omega t - \beta)} \end{aligned} \quad (4.19)$$

in which $\overline{x} = x + iy$, $\overline{\phi} = \phi_{0y} + i\phi_{0x}$.

Where m, J are the mass of the working body and moment of inertia about coordinate axis Oy and Ox respectively; f and f_ϕ are the resistance force and moment coefficients; k and k_ϕ are the spring stiffness along the horizontal direction and swing spring stiffness; x, y are the displacements of the mass center along x and y directions; ϕ_{0y}, ϕ_{0x} are the angular displacements of the body around Oy and Ox axis; and $\overline{x}$, $\overline{\phi}$ are the complex forms of the displacement and angular displacement.

As a matter of fact, Eq. (4.19) represents four vibration equations. The springs are installed symmetrically to the z-axis, the two equations above are independent. The stable state vibration amplitude can be found from the first equation:

$$\lambda = \frac{\Sigma m_0 \omega^2 r \cos \frac{\gamma}{2}}{\sqrt{(k - m\omega^2)^2 + f^2 \omega^2}} \quad (4.20)$$

From the second equation, the vibration angular amplitudes around the Oy and Ox axes can be found as:

$$\theta_{ox} = \theta_{oy} = \frac{\Sigma m_0 \omega^2 r L}{\sqrt{(k_\phi - J\omega^2)^2 + f_\phi^2 \omega^2}} \quad (4.21)$$

When the damping is ignored, and introducing a frequency ratio $z_0 = \omega/\omega_0$ and $z_{0\phi} = \omega/\omega_{0\phi}$ (the natural frequency in horizontal direction $\omega_0 = \sqrt{k/m}$, the natural frequency in swing direction $\omega_{0\phi} = \sqrt{k_\phi /J}$,) one obtains the expressions for amplitude λ and vibration amplitude phase angle θ:

$$\lambda \approx \frac{\Sigma m_0 r \cos \frac{\gamma}{2}}{m(\frac{1}{z_0^2} - 1)} \quad \theta \approx \frac{\Sigma m_0 \omega^2 r L}{J(\frac{1}{z_{0\phi}^2} - 1)} \quad (4.22)$$

In order to the increase the finisher working efficiency, the angle for the eccentric γ should be selected reasonably. The test results show that when $\gamma = 90°$, the systems have large vibrations both in horizontal and in the swing direction. The complex vibration can make the part be grinded better. The frequency ratios for the machines are selected from the non-resonant vibration machines.

4.3 Dynamics of Non-resonant Vibration Machines of Double-Axis Inertial Type

4.3.1 Dynamics of Non-resonant Vibrating Machines of Planer Double-Axis Inertial Type

Taking the double-axle inertial type of vibration screen as an example to analyze this type of machines.

In the double-axle inertial vibration screen, the spring stiffness has little effect on the vibration amplitudes, it can be taken as zero in the approximate calculations. Besides the swing vibration of the machines is small, it can be ignored in the approximation calculations as well. As shown in Fig. 4.8 the forces acting on the vibration body are only the inertial forces $m\omega^2\lambda\sin\omega t$ due to the body vibration and the inertial force of the eccentric $2m_0\omega^2 r\sin\omega t$, they should be balanced by each other:

$$m\omega^2\lambda \approx 2m_0\omega^2 rt \text{ or } m\lambda \approx 2m_0 r \qquad (4.23)$$

in which m is the calculated mass of the body(including the eccentric); λ is the vibration amplitude in the vibration direction; m_0 is the eccentric mass and r is the distance between the eccentric mass center and its rotation axis.

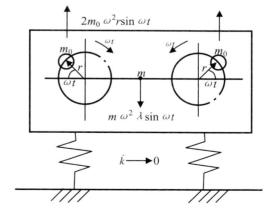

Fig. 4.8 The approximate diagram for the double axle inertial vibration screen

(1) Approximate Solution

When the eccentric mass m_0, eccentricity r and body mass m are all known, the approximate value for the amplitude is:

$$\lambda \approx \frac{2m_0 r}{m} \qquad (4.24)$$

If the single vibration amplitude λ and the mass of the body m are known, the mass moment needed for the eccentric can be calculated as follows:

$$m_0 r \approx \frac{m\lambda}{2} \qquad (4.25)$$

in which $m_0 r$ is the eccentric mass moment of every axles.

(2) The Exact Solutions

In order to calculate accurately for the double axle inertial vibration screen, the equations for the movements of the body can be derived and the solutions can be found from the equations.

The basic difference between the differential equations for the double axle inertial vibration screen and for the single one is the different excitation form. As shown in Fig. 4.9, the resultant inertial force generated by the eccentrics on the double axles is:

$$F = 2m_0 \omega^2 r \sin \omega t \qquad (4.26)$$

The decompositions of the force on the x and y axis are:

$$\begin{aligned} F_y &= 2m_0 \omega^2 r \sin \beta_0 \sin \omega t \\ F_x &= 2m_0 \omega^2 r \cos \beta_0 \sin \omega t \end{aligned} \qquad (4.27)$$

Fig. 4.9 The double axle inertial type of vibration screen

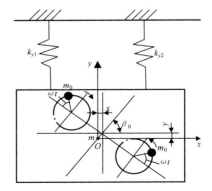

4.3 Dynamics of Non-resonant Vibration Machines ...

in which F_y and F_x are the components of the resultant inertial forces in y and x axis respectively; β_0 is the angle of the resultant inertial force and the x-axis.

The inertial force of the absolute movement for the eccentric should be the summation of the relative movement inertial force (the one generated by rotating around its axle-line: $2m_0\omega^2 r \sin\omega t$ and the inertial force):

$$F_y = -2m_0(\ddot{y} - \omega^2 r \sin\beta_0 \sin\omega t) \\ F_x = -2m_0(\ddot{x} - \omega^2 r \cos\beta_0 \sin\omega t) \tag{4.28}$$

In addition to the inertial force of the absolute movement of the eccentric, there are inertial forces of the body $F_{my} = -my$, $F_{mx} = -mx$, damping forces $F_{fy} = -f\dot{y}$, $F_{fx} = -f\dot{x}$ and spring forces $F_{ky} = -ky$, $F_{kx} = 0$. The sum of all the forces should be zero:

$$\text{In } y \text{ direction: } -m\ddot{y} - f\dot{y} - ky - 2m_0(\ddot{y} - \omega^2 r \sin\beta_0 \sin\omega t) = 0 \\ \text{In } x \text{ direction: } -m\ddot{x} - f\dot{x} - 2m_0(\ddot{x} - \omega^2 r \cos\beta_0 \sin\omega t) = 0 \tag{4.29}$$

in which $m = m_p + K_m m_m$.

where m_p is the real mass of the vibration body; K_m is the combined coefficient of the material; m_m is the material mass; f is the equivalent damping coefficient; k is the stiffness of the vibration isolation springs in their center lines, $k = k_{y1} + k_{y2}$; y, $\dot{y}$, $\ddot{y}$ and x, $\dot{x}$, $\ddot{x}$ are the displacement, velocity and acceleration of the vibration body in x and y direction respectively. Equation (4.29) can also be rewritten as:

$$(m + 2m_0)\ddot{y} + f\dot{y} + ky = 2m_0\omega^2 r \sin\beta_0 \sin\omega t \\ (m + 2m_0)\ddot{x} + f\dot{x} = 2m_0\omega^2 r \cos\beta_0 \sin\omega t \tag{4.30}$$

Equations (4.30) are the differential equations for the double axle inertial vibration screen in x and y directions.

Assume the displacements in x and y direction are:

$$y = \lambda_y \sin(\omega t - a_y) \qquad x = \lambda_x \sin(\omega t - a_x) \tag{4.31}$$

In which λ_y and λ_x are the amplitudes in x and y directions; a_y and a_x are the phase angle of the excitation force versus the displacement:

The velocity and acceleration are:

$$\dot{y} = \lambda_y \omega \cos(\omega t - a_y) \\ \ddot{y} = -\lambda_y \omega^2 \sin(\omega t - a_y) \\ \dot{x} = \lambda_x \omega \cos(\omega t - a_x) \\ \ddot{x} = -\lambda_x \omega^2 \sin(\omega t - a_x) \tag{4.32}$$

Substituting Eq. (4.32) into Eq. (4.30) and one can obtain:

$$\begin{aligned}-(m+2m_0)\omega^2\lambda_y + k\lambda_y &= 2m_0\omega^2 r \sin\beta_0 \cos a_y \\ f\omega\lambda_y &= 2m_0\omega^2 r \sin\beta_0 \sin a_y \\ -(m+2m_0)\omega^2\lambda_x &= 2m_0\omega^2 r \cos\beta_0 \cos a_x \\ f\omega\lambda_x &= 2m_0\omega^2 r \cos\beta_0 \sin a_x\end{aligned} \quad (4.33)$$

The amplitudes and phase angles can be found as follows:

$$\begin{aligned}\lambda_y &= \frac{2m_0\omega^2 r \sin\beta_0 \cos a_y}{k-(m+2m_0)\omega^2} = -\frac{2m_0 r \sin\beta_0 \cos \alpha_y}{m'_y} \\ \lambda_x &= \frac{2m_0 r \cos\beta_0 \cos a_x}{-(m+2m_0)} = -\frac{2m_0 r \cos\beta_0 \cos a_x}{m'_x} \\ a_y &= \arctan\frac{f\omega}{k-(m+2m_0)\omega^2} = \arctan\frac{-f}{m'_y\omega} \\ a_x &= \arctan\frac{-f}{(m+2m_0)\omega} = \arctan\frac{-f}{m'_x\omega}\end{aligned} \quad (4.34)$$

in which $m'_y = m+2m_0 - \frac{k}{\omega^2}$, $m'_x = m+2m_0$, m'_y and m'_x are the computed masses.

Since the stiffness is different in x and y directions, the resultant vibration direction and the excitation force direction are different. The resultant amplitude in x and y directions can be taken as the amplitude in the vibration direction (due to small damping, we have $a_y \approx a_x$):

$$\lambda = \sqrt{\lambda_x^2 + \lambda_y^2} = \frac{2m_0 r}{m'_y} \cos a_y \sqrt{(\frac{m'_x}{m'_y}\sin\beta_0)^2 + \cos^2\beta_0} \quad (4.35)$$

The real vibration angle β is:

$$\beta = \arctan\frac{\lambda_y}{\lambda_x} = \arctan(\frac{m'_x}{m'_y}\tan\beta_0) \quad (4.36)$$

Since the spring in y direction has curtain stiffness but the spring in x direction has no stiffness, m'_y in general is smaller than m'_x, the real vibration angle β is larger a little bit than the resultant inertial force angle β_0. However for the situation in which vibration frequency is far more larger than the resonance frequency ($k << (m+2m_0)\omega^2$, $m'_x = m'_y$, $a_x = a_y$, the resultant amplitude is:

$$\lambda = \frac{2m_0 r}{m+2m_0} \quad (4.37)$$

4.3 Dynamics of Non-resonant Vibration Machines … 157

The vibration angle is:

$$\beta = \beta_0$$

The results from Eq. (4.37) are consistent with those in the aforementioned approximate solutions.

4.3.2 Dynamics of Non-resonant Vibration Machines of Spatial Double-Axis Inertial Type

We will analyze its dynamics of this type of machines using the spiral vertical vibration conveyers as an example.

Figure 4.10a is the schematic of a parallel axle type of vertical vibration conveyers. It can be seen from the figure that the two eccentrics on the end of the parallel axle type of the inertial exciters are installed with an angle. When they are rotating anticlockwise the excitation forces are generated in the vertical direction and the forces make the trough body vibrate in the vertical direction, at the same time the forces form a force couple to make the trough rotationally vibrate. The combination of the two vibrations makes every points on the working body vibrate in angle between the body and the trough bottom surface.

Referred to Fig. 4.10b the excitation force and excitation force couple can be written as:

$$\begin{aligned} F(t) &= \Sigma m_0 r \omega^2 \sin \gamma \sin \omega t \\ M(t) &= \Sigma m_0 r \omega^2 a \cos \gamma \sin \omega t \end{aligned} \tag{4.38}$$

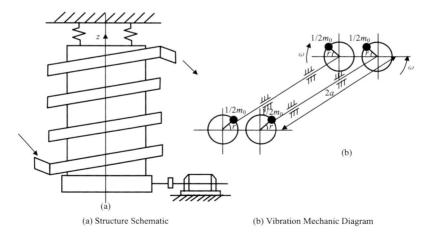

(a) Structure Schematic   (b) Vibration Mechanic Diagram

Fig. 4.10 The schematic of a parallel axle type of vertical vibration conveyers

in which γ is the angle between the horizontal plane and the position of the eccentric as shown in Fig. 4.10 and a is the half distance between the two eccentrics.

The vibration equations of the machines in z axis and rotation around z-axis are:

$$m\ddot{z} + f\dot{z} + k_z z = F(t)$$
$$J_z\ddot{\phi} + f_\phi\dot{\phi} + k_\phi\phi = M(t) \qquad (4.39)$$

Substituting Eq. (4.38) into Eq. (4.39) one can obtain the amplitude and vibration phase angle for the vibrations:

$$\lambda_z = \frac{\Sigma m_0 r \omega^2 \sin\gamma}{\sqrt{(k_z - m\omega^2)^2 + f^2\omega^2}} \approx -\frac{\Sigma m_0 r \sin\gamma}{m}$$
$$\theta_z = \frac{\Sigma m_0 r \omega^2 a \cos\gamma}{\sqrt{(k_\phi - J_z\omega^2)^2 + f_\phi^2\omega^2}} \approx -\frac{\Sigma m_0 r a \cos\gamma}{J_z} \qquad (4.40)$$

The vibration machine parameters can be computed approximately for the non-resonance condition. When approximating, the effects of spring stiffness and damping can be ignored. The resultant vibration amplitude and vibration phase angle for the point whose distance to the vertical axle is ρ are:

$$\lambda = \sqrt{\lambda_z^2 + \theta_z^2 \rho^2} \qquad \beta = \arctan\frac{\lambda_z}{\theta_z \rho} \qquad (4.41)$$

The angle between the vibration direction of spiral vertical vibration conveyers and the horizontal place β is in general 20°–35° larger than the spiral angle and the materials move upward while the vibration angle for the tower type of vibration cooling machines is equal or a little bit larger than 90°, the materials move downward.

Besides the forced synchronous parallel axle type of spiral vertical vibration conveyers, the self-synchronous cross axle type of spiral vertical vibration conveyers has been widely used. The angle of the two eccentrics on the end of axles of the vibration machine is zero while the angle between the axis and vertical line is γ. The machine can obtain the same trajectory of motion as those of the parallel axle type of vibration machines.

4.4 Dynamics of Non-resonant Vibration Machines of Multi-axis Inertial Type

The 4-axle inertial vibration table is a typical multi-axle inertial vibration machines. Figure 4.11 is a schematic of the vibration table. It can be seen from the Figure that the 4-axle exciters include a pair of high-speed shaft with the same velocity but the opposite rotation and a pair of low speed shaft with the same velocity but the

4.4 Dynamics of Non-resonant Vibration Machines ...

Fig. 4.11 Suspended type of 4-axle inertial vibration table

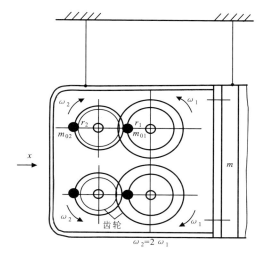

opposite rotation. The frequency ratio of the two major pair shafts is selected as 1:2. The combined harmonic wave vibration with frequency ratio 1:2 can make the table obtain the differential characteristics of the motions.

4.4.1 General Pattern of Planer Movement

The displacement of the vibration machines is composed of low and high frequency displacements:

$$x = x_1 + x_2 \quad (4.42)$$

The low frequency displacement is:

$$x_1 = \lambda_1 \sin \omega t \quad \lambda_1 = -\frac{\Sigma m_{01} r_1}{m} \quad (4.43)$$

in which Σm_{01} is the mass of low frequency eccentric; r_1 is the eccentricity of the low frequency eccentric; m is the mass of the vibration body (including the eccentric mass); ω and t are the low frequency vibration frequency and time respectively.

The negative symbol in Eq. (4.43) indicates the direction of the movement of the eccentric is opposite to the displacement direction.

The high frequency displacement can be expressed as:

$$x_2 = \lambda_2 \sin(2\omega t + \theta_2), \lambda_2 = -\frac{\Sigma m_{02} r_2}{m} \quad (4.44)$$

in which Σm_{02} is the mass of high frequency eccentric; r_2 is the eccentricity of the high frequency eccentric; θ_2 is the initial phase angle of the high frequency vibration.

As before the negative symbol in Eq. (4.44) indicates the direction of the movement of the eccentric is opposite to the displacement direction.

Substituting Eqs. (4.43) and (4.44) into Eq. (4.42) one obtains:

$$x = -\frac{1}{m}[m_{01}r_1 \sin \omega t + m_{02}r_2 \sin(2\omega t + \theta_2)]$$

or

$$x = -\frac{m_{01}r_1}{m}[\sin \omega t + \frac{m_{02}r_2}{m_{01}r_1} \sin(2\omega t + \theta_2)] \quad (4.45)$$

The velocity and acceleration are:

$$\dot{x} = -\frac{m_{01}r_1\omega}{m}[\cos \omega t + \frac{2m_{02}r_2}{m_{01}r_1} \cos(2\omega t + \theta_2)]$$
$$\ddot{x} = \frac{m_{01}r_1\omega^2}{m}[\sin \omega t + \frac{4m_{02}r_2}{m_{01}r_1} \sin(2\omega t + \theta_2)] \quad (4.46)$$

It can be seen from Eq. (4.46) that the displacements, velocity and acceleration are composed by a fundamental and a second order harmonics, their vibration amplitudes are directly proportional to the mass moment $m_{01}r_1$ or $m_{02}r_2$ and inversely proportional to the vibration mass. The excitation forces of the low and high eccentrics which generate the relative movements are:

$$F_1 = m_{01}\omega^2 r_1 \sin \omega t$$
$$F_2 = 4m_{02}\omega^2 r_2 \sin(2\omega t + \theta_2) \quad (4.47)$$

The resultant excitation force is:

$$F = F_1 + F_2 = m_{01}\omega^2 r_1[\sin \omega t + \frac{4m_{02}r_2}{m_{01}r_1} \sin(2\omega t + \theta_2)] \quad (4.48)$$

The excitation forces are always in opposite direction with the table inertial force.

4.4.2 Values of Displacement, Velocity and Acceleration Curves and Differential Coefficients When θ_2 is Equal to $\Pi/2$

Figure 4.12 shows the curves of displacements $\frac{mx}{m_{01}r_1}$, velocity $\frac{m\dot{x}}{m_{01}r_1\omega}$ and acceleration $\frac{m\ddot{x}}{m_{01}r_1\omega^2}$ when $\theta_2 = \frac{\pi}{2}$ and $\varepsilon = \frac{m_{02}r_2}{m_{01}r_1} = 0.15, 0.35$. It can be seen from the Figure

4.4 Dynamics of Non-resonant Vibration Machines ...

Fig. 4.12 The displacement, velocity and acceleration curves when $\theta_2 = \pi/2$

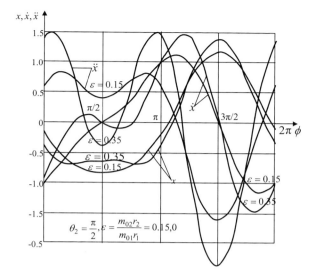

that when $\theta_2 = \frac{\pi}{2}$ the displacement and velocity curves are symmetric to a certain vertical axis line.

The maximum and minimum values can be found in the Figure and also can be found by Eq. (4.48). The corresponding angles to the extreme values can be found by the following:

$$\frac{dx}{d\phi} = -\frac{m_{01}r_1}{m}\left[\cos\phi + \frac{m_{02}r_2}{m_{01}r_1}2\cos(2\phi + \frac{\pi}{2})\right] = 0 \quad (4.49)$$

i.e.

$$\cos\phi\left(1 - 4\frac{m_{02}r_2}{m_{01}r_1}\sin\phi\right) = 0$$

The two conditions can be drawn from the above equations:

$$\cos\phi = 0, \phi_1 = -\frac{\pi}{2}, \phi_2 = \frac{\pi}{2}$$

$$1 - 4\frac{m_{02}r_2}{m_{01}r_1}\sin\phi = 0$$

$$\phi_3 = \arcsin\frac{m_{01}r_1}{4m_{02}r_2}, \phi_4 = \pi - \arcsin\frac{m_{01}r_1}{4m_{02}r_2} \quad (4.50)$$

The maximum occurs when $\phi = -\frac{\pi}{2}$, i.e.:

$$x_{\max} = \frac{m_{01}r_1}{m}\left(1 + \frac{m_{02}r_2}{m_{01}r_1}\right) \tag{4.51}$$

The minimum occurs when $\phi = \frac{\pi}{2}$, i.e.:

$$x_{\min} = -\frac{m_{01}r_1}{m}\left(1 - \frac{m_{02}r_2}{m_{01}r_1}\right) \tag{4.52}$$

It can be seen from Eq. (4.50) that when $\frac{m_{01}r_1}{4m_{02}r_2} > 1$, ϕ_3 and ϕ_4 have no solutions; only when $\frac{m_{01}r_1}{4m_{02}r_2} < 1$, ϕ_3 and ϕ_4 have solutions. The corresponding minimum displacements are:

$$x_{\min} = -\frac{m_{01}r_1}{m}\left[\frac{m_{02}r_2}{m_{01}r_1} + \frac{m_{01}r_1}{8m_{02}r_2}\right] \tag{4.53}$$

The maximum and minimum values of the acceleration can be found by the same method. The angle corresponding to the maximum is determined by:

$$\frac{d\ddot{x}}{d\phi} = \frac{\omega^2 m_{01}r_1}{m}[\cos\phi + \frac{8m_{02}r_2}{m_{01}r_1}\cos(2\omega t + \frac{\pi}{2})] = 0 \tag{4.54}$$

or

$$\cos\phi + \frac{8m_{02}r_2}{m_{01}r_1}\cos(2\phi + \frac{\pi}{2}) = \cos\phi(1 - \frac{8m_{02}r_2}{m_{01}r_1}\sin\phi) = 0$$

One can obtain:

$$\cos\phi = 0, \phi_1 = -\frac{\pi}{2}, \phi_2 = +\frac{\pi}{2} \tag{4.55}$$

$$1 - \frac{8m_{02}r_2}{m_{01}r_1}\sin\phi = 0, \phi_3 = \arcsin\frac{8m_{02}r_2}{m_{01}r_1}, \phi_4 = \pi - \arcsin\frac{8m_{02}r_2}{m_{01}r_1}$$

The minimum accelerations occur when $\phi = -\frac{\pi}{2}$, and its value is:

$$\ddot{x}_{\min} = -\frac{m_{01}r_1\omega^2}{m}(1 + \frac{4m_{02}r_2}{m_{01}r_1}) \tag{4.56a}$$

The maximum accelerations are:

$$\ddot{x}_{\max 1} = \frac{m_{01}r_1\omega^2}{m}(1 - \frac{4m_{02}r_2}{m_{01}r_1}) \tag{4.56b}$$

$$\ddot{x}_{\max 2,3} = \frac{m_{01}r_1\omega^2}{m}(4\frac{m_{02}r_2}{m_{01}r_1} + \frac{m_{01}r_1}{32m_{02}r_2}) \tag{4.57}$$

The differential coefficient for the table is the ratio of the absolute values of the minimum to maximum accelerations, i.e.:

$$z = \left|\frac{\ddot{x}_{\min}}{\ddot{x}_{\max}}\right| = \frac{m_{01}r_1 + 4m_{02}r_2}{[4m_{02}r_2 + \frac{(m_{01}r_1)^2}{32m_{02}r_2}]} \tag{4.58}$$

When $\frac{m_{02}r_2}{m_{01}r_1} = 0.15 - 0.35$, the differential coefficient $z = 1.98 - 1.63$. Thus $\frac{m_{02}r_2}{m_{01}r_1}$ should not be large. To increase the differential coefficient to change the Table selection characteristics, it is generally to change the angle θ_2.

4.5 Dynamics of Inertial Near-Resonant Type of Vibration Machines

4.5.1 Dynamics of Single Body Near-Resonant Vibration Machines

For the inertial type of near resonant vibration screens and near resonant vibration feeders used in industries when their main spring stiffness is equal to or approximate a constant, they can be treated as a linear vibration machines and the calculation errors are not significant. For example, the vibration machines using leaf springs and cylinder spiral springs as their main vibration springs, are all in this category. The vibration machines using the shear rubber springs and preloaded constant-area compression rubber spring as its main vibration springs can also be treated approximately the linear vibration machines.

The single body inertial type of near-resonant vibration machines is generally small machines. Large single body near resonant vibration machines will generate large unbalanced dynamic load and transmit to the base. This vibration force is harmful to the base and the buildings.

The single body inertial type of near-resonant vibration machines can be driven by single-axle inertial exciter, or a double-axle inertial exciter. The former is simple in structure while its guided rod (i.e. leaf spring or rubber connecting type of rod) is subject to some of the inertial force that can not be balanced out. The latter structure is complex while its exciter's inertial forces in the guided rod direction are cancelled out each other. However there is no significant difference between the two from the viewpoint of dynamics (Fig. 4.13).

This type of vibration machines generates vibration only in direction vertical to the centerline of the guided rod and its vibration equations have little different from those for the double axle line inertial type of vibration machines. If the term $\sum m_0 \omega^2 r \sin \omega t$ denotes the inertial force of the relative movement in vibration direction by the exciter, the vibration equations of the vibration body in the vibration direction can be written as:

Fig. 4.13 The schematic of a single body near resonant vibration machine

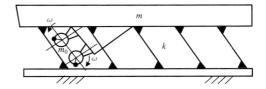

$$(m + \Sigma m_0)\ddot{s} + f\dot{s} + ks = \Sigma m_0 \omega^2 r \sin \omega t \quad (4.59)$$

where m is the computed mass of the body, $\sum m_0$ is the total mass of the eccentric; f is the equivalent damping coefficient; k is the stiffness of the main vibration spring; $s, \dot{s}, \ddot{s}$ are the displacement, velocity and acceleration of the vibration body; r is the distance from the combined mass center of the eccentric to the rotation axle line; ω is the angular velocity and t is the time.

The particular solutions for the above equations are:

$$s = \lambda \sin(\omega t - \alpha) \quad (4.60)$$

in which the amplitude λ and the phase angle α are:

$$\begin{aligned} \lambda &= \frac{\Sigma m_0 \omega^2 r \cos a}{k - m'\omega^2} = \frac{\Sigma m_0 \omega^2 r \cos a}{k(1 - z_0^2)} = \frac{z_0^2 \Sigma m_0 r \cos a}{m'(1 - z_0^2)} \\ a &= \arctan \frac{f\omega}{k - m'\omega^2} = \arctan \frac{2bz_0}{1 - z_0^2}, m' = m + \Sigma m_0 \end{aligned} \quad (4.61)$$

where z_0 is the frequency ratio ω/ω_0 and b is the damping ratio.

The relationship between the amplitude and the frequency ratio is shown in Fig. 4.2. It is worth noticing that the frequency ratio is selected to be close to 1 for this type of machines for making use of a series of advantages of the resonance (small excitation force, transmission portion of the machine is compact, less energy consuming and the durability). In general it is taken as:

$$z_0 = 0.75 \sim 0.95$$

Using the selected frequency ratio, one can compute the mass moment needed for the eccentric

$$\Sigma m_0 r = \frac{m'\lambda(1 - z_0^2)}{z_0^2 \cos a} \quad (4.62)$$

The phase angle can be computed from Eq. (4.61). In general the damping ratio is less than 0.05–0.07.

The stiffness of the main vibration springs is:

$$k = \frac{1}{z_0^2} m' \omega^2 \qquad (4.63)$$

4.5.2 Dynamics of Double Body Near-Resonant Vibration Machines

In recent decades, the double body inertial resonance vibration screens, resonance conveyers and resonance feeders have gradually gained momentum in its applications in industries. A typical structure of the machines is shown in Fig. 4.14a and its mechanics diagram is shown in Fig. 4.14b. In the following equation derivations, the damping of the absolute movements is considered to be small and ignored.

The forces acting on the body 1 include the inertial force, relative movement damping force and spring force of the main vibration springs and the vibration isolation spring force, and their sum should be zero:

$$-m_1 \ddot{x}_1 - f(\dot{x}_1 - \dot{x}_2) - k(x_1 - x_2) - k_{1x} x_1 = 0 \qquad (4.64)$$

The forces subjected by the body 2 include the body 2 inertial force, relative movement damping force, the spring force of the main vibration springs, the vibration isolation spring force and the inertial force generated by the transmission shaft eccentric, and those forces should be all balanced:

$$-m_2 \ddot{x}_2 - f(\dot{x}_2 - \dot{x}_1) - k(x_2 - x_1) - k_{2x} x_2 - \Sigma m_0 (\ddot{x} - \omega^2 r \sin \omega t) = 0 \qquad (4.65)$$

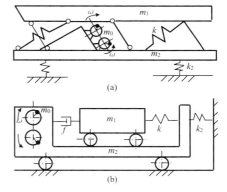

Fig. 4.14 The double body near the resonance vibration machines. **a** a typical structure of the machines, **b** mechanics diagram

where m_2 is the mass of the body 2; Σm_0 is the total mass of the eccentric block; ω is the axle rotation angular velocity; r is the distance between the eccentric block and rotational axle; t is the time.

In the linear vibration theory, the displacement has a relation with the acceleration as follows:

$$\ddot{x}_1 = -\omega^2 x_1, \ddot{x}_2 = -\omega^2 x_2 \tag{4.66}$$

Substituting Eq. (4.66) into Eqs. (4.64) and (4.65) and simplifying them, one can obtain the equations of the movement of the body 1 and body 2:

$$\begin{aligned} m'_1 \ddot{x}_1 + f(\dot{x}_1 - \dot{x}_2) + k(x_1 - x_2) = 0 \\ m'_2 \ddot{x}_2 - f(\dot{x}_1 - \dot{x}_2) - k(x_1 - x_2) = \Sigma m_0 \omega^2 r \sin \omega t \end{aligned} \tag{4.67}$$

where

$$m'_1 = m_1 - \frac{k_{1x}}{\omega^2} \approx m_1, m'_2 = m_2 + \Sigma m_0 - \frac{k_{2x}}{\omega^2} \approx m_2 + \Sigma m_0 \tag{4.68}$$

In the resonance screen and conveyers, the term $\frac{k_{1x}}{\omega^2}$ is much less than m_1 and $\frac{k_{2x}}{\omega^2}$ is much less than m_2, so the stiffness of the vibration isolation spring is combined by the computed masses m'_1 and m'_2.

In order to find the solutions in above vibration equations, it is convenient to the equation into the ones for the relative displacement, velocity and accelerations. Multiplying the first Equation in (4.67) with $\frac{m'_2}{m'_1+m'_2}$ and subtracting it from the second equation in (4.67) multiplied by $\frac{m'_1}{m'_1+m'_2}$, one has:

$$m\ddot{x} + f\dot{x} + kx = -\frac{m'_1}{m'_1 + m'_2} \Sigma m_0 \omega^2 r \sin \omega t \tag{4.69}$$

where m is the induced mass, $m = \frac{m'_1 m'_2}{m'_1+m'_2}$; $x, \dot{x}, \ddot{x}$ are the relative displacement, relative velocity and relative acceleration between the body 1 and body 2 respectively.

Due to the existence of damping, the free vibrations decay to zero with time and it will not be considered. So the particular solutions to the equation are in the form of:

$$x = \lambda \sin(\omega t - \alpha) \tag{4.70}$$

where λ is the relative amplitude and α is the lead angle difference of exciting force to the relative displacement.

Substituting Eq. (4.70) into Eq. (4.69) leads to:

4.5 Dynamics of Inertial Near-Resonant Type of Vibration Machines

$$\lambda = -\frac{m}{m_2'} \frac{\Sigma m_0 \omega^2 r \cos a}{k - m\omega^2} = -\frac{1}{m_2'} \frac{z_0^2 \Sigma m_0 r \cos a}{(1 - z_0^2)}$$

$$a = \arctan \frac{2bz_0}{1 - z_0^2} \quad (4.71)$$

The relative vibration amplitudes have been found for the body 1 and 2. Now we will find the absolute amplitudes and displacements x_1 and x_2.

It is obvious that the absolute displacements x_1 and x_2 have the following forms:

$$x_1 = \lambda_1 \sin(\omega t - a_1) \qquad x_2 = \lambda_2 \sin(\omega t - a_2) \quad (4.72)$$

Using Eq. (4.67) and Eq. (4.69) one has the absolute amplitudes:

$$\lambda_1 = \frac{k}{m_1' \omega^2} \frac{\lambda}{\cos \gamma_1} = -\frac{\Sigma m_0 r \cos a}{(m_1' + m_2')(1 - z_0^2) \cos \gamma_1} =$$

$$-\frac{\Sigma m_0 r \sqrt{1 + 4b^2 z_0^2}}{(m_1' + m_2')\sqrt{(1 - z_0^2)^2 + 4b^2 z_0^2}}$$

$$\lambda_2 = (\frac{k}{m_1' \omega^2} - 1)\frac{\lambda}{\cos \gamma_2} = (\frac{z_0^2}{m_2'} - \frac{1}{m_1' + m_2'})\frac{\Sigma m_0 r}{1 - z_0^2} \frac{\cos \alpha}{\cos \gamma_2} =$$

$$\frac{\Sigma m_0 r \sqrt{(1 - \frac{m_1'}{m} z_0^2)^2 + 4b^2 z_0^2}}{(m_1' + m_2')\sqrt{(1 - z_0^2)^2 + 4b^2 z_0^2}}$$

(4.73)

The phase angles α_1 and α_2 are respectively:

$$\alpha_1 = \alpha + \gamma_1, \alpha_2 = \alpha + \gamma_2$$

in which

$$\gamma_1 = \arctan 2bz_0, \gamma_2 = \arctan \frac{2bz_0}{1 - \frac{m_1'}{m} z_0^2} \quad (4.74)$$

$$b = \frac{f}{2m\omega_0}$$

It can be seen from Eq. (4.73) that the conditions that the inertial near-resonance vibration machine body acquires its maximum amplitudes (assume that damping is negligible) are:

$$m_1' + m_2' = 0 \text{ or } 1 - z_0^2 = 0 \quad (4.75)$$

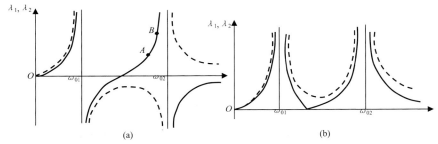

(a) λ_1, λ_2 have not been taken as absolute values, (b) λ_1, λ_2 have been taken as absolute values

Fig. 4.15 Double body near resonance vibration machine frequency-amplitude curves

The low natural frequency and high natural frequencies can be found approximately to be:

$$\omega_{0d} = \sqrt{\frac{k_{1x} + k_{2x}}{m_1 + m_2}} \qquad \omega_{0g} = \sqrt{\frac{k(m'_1 + m'_2)}{m'_1 m'_2}} \qquad (4.76)$$

The resonance curves for the double body inertial type of near resonance machines are shown in Fig. 4.15a. For the machines to have a stable amplitude and small excitation force, the main resonance frequencies are selected in the range of 0.75–0.95, i.e. the region AB as shown in Fig. 4.15a.

4.6 Dynamics of Single Body Elastic Connecting Rod Type of Near Resonance Vibration Machines

The elastic connecting rod type of vibration machines has been used for conveyance, screening, selecting and cooling, etc. Their structures are simple, easily manufactured, and subject to small forces when working. The body balancing characteristics are excellent (for double and multiple bodies).

The machines can be categorized by their number of vibration bodies: single-body type, double-body type and multiple-body type. We will analyze their dynamics one by one.

This kind of machines is installed either horizontally or in a small angle with the horizontal. Figure 4.16 shows its working mechanism and mechanics diagram. We can see that it has only one body. The body and bases are connected by main vibration springs; the eccentric drive shaft makes the connecting rod move reciprocally and the body is forced to vibrate by the end of the connecting rod via the driving springs.

Referred to Fig. 4.16b the equation of motion for the vibration body can be derived as:

4.6 Dynamics of Single Body Elastic Connecting Rod Type …

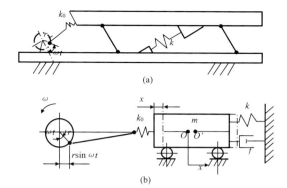

Fig. 4.16 Eccentric connecting rod type of vibration conveyer. **a** mechanism, **b** mechanics diagram

$$-m\ddot{x} - f\dot{x} - kx - k_0(x - r\sin\omega t) = 0 \tag{4.77}$$

in which x, $\dot{x}$, $\ddot{x}$ are the body displacement, velocity and acceleration respectively; f is the damping coefficient; m is the vibration mass; k is the spring stiffness of the main vibration springs; k_0 is the spring stiffness of the connecting rod; r is the eccentricity of the drive shaft; ω is the angle frequency; t is the time.

Simplifying Eq. (4.78) one can get:

$$m\ddot{x} + f\dot{x} + (k + k_0)x = k_0 r \sin\omega t \tag{4.78}$$

Equation (4.78) indicates that the system is a single degree of freedom forced vibration system, its excitation force amplitude is $k_0 r$, while the spring stiffness of the system is the sum of the stiffness of the main vibration spring and the connecting rod spring. The solutions of the equation include the particular solution for the equation and the general solutions of the corresponding free vibration equation. Due to the existence of the damping, the free vibration decays to zero and only the forced vibration remains in the operation over time. We only study the forced vibration of the machines.

The particular solution for the equation has the following form:

$$x = \lambda \sin(\omega t - \alpha) \tag{4.79}$$

where λ is the amplitude of the vibration body; α is the lag phase angle of the displacement relative to the nominal excitation $k_0 \sin \omega t$.

The velocity and acceleration are respectively:

$$\begin{aligned} \dot{x} &= \lambda\omega \cos(\omega t - \alpha) \\ \ddot{x} &= -\lambda\omega^2 \sin(\omega t - \alpha) \end{aligned} \tag{4.80}$$

Substituting Eq. (4.80) into Eq. (4.78) results in:

$$-m\omega^2 \lambda \sin(\omega t - \alpha) + f\omega\lambda \cos(\omega t - \alpha) + (k + k_0)\lambda \sin(\omega t - \alpha) = k_0 r \sin \omega t \quad (4.81)$$

The triangular equation is as follows:

$$k_0 r \sin \omega t = k_0 r \sin(\omega t - \alpha + \alpha)$$

$$= k_0 r \cos \alpha \sin(\omega t - \alpha) + k_0 r \sin \alpha \cos(\omega t - \alpha) \quad (4.82)$$

Substituting Eq. (4.82) into Eq. (4.81) and equating the coefficients of the triangular terms on both sides one obtains:

$$-m\omega^2 \lambda + (k + k_0)\lambda = k_0 r \cos \alpha$$

$$f\omega\lambda = k_0 r \sin \alpha \quad (4.83)$$

The physical meanings behind the Eq. (4.83) are: the cosine component of the nominal excitation amplitude is balanced by the difference between the inertial force and the spring force amplitude; its sine component of the excitation is balanced by the damping force (the vector diagram is shown in Fig. 4.17). The vibration amplitude of the body and the lag phase angle relative to the nominal excitation force can be found by Eq. (4.83):

$$\lambda = \frac{k_0 r \cos \alpha}{k + k_0 - m\omega^2} = \frac{k_0 r \cos \alpha}{(k + k_0)(1 - z_0^2)} = \frac{k_0 r z_0^2 \cos \alpha}{m\omega^2 (1 - z_0^2)} = \frac{k_0 r}{(k + k_0)\sqrt{(1 - z_0^2)^2 + 4b^2 z_0^2}} \quad (4.84)$$

$$\alpha = \arcsin \frac{f\omega\lambda}{k_0 r} = \arctan \frac{f\omega}{(k + k_0) - m\omega^2} = \arctan \frac{2bz_0}{1 - z_0^2}$$

Fig. 4.17 Vector relation of nominal exciting force amplitude and system forces

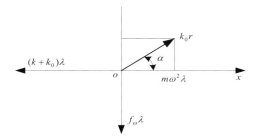

where the damping ratio is $b = \frac{f}{2m\omega_0}$ and its values are in the range of 0.03–0.07.

For this kind of machines, the frequency ratio is taken as $z_0 = 0.03 - 0.07$, i.e. the machines are working under the low critical and near resonant dynamic conditions and thedamping coefficients have large effect on the vibration of the systems.

4.7 Dynamics of Double Body Elastic Connecting Rod Type of Near Resonance Vibration Machines

The structures of the double body elastic connecting rod type of vibration machines are more complex than those of the single body counterpart, and their dynamics are obvious different from the single body. Now we will discuss two types of the double body elastic connecting rod vibration machines: i.e. the balanced and unbalanced.

4.7.1 Balanced Type of Vibration Machines with Double Body Elastically Connecting Rod

We know from the dynamic analysis of the single body elastic connecting rod type of vibration machines that it has the shortcomings of dynamic unbalanced forces, and the forces could be all transferred to the base and thus cause the base and the buildings to vibrate. The balanced type of vibration machines is a kind of the double-body working mechanism which can reduce the vibration effect on the base. Figure 4.18 shows its mechanics diagram. The rubber hinge joint type of the connecting rod connects the two bodies while the whole machine is fixed on to the bottom frame via the hinge joints at the middle of the rods and the supports. There are the elastic connecting rod types of driving device and the main vibration springs. When working the inertial forces of the two bodies are in the opposite directions as the bodies roll around the center of the guided rods. When the masses of the two bodies are equal, their inertial forces are balanced. Actually the masses of the materials in the trough body and trough themselves are not completely equal, the unbalance part of the inertial forces are transferred to the base. However, compared to the machine mass, the inertial force can be neglected in the analysis.

Fig. 4.18 Mechanical diagram of the vibration machines

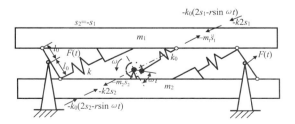

Referred to Fig. 4.18 the roll vibration equation around the support hinge joint point O of the two bodies (i.e. the rotation movement) and the balanced equation along the vibration direction can be derived:

$$-(m_1 + m_2)\ddot{s}_1 l_0 - (f_1 + f_2)\dot{s}_1 l_0 - k \times 2s_1 \times 2l_0 - k_0(2s_1 - r\sin\omega t) \times 2l_0 = 0$$

$$m_1 \ddot{s}_1 + m_2 \ddot{s}_2 = (m_1 - m_2)\ddot{s}_1 = F(t) \tag{4.85}$$

where k, k_0 are the main vibration and connecting rod spring stiffness respectively; m_1, m_2 are the masses of the body 1 and 2 respectively; f_1, f_2 are the damping coefficients between the body 1 and body 2; $s_1, \dot{s}_1$ and $\ddot{s}_1$ are the displacement velocity and acceleration along the vibration direction of the body 1; l_0 is the distance from the hinge joint center of the guided rod to the center hinge joint; and $F(t)$ is the acting force on the support hinge joint.

The first in Eq. (4.85) can be simplified as:

$$\frac{1}{4}(m_1 + m_2)\ddot{s}_1 + \frac{1}{4}(f_1 + f_2)\dot{s}_1 + (k + k_0)s_1 = \frac{1}{2}k_0 r \sin \omega t \tag{4.86}$$

The particular solution to the Eq. (4.86) can be found:

$$s_1 = \lambda_1 \sin(\omega t - \alpha_1) \tag{4.87}$$

The amplitudes of the body 1 and 2 are:

$$\lambda_1 = \frac{1}{2} \times \frac{k_0 r \cos\alpha_1}{k + k_0 - m\omega^2} = \frac{1}{2} \times \frac{k_0 r \cos\alpha_1}{(k+k_0)(1 - z_0^2)} =$$

$$\frac{1}{2} \times \frac{k_0 r z_0^2 \cos\alpha_1}{m\omega^2 (1 - z_0^2)} = \frac{1}{2} \times \frac{k_0 r}{(k+k_0)\sqrt{(1-z_0^2)^2 + 4b^2 z_0^2}} \tag{4.88}$$

The phase angle is:

$$\alpha_1 = \arctan \frac{f\omega}{k + k_0 - m\omega^2} = \arctan \frac{2b z_0}{1 - z_0^2} = \arccos \frac{1 - z_0^2}{\sqrt{(1-z_0^2)^2 + 4b^2 z_0^2}} \tag{4.89}$$

where m and f are the induced mass and induced damping coefficient:

$$m = \frac{m_1 + m_2}{4} \qquad f = \frac{f_1 + f_2}{4} \tag{4.90}$$

The natural frequency and frequency ratio are:

4.7 Dynamics of Double Body Elastic Connecting Rod Type ...

$$\omega_0 = \sqrt{\frac{(k+k_0) \times 4}{m_1 + m_2}}$$

$$z_0 = \frac{\omega}{\omega_0} = \omega \sqrt{\frac{m_1 + m_2}{4(k+k_0)}} \quad (4.91)$$

The vibration relative amplitude is:

$$\lambda = \lambda_1 - \lambda_2 = \lambda_1 + |\lambda_2| = 2\lambda_1 = \frac{k_0 r}{(k+k_0)\sqrt{(1-z_0^2)^2 + 4b^2 z_0^2}} \quad (4.92)$$

The relative vibration amplitudes of the support type of vibration machines are 2 time as much as that of the body 1 or body 2.

It can be seen from Eqs. (4.85) and (4.87) that the dynamic load amplitude the machines transferred to the base is:

$$F_d = (m_1 - m_2)\omega^2 \lambda_1 \quad (4.93)$$

The dynamic load amplitude the machines transferred to the base is the difference between the inertial forces of the two bodies. Thus it can be seen from Eq. (4.93) that reducing the difference between the two body masses can reduce the dynamic load transferred to the base.

4.7.2 Non-balance Double Body Type of Elastically Connecting Rod Vibration Machines

Figure 4.19b is the mechanics diagram of the "unbalanced type" of double-body vibration machines. The two bodies for this kind of machines are installed upper and

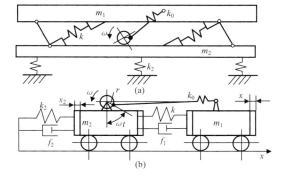

Fig. 4.19 Unbalanced type of double- body elastic connecting rod vibration machine. **a** mechanism, **b** mechanics diagram

lower positions. They are connected by the guided rod, the main vibration springs and elastic connecting rod type of exciter. The vibration isolation springs are installed under the lower body to reduce the impact to the base.

Unlike the "balanced type" of vibration machine the "unbalanced type" of machines uses the vibration isolation springs to reduce its impact on the base. Next we will analyze the systems with two degrees of freedom.

Assume that the stiffness of the vibration isolation springs in the vertical and vibration direction are all equal and the equivalent damping are approximately directly proportional to the relative velocity. The equations of the movements for the body 1 and body 2 are:

$$-m_1\ddot{x}_1 - f_1\dot{x}_1 - f\dot{x} - kx - k_0(x - r\sin\omega t) = 0 \\ -m_2\ddot{x}_2 - f_2\dot{x}_2 + f\dot{x} + kx - k_2 x_2 + k_0(x - r\sin\omega t) = 0 \quad (4.94)$$

where $x = x_1 - x_2$, $\dot{x} = \dot{x}_1 - \dot{x}_2$, $\ddot{x} = \ddot{x}_1 - \ddot{x}_2$.

$x_1, x_2, \dot{x}_1, \dot{x}_2, \ddot{x}_1, \ddot{x}_2$ are the displacement, velocity and acceleration of the body 1 and 2 respectively; f is the relative damping coefficient between the two bodies, f_1 and f_2 is the absolute damping coefficients of the body 1 and 2 respectively; m_1 and m_2 are the computed masses of the body 1 and 2; k, k_0 and k_2 are the stiffness of the main vibration spring, connecting rod spring and vibration isolation spring and r is the eccentricity of the shaft.

Simplifying the equations above, one has:

$$\begin{cases} m_1\ddot{x}_1 + f_1\dot{x}_1 + f\dot{x} + (k + k_0)x = k_0 r\sin\omega t \\ m_2\ddot{x}_2 + f_2\dot{x}_2 - f\dot{x} - (k + k_0)x + k_2 x_2 = -k_0 r\sin\omega t \end{cases} \quad (4.95)$$

Under the normal working conditions, the spring force $-k_2 x_2$ of the vibration isolation spring is much smaller than the inertial force $-m_2\ddot{x}_2$ of the body 2, and the spring force can be expressed as a function of the acceleration in the linear systems under the harmonic excitations:

$$k_2 x_2 = -\frac{k_2}{\omega^2}\ddot{x}_2 \quad (4.96)$$

For the sake of computation convenience, the spring force of the vibration isolation spring is combined into the computed inertial force of the body 2:

$$m_2\ddot{x}_2 + k_2 x_2 = (m_2 - \frac{k_2}{\omega^2})\ddot{x}_2 = m'_2\ddot{x}_2 \quad (4.97)$$

where m'_2 is the computed mass of the body 2.

$$m'_2 = m_2 - \frac{k_2}{\omega^2} \quad (4.98)$$

4.7 Dynamics of Double Body Elastic Connecting Rod Type ...

Thus Eq. (4.95) can be written as a simpler form:

$$\begin{cases} m_1\ddot{x}_1 + f_1\dot{x}_1 + f\dot{x} + (k+k_0)x = k_0 r \sin \omega t \\ m_2'\ddot{x}_2 + f_2\dot{x}_2 - f\dot{x} - (k+k_0)x = -k_0 r \sin \omega t \end{cases} \quad (4.99)$$

Multiplying the first equation in Eq. (4.99) by $\frac{m_2'}{m_1+m_2'}$ and the second by $\frac{m_1}{m_1+m_2'}$ and subtracting each other, one obtains the following vibration equation for the relative displacement, velocity and acceleration:

$$m\ddot{x} + (f_{1m} + f)\dot{x} + (k+k_0)x = k_0 r \sin \omega t \quad (4.100)$$

in which

$$f_{1m} \approx \frac{m_2' f_1}{m_1 + m_2'} \approx \frac{m_1 f_2}{m_1 + m_2'}$$

$$m = \frac{m_1 m_2'}{m_1 + m_2'} \quad (4.101)$$

It can be seen from Eq. (4.101) that this equation is the same as that for the single body vibration machine, that is to say that the vibration system of the double-body vibration machine can be converted into an equivalent single body vibration machine for computations. However m is the induced mass and x is the relative displacement, $\dot{x}$ is the relative velocity, and $\ddot{x}$ is the relative acceleration, $k + k_0$ is the induced stiffness.

The particular solution for the equation Eq. (4.100) has the following form:

$$x = \lambda \sin(\omega t - \alpha) \quad (4.102)$$

Substituted it into Eq. (4.100) the relative amplitude is:

$$\lambda = \frac{k_0 r \cos \alpha}{k + k_0 - m\omega^2} = \frac{k_0 r \cos \alpha}{(k + k_0)(1 - z_0^2)} \quad (4.103)$$

or

$$\lambda = \frac{k_0 r}{(k + k_0)\sqrt{(1 - z_0^2)^2 + 4b^2 z_0^2}}$$

The phase angle is:

$$\alpha = \arctan \frac{f\omega}{k + k_0 - m\omega^2} = \arctan \frac{2bz_0}{1 - z_0^2} \quad (4.104)$$

in which

$$z_0 = \frac{\omega}{\omega_0} \qquad \omega_0 = \sqrt{\frac{k+k_0}{m}} = \sqrt{(k+k_0)\frac{m_1+m_2'}{m_1 m_2'}} \qquad b = \frac{f}{2m\omega_0}$$

in which z_0 is the frequency ratio, ω_0 is the natural frequency and b is the relative damping ratio.

The absolute displacements for the body 1 and 2 can be found as follows. Adding the first and second equation together in Eq. (4.99) one has:

$$m_1\ddot{x}_1 = -m_2\ddot{x}_2 \qquad (4.105)$$

The absolute displacements for the body 1 and 2 have the following forms:

$$\begin{cases} x_1 = \lambda_1 \sin(\omega t - \alpha) \\ x_2 = \lambda_2 \sin(\omega t - \alpha) \end{cases} \qquad (4.106)$$

Substituting them into Eq. (4.105) one gets:

$$m_1 \lambda_1 = -m_2' \lambda_2$$

i.e.

$$\left|\frac{\lambda_1}{\lambda_2}\right| = \frac{m_2'}{m_1} \qquad (4.107)$$

The relationship among the absolute and the relative amplitudes can be found:

$$\begin{cases} \lambda_1 = \frac{m_2'}{m_1+m_2'}(\lambda_1 - \lambda_2)\frac{m}{m_1}\lambda \\ \lambda_2 = \frac{-m_1}{m_1+m_2'}(\lambda_1 - \lambda_2) = -\frac{m}{m_2'}\lambda \end{cases} \qquad (4.108)$$

Substituting the relative amplitude Eq. (4.103) into Eq. (4.108) one obtains the amplitudes of the body 1 and body 2:

$$\begin{cases} \lambda_1 = \frac{m}{m_1}\frac{k_0 r \cos\alpha}{k+k_0-m\omega^2} = \frac{m}{m_1}\frac{k_0 r \cos\alpha}{(k+k_0)(1-z_0^2)} \\ \lambda_2 = -\frac{m}{m_2'}\frac{k_0 r \cos\alpha}{k+k_0-m\omega^2} = -\frac{m}{m_2'}\frac{k_0 r \cos\alpha}{(k+k_0)(1-z_0^2)} \end{cases} \qquad (4.109)$$

The formula for the phase angle is the same as Eq. (4.104).

$$\alpha = \arctan \frac{f\omega}{k + k_0 - m\omega^2} = \arctan \frac{2bz_0}{1 - z_0^2}$$

4.8 Multi-body Elastic-Connecting Rod Type of Near-Resonant Vibration Machines

There are multi-body vibration machines with more than three vibration bodies in industries. As the number of vibration bodies increases, the degrees of freedom for the machines will increase correspondingly. Their analyses for the multi-body vibration machines would be much more complex. The number of equations describing the motions is equal to the degrees of the freedom while the number of natural frequencies would be the same as that of the degrees of freedom and there are as many peaks in the frequency amplitude responses of the system as the d.o.f. Since the response characteristics in the working frequency regions are the most important portion, we should emphasize on the response curve changes and other properties in these regions when we conduct analytical analysis on the vibration machine dynamics while we can ignore other portions of the curves.

For the aforementioned balanced type of double body vibration machines, the portion of the unbalanced inertial forces would be transferred to the base when the two masses are not equal. While the vibration isolation spring installed underneath the chassis can reduce the inertial force transferred dramatically. In this case, the vibration machines have three bodies, hence this machine can be called the spring vibration isolation balanced type of three-body vibration machine as shown in Fig. 4.20. To simplify the analysis process the damping is ignored. The equations of the motion for this machine is:

$$\sum M_0 = (m_1\omega^2\lambda_1 - m_2\omega^2\lambda_2)l_0 - k\lambda \times 2l_0 - k_0(\lambda - r) \times 2l_0 = 0$$

$$\sum P = m_1\omega^2\lambda_1 + m_2\omega^2\lambda_2 + m_3'\omega^2\lambda_3 = 0 \quad (4.110)$$

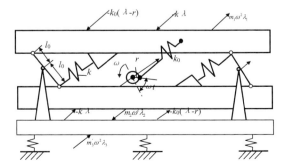

Fig. 4.20 The mechanics diagram for the three-body type of vibration conveyer

in which m_1, m_2 and m'_3 are the computed masses for trough body 1, 2 and chassis respectively and $m'_3 = m - \dfrac{k_3}{\omega_2}$; m_3 is the real mass of the chassis; k_3 is the stiffness of the vibration isolation spring in the vibration direction; $\lambda_1, \lambda_2, \lambda_3$ are the vibration amplitudes of trough body 1, 2 and chassis in the vibration directions respectively, λ is the relative amplitude between the body 1 and 2.

According to the geometric relationship $\lambda_1, \lambda_2, \lambda_3$ and λ have the following relations:

$$\lambda_1 - \lambda_2 = \lambda_1 \lambda_1 + \lambda_2 = 2\lambda_3$$

i.e.

$$\lambda_1 - \lambda_3 = -(\lambda_2 - \lambda_3)$$

or

$$\begin{cases} \lambda_1 = \dfrac{1}{2}\lambda + \lambda_3 \\ \lambda_2 = -\dfrac{1}{2}\lambda + \lambda_3 \end{cases} \quad (4.111)$$

Substituting Eq. (4.111) into Eq. (4.110) one gets:

$$\lambda_3 = \dfrac{-(m_1 - m_2)\lambda}{2(m_1 + m_2 + m'_3)} \quad (4.112)$$

Substituting Eq. (4.112) into Eq. (4.111) one obtains:

$$\begin{cases} \lambda_1 = \dfrac{1}{2}\lambda + \lambda_3 = \dfrac{(2m_2 + m'_3)\lambda}{2(m_1 + m_2 + m'_3)} \\ \lambda_2 = -\dfrac{1}{2}\lambda + \lambda_3 = \dfrac{-(2m_1 + m'_3)\lambda}{2(m_1 + m_2 + m'_3)} \end{cases} \quad (4.113)$$

The first equation in Eq. (4.110) can be rewritten as:

$$\begin{aligned} m_1\omega^2\lambda_1 - m_2\omega^2\lambda_2 &= \dfrac{1}{2}(m_1 + m_2)\omega^2\lambda + (m_1 - m_2)\omega^2\lambda_3 \\ &= \dfrac{1}{2}[m_1 + m_2 - \dfrac{(m_1 - m_2)^2}{m_1 + m_2 + m'_3}]\omega^2\lambda \end{aligned} \quad (4.114)$$

This equation can be rewritten as:

$$-\dfrac{1}{4}[m_1 + m_2 - \dfrac{(m_1 - m_2)^2}{m_1 + m_2 + m'_3}]\omega^2\lambda + (k + k_0)\lambda = k_0 r \quad (4.115)$$

Hence the relative vibration amplitude is:

$$\lambda = \frac{k_0 r}{k + k_0 - \frac{1}{4}[m_1 + m_2 - \frac{(m_1-m_2)^2}{m_1+m_2+m_3'}]\omega^2} = \frac{k_0 r}{(k+k_0)(1-z_0^2)} \quad (4.116)$$

in which z_0 is the frequency ratio; ω_0 is the natural frequency $\omega_0 = \sqrt{\frac{k+k_0}{m}}$; m is the induced mass, $m = \frac{1}{4}[m_1 + m_2 - \frac{(m_1-m_2)^2}{m_1+m_2+m_3'}]$.

The relative vibration amplitude can be found from Eq. (4.116) and then substituted into Eq. (4.113) the amplitude for the body 1 and 2 can be found to be:

$$\begin{cases} \lambda_1 = \frac{(2m_2+m_3')}{2(m_1+m_2+m_3')} \frac{k_0 r}{k+k_0-\frac{1}{4}[m_1+m_2-\frac{(m_1-m_2)^2}{m_1+m_2+m_3'}]\omega^2} \\ \lambda_2 = \frac{-(2m_1+m_3')}{2(m_1+m_2+m_3')} \frac{k_0 r}{k+k_0-\frac{1}{4}[m_1+m_2-\frac{(m_1-m_2)^2}{m_1+m_2+m_3'}]\omega^2} \\ \lambda_3 = \frac{(m_1-m_2)}{2(m_1+m_2+m_3')} \frac{k_0 r}{k+k_0-\frac{1}{4}[m_1+m_2-\frac{(m_1-m_2)^2}{m_1+m_2+m_3'}]\omega^2} \end{cases} \quad (4.117)$$

The dynamic load transferred to the base by a double trough balanced type of vibration conveyer with vibration isolation springs can be found by directly multiplying the vibration isolation spring stiffness with the vibration amplitude of the chassis.

The dynamic loads in the vertical and horizontal directions are respectively:

$$F_c = k_{gc}\lambda_3 \sin\delta \qquad F_s = k_{gs}\lambda_3 \cos\delta \quad (4.118)$$

in which δ is the angel between the vibration direction and horizontal plane; k_{gc}, k_{gs} are the stiffness of the isolation spring in horizontal and vertical directions respectively.

The resultant dynamic load is:

$$F_d = \sqrt{F_c^2 + F_s^2} \quad (4.119)$$

The practice indicates that the dynamic load to the base is reduced dramatically after the isolation springs are used.

4.9 Dynamics of Electric–Magnetic Resonant Type of Vibrating Machines with Harmonic Electric–Magnetic Force

4.9.1 Basic Categories of Electric–Magnetic Forces of Electric–Magnetic Vibration Machines

The electric–magnetic type of vibration machines, short for EMTVM, can be categorized as the followings by their excitation forces:

(1) The linear EMTVM with the harmonic form of the electric force. The elastic forces are linear, the whole vibration system is linear. This type of EMTVM includes the alternative magnetization EMTVM; an EMTVM with small magnetism leak magnets, and one-half period rectification of negligible internal resistance and an EMTVM with a one-half period + one period rectifications.
(2) The linear EMTVM with the non-harmonic form of the electric–magnetic force. It includes the EMTVM with controllable one-half period rectification; the EMTVM with one-half period or controllable one period rectification; and an EMTVM with a one-half period + one period rectifications of negligible internal resistance.
(3) The EMTVM with the pseudo or non-linear elastic force. It includes the EMTVM with shear rubber spring or compressive rubber springs; the EMTVM with installation gaps of rubber spring and the ones with curved leaf springs.
(4) Impact EMTVM working by the impact principles. Such as impact electric–magnetic vibration sand shaker, etc.

The purposes for analyzing and studying EMTVM dynamic characteristics are:

(1) Select an appropriate working point for good stability;
(2) Propose the correct computation methods for EMTVM's dynamic parameters (such as vibration amplitude, spring stiffness, excitation forces etc.)
(3) Reveal the effect of the linearity and/or non-linearity of the vibration system upon the body vibrations and thus choose the proper dynamic characteristics by its specific working requirements.

4.9.2 Dynamics of Electric–Magnetic Type of Vibrating Machines with Harmonic Electric–Magnetic Force

EMTVM's explained in Sect. 4.9.1 are mostly the double-body vibration systems and their working principle and mechanics diagram are shown in Fig. 4.21.

The vibration differential equations for the linear EMTVM with the harmonic electric–magnetic force:

4.9 Dynamics of Electric–Magnetic Resonant Type ...

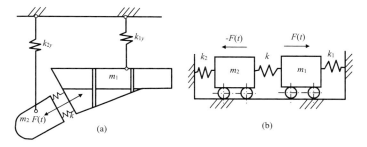

(a) EMTVM working principle   (b) Mechanics Diagram

Fig. 4.21 The working principle and mechanics diagram for EMTVM

$$\begin{cases} m_1\ddot{x}_1 + f_1\dot{x}_1 + k_1 x_1 + f(\dot{x}_1 - \dot{x}_2) + k(x_1 - x_2) = F'_0 + F'_1 \sin \omega t_1 + F'_2 \sin 2\omega t_2 \\ m_2\ddot{x}_2 + f_2\dot{x}_2 + k_2 x_2 - f(\dot{x}_1 - \dot{x}_2) - k(x_1 - x_2) = -(F'_0 + F'_1 \sin \omega t_1 + F'_2 \sin 2\omega t_2) \end{cases} \quad (4.120)$$

where the terms on the right-hand side are the harmonic electric–magnetic excitation. m_1, m_2 are the masses of the body 1 and 2; f_1 and f_2 are the damping coefficients of the absolute movements; f is the damping coefficient of the relative movement between the body 1 and 2; k_1, k_2 are the stiffness of the vibration isolation along the vibration direction; k is the stiffness of the main vibration spring; x_1, x_2 are the displacements of the body 1 and 2 along the vibration direction; $\dot{x}_1$, $\dot{x}_2$ are the velocities of the body 1 and 2 along the vibration direction; $\ddot{x}_1$, $\ddot{x}_2$ are the accelerations of the body 1 and 2 along the vibration direction; F'_0 is the averaged electric–magnetic force; F'_1, F'_2 are the first and the second harmonic excitation force amplitudes; ω is the electric power angular frequency; t_1, t_2 are the time periods of the first and the second harmonic excitation force.

Since the EMTVM is in generally working near the resonance region of the main harmonic excitation force, the following analysis is concentrated on the vibrations near the main resonance region. The spring force of the vibration isolation spring of the EMTVM is in general much less than those of the inertial forces of the body 1 and 2 and of the spring force of the main vibration springs, it is neglected for approximate calculations. In the accurate calculations, the stiffness k_1 and k_2 of the vibration isolation spring is combined into the masses m_1 and m_2.

Replacing $m_1\ddot{x}_1 + k_1 x_1$ with $m'_1\ddot{x}_1$ one has:

$$m'_1 = m_1 - \frac{k_1}{\omega'^2}$$

Replacing $m_2\ddot{x}_2 + k_2 x_2$ with $m'_2\ddot{x}_2$ one has:

$$m'_2 = m_2 - \frac{k_2}{\omega'^2}$$

where m_1', m_2' are the computed masses of body 1 and 2; ω is the forced vibration angular frequency.

For most of the EMTVM's the computed mass $m_1' \approx m_2'$, the effect of the absolute movement damping force on the body vibration is not obvious. To approximate its effects, it can be taken as the following:

$$\frac{f_1 m_2'}{m_1' + m_2'} \approx \frac{f_2 m_1'}{m_1' + m_2'}$$

Using the above conditions, multiplying the first equation in Eq. (4.120) with $\frac{m_2'}{m_1'+m_2'}$ and the second one with $\frac{m_1'}{m_1'+m_2'}$, subtracting one from another, one obtains the vibration equation for the relative displacement, velocity and acceleration:

$$m_u \ddot{x} + f_u \dot{x} + kx = F_0' + F_1' \sin \omega t_1 + F_2' \sin 2\omega t_2 \quad (4.121)$$

where
$$m_u = \frac{m_1' m_2'}{m_1' + m_2'} \qquad f_u = f + \frac{f_1 m_2'}{m_1' + m_2'}$$
$$x = x_1 - x_2 \qquad \dot{x} = \dot{x}_1 - \dot{x}_2 \qquad \ddot{x} = \ddot{x}_1 - \ddot{x}_2$$

where m_u is the induced mass; f_u is the induced damping coefficient; $x, \dot{x}, \ddot{x}$ are the relative displacement, velocity and acceleration of the body 1 relative to the body 2. Adding the first and the second equations of Eqs. (4.120) one gets:

$$m_1' \ddot{x}_1 + m_2' \ddot{x}_2 + f_1 \dot{x}_1 + f_2 \dot{x}_2 = 0$$

The above equation can be used to find the relationship between the relative and absolute displacements of the body 1 and 2:

$$x_1 = \frac{m_2'}{m_1'} x_2 = \frac{m_2'}{m_1' + m_2'} x$$
$$x_2 = -\frac{m_1'}{m_1' + m_2'} x$$

When the EMTVM is working in a normal state, the free vibration disappears quickly, the remainings are the forced vibration of the machine. So we only study the particular solution of Eq. (4.121).

The particular solution of Eq. (4.121) has the following form:

$$x = x_1 - x_2 = \Delta + \lambda \sin(\omega t_1 - a') + \xi \sin(2\omega t_2 - \theta') \quad (4.122)$$

Substituting Eq. (4.122) into Eq. (4.121) and simplifying it one can get the relative static displacements Δ of the body 1 and 2, and the absolute static displacements Δ_1 and Δ_2 of the body 1 and 2 under the averaged electric–magnetic force:

4.9 Dynamics of Electric–Magnetic Resonant Type …

$$\begin{cases} \Delta = \dfrac{F'_0}{k} = \dfrac{(\frac{1}{2} + A^2)}{k} F_a \\ \Delta_1 = \dfrac{k_2}{k_1 + k_2} \Delta = \dfrac{k_2}{k_1 + k_2} \dfrac{(\frac{1}{2} + A^2)}{k} F_a \\ \Delta_2 = -\dfrac{k_1}{k_1 + k_2} \Delta = -\dfrac{k_1}{k_1 + k_2} \dfrac{(\frac{1}{2} + A^2)}{k} F_a \end{cases} \tag{4.123}$$

where

$$F_a = \frac{2B_a^2 S'}{\mu_0} \qquad B_a = \frac{\sqrt{2}u_1(1-\sigma_a)}{w\omega S'} \sin\phi \qquad \sigma_a = \frac{L_2}{L_0}$$

in which A is the character number of the EMTVM; F_a is the basic electric–magnetic force; B_a is the basic magnetic flux density considering electric induction coefficient; σ_a is a constant; μ_0 is the air magnetic conductivity, $\mu_0 = 4\pi \times 10^7$ H/m; ϕ is the phase angle of the alternative magnetic density relative to the alternative voltage; w is the number of coils; L_2, L_0 are the induction leak and total electric induction in the circuit for the averaged working air gap.

The relative amplitude λ and phase angle a' and the absolute amplitudes λ_1 and λ_2 of the body 1 and 2 due to the first harmonic excitation force are respectively:

$$\begin{cases} \lambda = \dfrac{F'_1 \cos a'}{k - m_u \omega^2} = \dfrac{2A F_a \cos a'}{k - m_u \omega^2} \\ \lambda_1 = \dfrac{m_u}{m'_1} \lambda = \dfrac{m_u}{m'_1} \dfrac{2A F_a \cos a'}{k - m_u \omega^2} \\ \lambda_2 = -\dfrac{m_u}{m'_2} \lambda = -\dfrac{m_u}{m'_2} \dfrac{2A F_a \cos a'}{k - m_u \omega^2} \end{cases} \tag{4.124}$$

$$a' = \arctan \frac{f_u \omega}{k - m_u \omega^2} \tag{4.125}$$

The relative amplitude ξ and phase angle θ' and the absolute amplitudes ξ_1 and ξ_2 of the body 1 and 2 due to the first harmonic excitation forces are respectively:

$$\begin{cases} \xi = \dfrac{F'_2 \cos \theta'}{k - 4m_u \omega^2} = \dfrac{1}{2} \dfrac{F_a \cos \theta'}{k - 4m_u \omega^2} \\ \xi_1 = \dfrac{1}{2} \dfrac{m_u}{m'_1} \dfrac{F_a \cos \theta'}{k - 4m_u \omega^2} \\ \xi_2 = -\dfrac{1}{2} \dfrac{m_u}{m'_2} \dfrac{F_a \cos \theta'}{k - 4m_u \omega^2} \\ \theta' = \arctan \dfrac{2f_u \omega}{k - 4m_u \omega^2} \end{cases} \tag{4.126}$$

4.9.3 Amplitudes and Phase Angle Differentials of One-Half-Period Rectification EMTVM

The character number of EMTVM with the one-half period rectification $A \approx 1$ and the electric–magnetic excitation force is:

$$\begin{aligned} F(t) &= (\frac{1}{2} + A^2)F_a + 2AF_a \sin \omega t_1 + \frac{1}{2}F_a \sin 2\omega t_2 \\ &= \frac{3}{2}F_a + 2F_a \sin \omega t_1 + \frac{1}{2}F_a \sin 2\omega t_2 \end{aligned} \quad (4.127)$$

The electric–magnetic excitation force of the one-half-period rectification EMTVM is composed of three parts: the averaged electric–magnetic force $\frac{3}{2}F_a$, the first harmonic excitation force $2F_a \sin \omega t_1$ and the second harmonic excitation force $\frac{1}{2}F_a \sin 2\omega t_2$. The relative and absolute displacements for the body 1 and 2 due to the averaged electric–magnetic force $\frac{3}{2}F_a$ are respectively:

$$\begin{cases} \Delta = \dfrac{3}{2}\dfrac{F_a}{k} \\ \Delta_1 = \dfrac{k_2}{k_1 + k_2} \times \dfrac{3}{2}\dfrac{F_a}{k} \\ \Delta_2 = -\dfrac{k_1}{k_1 + k_2} \times \dfrac{3}{2}\dfrac{F_a}{k} \end{cases} \quad (4.128)$$

The relative, absolute displacements and phase angle for the body 1 and 2 due to the first harmonic excitation force $2F_a \sin \omega t_1$ are respectively:

$$\begin{cases} \lambda = \dfrac{2F_a \cos a'}{k - m_u \omega^2} = \dfrac{2F_a z_0^2 \cos a'}{m_u \omega^2 (1 - z_0^2)} \\ \lambda_1 = \dfrac{m_u}{m'_1}\dfrac{2F_a \cos a'}{k - m_u \omega^2} = \dfrac{2F_a z_0^2 \cos a'}{m'_1 \omega^2 (1 - z_0^2)} \\ \lambda_2 = -\dfrac{m_u}{m'_2}\dfrac{2F_a \cos a'}{k - m_u \omega^2} = -\dfrac{2F_a z_0^2 \cos a'}{m'_2 \omega^2 (1 - z_0^2)} \end{cases} \quad (4.129)$$

in which

$$a' = \arctan \frac{f_u \omega}{k - m_u \omega^2} = \arctan \frac{2bz_0}{1 - z_0^2}$$

After the amplitude of the body 1 is selected by the technological requirements, the basic electric–magnetic force F_a can be found by the following:

4.9 Dynamics of Electric–Magnetic Resonant Type ...

$$F_a = \frac{m_1' \omega^2 \lambda_1 (1 - z_0^2)}{2 z_0^2 \cos a'} \qquad (4.130)$$

Under the second harmonic excitation force, the relative, absolute amplitude and phase angle of the twice-alternative power frequency are respectively:

$$\begin{cases} \xi = \dfrac{1}{2} \dfrac{F_a \cos\theta'}{k - 4m_u\omega^2} = \dfrac{1}{8} \dfrac{F_a z_0'^2 \cos\theta'}{m_u\omega^2 (1 - z_0'^2)} \\ \xi_1 = \dfrac{1}{2} \dfrac{m_u}{m_1'} \dfrac{F_a \cos\theta'}{k - 4m_u\omega^2} = \dfrac{1}{8} \dfrac{F_a z_0'^2 \cos\theta'}{m_1'\omega^2 (1 - z_0'^2)} \\ \xi_2 = \dfrac{1}{2} \dfrac{m_u}{m_2'} \dfrac{F_a \cos\theta'}{k - 4m_u\omega^2} = -\dfrac{1}{8} \dfrac{F_a z_0'^2 \cos\theta'}{m_2'\omega^2 (1 - z_0'^2)} \end{cases} \qquad (4.131)$$

$$\theta' = \arctan \frac{2 f_u \omega}{k - 4m_u\omega^2} = \arctan \frac{2 b z_0'}{1 - z_0'^2}$$

where z_0' is the forced vibration frequency ratio for the second harmonic excitation, $z_0' = \frac{2\omega}{\omega_0}$.

The natural frequency ω_0 of the EVTVM main vibration system is in general selected in the neighborhood of the electric power frequency ω. It is usually taken as $z_0 = 0.85 \sim 0.95$. The amplitude ξ generated by the second harmonic excitation is much smaller than that by the first harmonic excitation, it is usually neglected.

4.9.4 Amplitudes and Phase Angle Differentials of One-Half-Period Plus One-Period Rectification EMTVM

The character number for the One-Half-Period Plus One-Period Rectification EMTVM with a Π-shaped and $|\perp|$-shaped core is:

$$A = 1 + \frac{B_0}{B_{a0}}$$

in which B_0 is the D.C. magnetic density without the effection of σ_a; B_{a0} is the basic magnetic density without the effection of σ_a.

For this type of EMTVM, it is usually taken $\frac{B_0}{B_{a0}} = 1$, so $A = 1 + \frac{B_0}{B_{a0}} = 2$. Its electric–magnetic force expression is:

$$F(t) = \left(\frac{1}{2} + A^2\right)F_a + 2AF_a \sin \omega t_1 + \frac{1}{2}F_a \sin 2\omega t_2 = \frac{9}{2}F_a + 4F_a \sin \omega t_1 + \frac{1}{2}F_a \sin 2\omega t_2 \quad (4.132)$$

Its relative and absolute displacements of the body 1 and 2 due to the averaged electric–magnetic force are respectively:

$$\begin{cases} \varDelta = \dfrac{9}{2}\dfrac{F_a}{k} \\ \varDelta_1 = \dfrac{k_2}{k_1 + k_2}\varDelta = \dfrac{k_2}{k_1 + k_2}\dfrac{9F_a}{2k} \\ \varDelta_2 = -\dfrac{k_1}{k_1 + k_2}\varDelta = -\dfrac{k_1}{k_1 + k_2}\dfrac{9F_a}{2k} \end{cases} \quad (4.133)$$

The relative, absolute displacements and phase angle for the body 1 and 2 due to the first harmonic excitation force $2F_a \sin \omega t_1$ are respectively:

$$\begin{cases} \lambda = \dfrac{4F_a \cos \alpha'}{k - m_u \omega^2} = \dfrac{4F_a z_0^2 \cos \alpha'}{m_u \omega^2 (1 - z_0^2)} \\ \lambda_1 = \dfrac{m_u}{m_1'}\dfrac{4F_a \cos \alpha'}{k - m_u \omega^2} = \dfrac{4F_a z_0^2 \cos \alpha'}{m_1' \omega^2 (1 - z_0^2)} \\ \lambda_2 = -\dfrac{m_u}{m_2'}\dfrac{4F_a \cos \alpha'}{k - m_u \omega^2} = -\dfrac{4F_a z_0^2 \cos \alpha'}{m_2' \omega^2 (1 - z_0^2)} \end{cases} \quad (4.134)$$

$$\alpha' = \arctan \dfrac{f_u \omega}{k - m_u \omega^2} \arctan \dfrac{2bz_0}{1 - z_0^2}$$

After the amplitude λ_1 is selected for the body 1, the following formula can be used to find the basic electric–magnetic force needed:

$$F_a = \dfrac{m_1' \omega^2 \lambda_1 (1 - z_0^2)}{4z_0^2 \cos \alpha'} \quad (4.135)$$

4.10 Dynamics of Electric–Magnetic Type of Near-Resonant Vibration Machines with Non-Harmonic Electric–Magnetic Force

For some of the vibration machines used in industries, such as EMTVM with thyristor rectification, some of the decreased frequency EMTVM, and EMTVM with large electric resistance in circuits, their expressions of the excitation forces are not in the

4.10 Dynamics of Electric–Magnetic Type of Near-Resonant ...

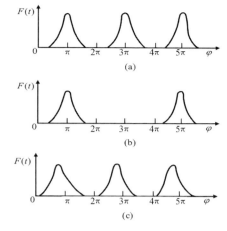

Fig. 4.22 The non-harmonic form of Electric–magnetic force of EMTVM. **a** controlled one-half period rectification EMTVM; **b** controlled decrease frequency EMTVM; **c** non-neglect electric resistance EMVTM

harmonic forms in the whole vibration periods (see Fig. 4.22). For those EMTVM's their vibration amplitudes can be calculated by two methods: i. The Fourier Series Expansion method, i.e., expanding the non-harmonic excitation force into harmonics in Fourier Series form and then conduct the EMTVM dynamic analysis; ii. the piece-wise integration method. Now we introduce the Fourier Series Method.

For any periodical functions, even their expressions are not in the harmonic forms, they can be expanded in Fourier Series:

$$F = F_0' + \sum_{n=1}^{\infty} F_n' \sin(n\omega t + \gamma_n) = F_0' + F_1' \sin(\omega t + \gamma_1) + F_2' \sin(2\omega t + \gamma_2) + \cdots = F_{2a}' \sin 2\omega t + F_{2b}' \sin \omega t + \cdots F_0' + F_{1a}' \sin \omega t + F_{1b}' \sin \omega t$$

where $n = 1, 2, 3, \cdots$, $F_n' = \sqrt{F_{na}^2 + F_{nb}^2}$, $\gamma_n = \arctan \frac{F_{nb}'}{F_{na}'}$.

4.10.1 Relationships Between Electric–Magnetic Force and Amplitudes of Controlled One-Half-Period Rectification EMTVM

For this type of EMTVM, its electric–magnetic force can be approximately expressed as:

$$F = \begin{cases} F_0 + F_1 \sin(\omega t - \frac{\pi}{2}) + F_2 \sin(2\omega t - \frac{3}{2}\pi) & \text{if } \omega t = \varepsilon \sim 2\pi - \varepsilon \\ 0 \quad \text{else } \omega t = 0 \sim \varepsilon; 2\pi - \varepsilon \sim 2\pi & \text{else} \end{cases}$$

(4.136)

where $F_0 = (\frac{1}{2} + A^2)F_a$, $F_1 = 2AF_a$, $F_2 = \frac{1}{2}F_a$, $A = \cos\varepsilon$ in which ε is the triggering angle for the thyristor.

It can be seen from Eq. (4.136) that the electric–magnetic force exists only within the time by the triggering angle $\varepsilon \sim (2\pi - \varepsilon)$ (here is the approximate, neglecting the electric resistance. When the internal resistance is considered, the stopping angle should be smaller than $2\pi - \varepsilon$). After the thyristor is conducted, the current in the coil of the magnet is increasing with the time. Since the magnet of the EMTVM belongs to the negative conductivity load, when the power voltage is reaching to the end of the first half period, the current is reaching its maximum; the magnet force is also approaching to its maximum. When the power voltage is entering the negative half period, the electric-magnet is discharging, and the current is becoming smaller. When ωt is approaching to $2\pi - \varepsilon$, the current is reaching to zero, the thyristor stops. Hence, when $\omega t = (2\pi - \varepsilon) \sim (2\pi + \varepsilon)$, the electric magnetic force are zero. The aforementioned methods can not be used to conduct the dynamic analysis for this type of the electric–magnetic force. The electric–magnetic force must be expanded, then the methods can be used.

The above expression is expanded as the follows, the coefficients are:

$$F'_a = \frac{1}{2\pi}\int_0^{2\pi} F(\omega t)d\omega t = \frac{1}{2\pi}\int_\varepsilon^{2\pi-\varepsilon} F_a\left[A + \sin\left(\omega t - \frac{\pi}{2}\right)\right]^2 d\omega t$$

$$= \frac{1}{2\pi}F_a[2(\pi - \varepsilon)(\frac{1}{2} + A^2) + 4A\sin\varepsilon - \frac{1}{2}\sin 2\varepsilon]$$

$$= \frac{1}{2\pi}F_a[2(\pi - \varepsilon)(\frac{1}{2} + \cos^2\varepsilon) + 1.5\sin 2\varepsilon] \quad (4.137)$$

$$F'_{1a} = \frac{1}{\pi}\int_0^{2\pi} F(\omega t)\sin\omega t\, d\omega t = 0$$

$$F'_{1b} = \frac{1}{\pi}\int_0^{2\pi} F(\omega t)\cos\omega t\, d\omega t = \frac{1}{\pi}\int_\varepsilon^{2\pi-\varepsilon} F_a\left[A + \sin\left(\omega t - \frac{\pi}{2}\right)\right]^2 \cos\omega t\, d\omega t$$

$$= -\frac{1}{\pi}F_a[2(\pi - \varepsilon)A + (1.5 + 2A^2)\sin\varepsilon - A\sin 2\varepsilon + \frac{1}{6}\sin 3\varepsilon]$$

$$= -\frac{1}{\pi}F_a[2(\pi - \varepsilon)\cos\varepsilon + 1.5\sin\varepsilon + \frac{1}{6}\sin 3\varepsilon]$$

$$F'_1 = \sqrt{F'^2_{1a} + F'^2_{1b}} = F'_{1b}$$

$$\gamma_1 = \arctan\frac{F'_{1b}}{F'_{1a}} = -\frac{\pi}{2}$$

$$F'_{2a} = \frac{1}{\pi}\int_0^{2\pi} F(\omega t)\sin 2\omega t\, d\omega t = 0$$

4.10 Dynamics of Electric–Magnetic Type of Near-Resonant ...

$$F'_{2b} = \frac{1}{\pi} \int_0^{2\pi} F(\omega t) \cos 2\omega t \, d\omega t = \frac{1}{\pi} \int_\varepsilon^{2\pi-\varepsilon} F_a \left[A + \sin\left(\omega t - \frac{\pi}{2}\right) \right]^2 \cos 2\omega t \, d\omega t$$

$$= \frac{1}{\pi} F_a \left[\frac{1}{2}(\pi - \varepsilon) + 2A \sin\varepsilon - \left(\frac{1}{2} + A^2\right) \sin 2\varepsilon + \frac{2}{3} A \sin 3\varepsilon - \frac{1}{8} \sin 4\varepsilon \right]$$

$$= \frac{1}{\pi} F_a \left[\frac{1}{2}(\pi - \varepsilon) + 0.5 \sin 2\varepsilon - \cos^2\varepsilon \sin 2\varepsilon + \frac{2}{3} \cos\varepsilon \sin 3\varepsilon - \frac{1}{8} \sin 4\varepsilon \right]$$

$$F'_2 = F'_{2b}, \quad \gamma_2 = \frac{\pi}{2}$$

Table 4.2 lists the relation between F'_0, F'_1, F'_2 and F_a for different triggering angle ε.

According to the data in Table 4.2 a relation curve between the first harmonic force amplitude F'_1 and the conducting angle $\theta \approx 2(\pi - \varepsilon)$ is shown in Fig. 4.23. It can be seen from Fig. 4.23 that when the triggering angle $\varepsilon > 120°$, the electric–magnetic force is very small and the amplitude can not be large. Hence the triggering angle

Table 4.2 The relation between the triggering angle of controlled one-half-rectification EMTVM and the excitation force amplitude

The triggering angle $\varepsilon/(°)$	180	150	120	90	60	30	0
The conducting angle $\theta/(°) \; \theta \approx 2(\pi - \varepsilon)$	0	60	120	180	240	300	360
The average electromagnetic force $F'_0 = K_0 F_a$	0	$0.002 F_a$	$0.042 F_a$	$0.25 F_a$	$0.71 F_a$	$1.255 F_a$	$1.5 F_a$
The first harmonic excitation $F'_1 = K_1 F_a$	0	$0.002 F_a$	$0.081 F_a$	$0.424 F_a$	$1.084 F_a$	$1.742 F_a$	$2 F_a$
The second harmonic excitation $F'_2 = K'_2 F_a$	0	$0.002 F_a$	$0.1 F_a$	$0.25 F_a$	$0.408 F_a$	$0.49 F_a$	$0.5 F_a$

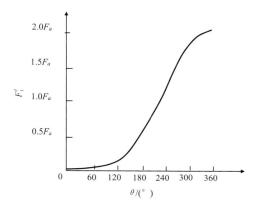

Fig. 4.23 The relation between the first harmonic force amplitude and the conducting angle θ

ε should be selected in the range of $0° \sim 120°$, i.e., the conducting angle θ can be adjusted in the range of $120° \sim 360°$.

For EMTVM with controlled one-half-period rectification, the maximum amplitude can be calculated approximately as that for the EMTVM with one-half-period rectification. i.e., the triggering angle for the thyristor tube can be selected as $0°$ and the conducting angle is close to $360°$ (should be less than $360°$). The character number $A = \cos\varepsilon \approx \cos\theta = 1$. Its expression for the electric–magnetic force is:

$$F(t) = (\frac{1}{2} + A^2)F_a + 2AF_a \sin\omega t_1 + \frac{1}{2}F_a \sin 2\omega t_2 = \frac{3}{2}F_a + 2F_a \sin\omega t_1 + \frac{1}{2}F_a \sin 2\omega t_2$$

The relative, absolute and phase angle for the body 1 and 2 are respectively:

$$\begin{cases} \lambda = \dfrac{2F_a z_0^2 \cos\alpha'}{m_u \omega^2(1-z_0^2)} \\ \lambda_1 = \dfrac{2F_a z_0^2 \cos\alpha'}{m_1' \omega^2(1-z_0^2)} \\ \lambda_2 = -\dfrac{2F_a z_0^2 \cos\alpha'}{m_2' \omega^2(1-z_0^2)} \end{cases} \quad (4.138)$$

$$\alpha' = \arctan\frac{2bz_0}{1-z_0^2}$$

From Eq. (4.138) the basic electric–magnetic force needed is:

$$F_a = \frac{m_1'\omega^2(1-z_0^2)\lambda_1}{2z_0^2 \cos\alpha'} \quad (4.139)$$

4.10.2 Relationships Between Electric–Magnetic Force and Amplitudes of the Decreased Frequency EMTVM

For the EMTVM with a Π-shaped and $|\bot|$-shaped magnetic core, one-half-period or controlled one-half-period rectification, its excitation forces are active only in the first half of the period, its working frequency ω' is the half of that of the power supplier and its excitation force expression is:

$$F = \begin{cases} F_a(A - \cos 2\omega't)^2 & \text{when } \omega't = \dfrac{\varepsilon}{2} \sim \pi - \dfrac{\varepsilon}{2} \\ 0 & \text{when } \omega't = 0 \sim \dfrac{\varepsilon}{2}, \pi - \dfrac{\varepsilon}{2} \sim 2\pi \end{cases} \quad (4.140)$$

4.10 Dynamics of Electric–Magnetic Type of Near-Resonant ...

Its Fourier series coefficients are:

$$\begin{cases} F_0' = \dfrac{1}{2\pi} \int_{\frac{\varepsilon}{2}}^{\pi-\frac{\varepsilon}{2}} F_a(A - \cos 2\omega' t)^2 d\omega' t = \\ \dfrac{1}{2\pi} F_a[(\pi - \varepsilon)(\dfrac{1}{2} + A^2) + 2A \sin \varepsilon - \dfrac{1}{4} \sin 2\varepsilon] \\ F_{1a}' = \dfrac{1}{\pi} \int_{\frac{\varepsilon}{2}}^{\pi-\frac{\varepsilon}{2}} F_a(A - \cos 2\omega' t)^2 \sin \omega' t d\omega' t = \dfrac{1}{\pi} F_a[(1 + 2A + 2A^2) \cos \dfrac{\varepsilon}{2} - \\ (\dfrac{1}{6} + \dfrac{2}{3} A) \cos \dfrac{3}{2}\varepsilon + \dfrac{1}{10} \cos \dfrac{5}{2}\varepsilon] \\ F_{1b}' = \dfrac{1}{\pi} \int_{\frac{\varepsilon}{2}}^{\pi-\frac{\varepsilon}{2}} F_a(A - \cos 2\omega' t)^2 \cos \omega' t d\omega' t = 0 \end{cases}$$

$$\begin{cases} F_1' = F_{1a}', \quad \gamma_1 = \arctan \dfrac{F_{1b}'}{F_{1a}'} = 0 \\ F_{2a}' = 0 \\ F_{2b}' = \dfrac{1}{\pi} \int_{\frac{\varepsilon}{2}}^{\pi-\frac{\varepsilon}{2}} F_a(A - \cos 2\omega' t)^2 \cos 2\omega' t d\omega' t = \dfrac{1}{\pi} F_a[-(\pi - \varepsilon)A - \\ (\dfrac{3}{4} + A^2) \sin \varepsilon + \dfrac{1}{2} A \sin \varepsilon - \dfrac{1}{12} \sin 3\varepsilon] \\ F_2' = F_{2b}', \quad \gamma_2 = -\dfrac{\pi}{2} \\ F_{3a}' = \dfrac{1}{\pi} \int_{\frac{\varepsilon}{2}}^{\pi-\frac{\varepsilon}{2}} F_a(A - \cos 2\omega' t)^2 \sin 3\omega' t d\omega' t = \dfrac{1}{\pi} F_a[-(\dfrac{1}{2} + 2A) \cos \dfrac{\varepsilon}{2} + \\ (\dfrac{1}{3} + \dfrac{2}{3} A^2) \cos \dfrac{3}{2}\varepsilon - \dfrac{2}{5} A \cos \dfrac{5}{2}\varepsilon + \dfrac{1}{14} \cos \dfrac{7}{2}\varepsilon] \\ F_{3b}' = 0, \quad F_3' = F_{3a}', \quad \gamma_3 = 0 \end{cases}$$

(4.141)

$$F = F_a(0.75 + 1.36 \sin \omega' t - 1 \cos 2\omega' t - 0.58 \sin 3\omega' t + \cdots)$$

For the decreasing frequency EMTVM with controllable one-half-period rectification, $A = \cos \varepsilon$; for the decreasing frequency EMTVM with one-half-period rectification $A = 1$. Substituting $A = 1$ s and $\varepsilon = 0$ into Eq. (4.141) one can obtain the electric–magnetic excitation force expanded in Fourier series for the decreasing frequency EMTVM with one-half-period rectification:

$$F = F_0' + F_1' \sin \omega' t + F_2' \sin(2\omega' t - \dfrac{\pi}{2}) + \cdots - F_0' - F_1' \sin(\omega' t - \pi)$$

$$-F_2' \sin(2\omega' t - \dfrac{5}{2}\pi) - \cdots = 2F_1' \sin \omega' t + 2F_3' \sin 3\omega' t + \cdots$$

$$= 2.72F'_a \sin\omega't + 1.162F_a \sin 3\omega't + \cdots \qquad (4.142)$$

Table 4.3 lists the relation between the triggering angle and the excitation force amplitude for a controllable decreasing frequency EMTVM. A curve for the relation between the first harmonic excitation and the conducting angle θ can be drawn by the data in Table 4.3, as shown in Fig. 4.24. It can be seen from Fig. 4.24 that when the conducting angle θ is less than $120°$, i.e., the triggering angle ε is larger than $120°$, the first harmonic force amplitude is very small. Hence the conducting angle

Table 4.3 Relation between the triggering angle and excitation force amplitude for a controllable decreasing frequency EMTVM

The triggering angle $\varepsilon/(°)$		180	150	120	90	60	30	0
The conducting angle $\theta/(°)\ \theta \approx 2(\pi - \varepsilon)$		0	60	120	180	240	300	360
Single beating decreasing frequency EMTVM	The average electromagnetic force $F'_0 = K'_0 F_0$	0	$0.001F_a$	$0.021F_a$	$0.125F_a$	$0.35F_a$	$0.627F_a$	$0.75F_a$
	The first harmonic excitation $F'_1 = K'_1 F_a$	0	$0.002F$	$0.042F_a$	$0.22F_a$	$0.665F_a$	$1.14F_a$	$1.36F_a$
	The second harmonic excitation $F'_2 = K'_2 F_a$	0	$0.0015F$	$0.035F_a$	$0.212F_a$	$0.55F_a$	$0.9F_a$	$1F_a$
	The third harmonic excitation $F'_3 = K'_3 F_a$	0						$0.58F_a$
Double beating decreasing frequency EMTVM	The average electromagnetic force $F'_0 = K'_0 F_0$	0	0	0	0	0	0	0
	The first harmonic excitation $F'_1 = K'_1 F_a$	0	$0.004F$	$0.084F_a$	$0.44F_a$	$1.3F_a$	$2.28F_a$	$2.72F_a$
	The second harmonic excitation $F'_2 = K'_2 F_a$	0	0	0	0	0	0	0
	The third harmonic excitation $F'_3 = K'_3 F_a$	0						$1.16F_a$

4.10 Dynamics of Electric–Magnetic Type of Near-Resonant ...

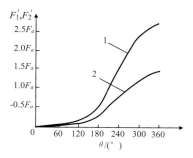

Fig. 4.24 Relation between the first harmonic force and the conducting angle θ. 1- Double beating decreasing frequency EMTVM; 2- Single beating decreasing frequency EMTVM

should be selected in the range of $120° \sim 360°$, i.e. the triggering angle should be in the range of $0° \sim 120°$ when designing the machines.

The maximum excitation amplitudes for the two types of EMTVM can be calculated approximately assuming the triggering angle is zero and its working frequency is half of that of the power supply.

For the single beating decreasing frequency EMTVM, the amplitudes for the body 1 and 2 are:

$$\begin{cases} \lambda_1 = \dfrac{1.36 F_a z_0^2 \cos \alpha'}{m_1' \omega'^2 (1 - z_0^2)} \\ \lambda_2 = -\dfrac{1.36 F_a z_0^2 \cos \alpha'}{m_2' \omega'^2 (1 - z_0^2)} \end{cases} \qquad (4.143)$$

Chapter 5
Utilization of Nonlinear Vibration

5.1 Introduction

The majority of the systems which are working by the vibration principles or the wave theories are the nonlinear ones. For the nonlinear vibration utilities, some of them are not active and intentional and the rest of them are because the technological processes can be achieved only by the nonlinear working states. The characteristics of the nonlinear vibration are made use of by these systems.

For the nonlinear vibration to be utilized effectively, the most common nonlinear vibration in the engineering, the differential equations of the motions and some of the features of the nonlinear vibration system solutions should be understood.

The most common nonlinear equations are:

(1) The Single, Double and Friction Pendulum Systems

  i. Single pendulum
     Without damping:

$$\frac{d^2 x}{d t^2} + k \sin x = 0 \quad k = \frac{g}{L} \tag{5.1}$$

   With damping:

$$\frac{d^2 x}{d t^2} + c\frac{d x}{d t} + k \sin x = 0 \tag{5.2}$$

  ii. Double pendulum

$$\frac{d^2 x}{d t^2} + k \sin x = 0 \quad k = \frac{GL}{J} \tag{5.3}$$

iii. Friction pendulum

$$\frac{d^2x}{dt^2} + F\operatorname{sgn}\frac{dx}{dt} + k\sin x = 0 \tag{5.4}$$

(2) Duffing System and Nonlinear System with Soft Spring

i. Without damping

$$\frac{d^2x}{dt^2} + kx + dx^3 = 0 \tag{5.5}$$

ii. With damping

$$\frac{d^2x}{dt^2} + c\frac{dx}{dt} + kx + dx^3 = 0 \tag{5.6}$$

iii. With squared term

$$\frac{d^2x}{dt^2} + c\frac{dx}{dt} + kx + bx^2 + dx^3 = 0 \tag{5.7}$$

iv. Nonlinear system with soft spring

$$\frac{d^2x}{dt^2} + c\frac{dx}{dt} + kx - dx^3 = 0 \tag{5.8}$$

(3) Symmetric and Non-Symmetric Piece-Wise Nonlinear Systems

$$\frac{d^2x}{dt^2} + f_c\left(\frac{dx}{dt}\right) + f_k(x) = 0 \tag{5.9}$$

i. Symmetric piece-wise nonlinear system ($e > 0$)

$$f_k(x) = \begin{cases} kx & -e \leq x \leq e \\ kx + \Delta k(x-e) & e \leq x \\ kx + \Delta k(x+e) & -e \geq x \end{cases} \tag{5.10}$$

$$f_c\left(\frac{dx}{dt}\right) = \begin{cases} ck\frac{dx}{dt} & -e \leq x \leq e \\ c(k+\Delta k)\frac{dx}{dt} & e \leq x,\ x \leq -e \end{cases}$$

ii. Asymmetric piece-wise nonlinear system

$$f_k(x) = \begin{cases} -k(a_0 - a - x) + \Delta k(x+a) & x \geq a \\ -k(a_0 - a - x) & x \leq -a \end{cases} \tag{5.11}$$

$$f_c\left(\frac{dx}{dt}\right) = \begin{cases} (f + \Delta f)\frac{dx}{dt} & x \geq a \\ f\frac{dx}{dt} & x \leq -a \end{cases}$$

5.1 Introduction

(4) Symmetric and Asymmetric Hysteresis System

$$\frac{d^2 x}{dt^2} + f_k(x) = 0 \qquad (5.12)$$

i. Symmetric hysteresis nonlinear system

$$f_k(x) = \begin{cases} k(A \sin \omega t - b) & 0 \le \omega t \le \pi/2 \\ k(A - b) & \pi/2 \le \omega t \le \pi - \varphi_e \\ k(A \sin \omega t + b) & \pi - \varphi_e \le \omega t \le \pi \end{cases} \qquad (5.13)$$

$$\phi_e = \arcsin\left(\frac{A - 2b}{A}\right)$$

ii. Asymmetric hysteresis nonlinear system (expression for one period)

$$f_k(x) = \begin{cases} k_1(A \sin \omega t - b) & -\varphi_e \le \omega t \le \pi/2 \\ k(A - b) & \pi/2 \le \omega t \le \pi - \varphi_e \\ k_2 A \sin \omega t & \pi - \varphi_e \le \omega t \le 2\pi - \varphi_e \end{cases} \qquad (5.14)$$

(5) Self-Excitation Systems

i. Van der Pol Equation

$$\frac{d^2 x}{dt^2} - \{1 - x^2\}\frac{dx}{dt} + kx = 0 \qquad (5.15)$$

ii. Rayleigh Equation

$$\frac{d^2 x}{dt^2} + \left\{-A + B\left(\frac{dx}{dt}\right)^2\right\}\frac{dx}{dt} + kx = 0 \qquad (5.16)$$

iii. Self-excitation vibration equation for cutting process

$$m\frac{d^2 x}{dt^2} + \varphi\left(\frac{dx}{dt}\right) + kx = 0$$

$$\varphi\left(\frac{dx}{dt}\right) = kS_0 - F\left(v_0 - \frac{dx}{dt}\right) = F(v_0) - F\left(v_0 - \frac{dx}{dt}\right) \qquad (5.17)$$

(6) Impact Nonlinear Vibration Systems

$$m\ddot{y} + ky = F_0 \sin \nu t \quad (y \le y_0) \qquad (5.18)$$

(7) Slow-changing Parameter Vibration Systems

$$\frac{d}{dt}\left[m(\tau)\frac{dx}{dt}\right] + k(\tau)x = f\left(x, \frac{dx}{dt}, \nu t, \varepsilon\right) \quad \tau = \varepsilon t \qquad (5.19)$$

(8) Parametric Excitation Vibration Systems

  i. Hill equations

$$\frac{d^2x}{dt^2} + \{-A + Kf(t)\}x = 0 \tag{5.20}$$

  ii. Mathieu equation

$$\frac{d^2x}{dt^2} + p^2\{1 + h\cos vt\}x = 0 \tag{5.21}$$

  iii. Parametric excitation vibration equation with non-linear terms

$$\frac{d^2x}{dt^2} + k(t)x + f\left(x, \frac{dx}{dt}\right) = 0 \tag{5.22}$$

Compared with the solution methods for linear vibration equations, the nonlinear solution methods have the following features:

(1) The natural frequencies are the function of the amplitude in the case of nonlinear restoring force

Duffing's equation, its restoring force term is the cubic term of the displacement, has the free vibration frequencies:

$$\omega = \sqrt{\frac{1}{M}\left(k + \frac{3}{4}bA^2\right)} \tag{5.23}$$

It can be seen from the above equation that for the hardening spring of the hardening nonlinear vibration systems, the natural frequencies will increase with the amplitude increases while for the soft spring nonlinear system the free natural frequency will decrease with the increase of the amplitudes.

(2) The resonant curves for nonlinear vibration systems are different from those for linear systems

The resonant curves for a nonlinear vibration system, i.e. the relation between amplitude and frequency (frequency response function FRF) and the relation of phase angle and the frequency (phase response function) have total different characteristics from those for linear vibration systems. For a nonlinear vibration system with hardening springs, its first order approximation solution is:

$$A = \frac{F_0 \cos \alpha}{k \pm \frac{3}{4}bA^2 - M\omega^2},$$
$$\alpha = \arctan \frac{c\omega}{k \pm \frac{3}{4}bA^2 - M\omega^2} \tag{5.24}$$

5.1 Introduction

The resonant curves for the nonlinear equation solutions can be drawn from the above equations, the hump portion of it for the hardening nonlinear system inclines to the right hand side and it inclines to the left-hand side for the hardening nonlinear system.

(3) The vibrations for the forced nonlinear vibration systems have lag and jump phenomena

For a nonlinear vibration system, if we keep the excitation force amplitude a constant, but increase the excitation frequency slowly, the vibration amplitude of the system will increase along the resonant curve direction. When the amplitude approaches its maximum value, the amplitude will jump and then decrease gradually. When excitation frequencies decrease slowly, the amplitudes increase gradually and will jump after reaching to the maximum value and amplitudes decrease thereafter. This jump phenomenon is impossible for a linear system under an ideal source condition.

The jump in the backward process always lags the jump in the forward process. This phenomenon is called the lag. The phenomenon will never occur in any linear systems.

(4) The resonant curves have stable and unstable regions

For a nonlinear vibration system under harmonic excitation, its resonant curves have stable and unstable regions. The line segment between the two jumps in the resonant curves is unstable, while other portions of the line segment are stable. For a linear vibration system under ideal power conditions, the vibration is in general stable when damping exists and the vibration in resonant condition is unstable when the damping is zero.

(5) The forced vibration system responses contain super and sub harmonics components

For a nonlinear vibration system under harmonic excitation, its forced vibration is not necessarily harmonic vibrations, and its responses contain a variety of sub-harmonic components. Those components have the same frequency harmonic as the excitation, and have additional sub-harmonics whose frequencies are a fraction of the excitation frequency, i.e. the sub-harmonic components of the responses with frequencies Ω/n and the super-harmonic responses with frequencies integer multiple of Ω ($m\Omega$). The sub-harmonic and super-harmonics have different features:

i. The super harmonic response components exist more or less in a nonlinear system, while the sub-harmonic components exist only in certain conditions.
ii. When damping exists, the damping has effect only on the super-harmonic amplitudes; while for sub-harmonic responses, the sub-harmonic will be killed only when damping is larger than a certain value.

Due to the existence of sub and super harmonics, the numbers of the resonant frequencies of the nonlinear systems will be larger than the degrees of freedom of the system.

When the excitation frequency is near the integral multiple of the natural frequency, for example, it's 3 times of natural frequency, the sub-harmonics resonance with the natural frequency and large amplitude will appear in the system responses; while the excitation frequency is a fraction of the natural frequency, say 1/3, then the super-harmonic resonance with the natural frequency and large amplitude will appear in the system responses.

(6) The combined vibrations under multiple harmonic excitations

As an example, a system is excited by two excitation forces, $F_1 \cos \Omega_1 t$ and $F_2 \cos \Omega_2 t$, then the system would have not only the vibrations of frequencies Ω_1, Ω_1, Ω_2, $2\Omega_1$, $2\Omega_2$, $3\Omega_1$, $3\Omega_2$, ..., but also the vibrations of the combinations of the differences and additions, i.e.: $|m\Omega_1 \pm n\Omega_2|$ (n, m are integers), such as $|\Omega_1 \pm \Omega_2|$, $|2\Omega_1 \pm \Omega_2|$, $|\Omega_1 \pm 2\Omega_2|$, etc. In some cases, the number of the combined frequency vibrations is much larger than that of the other frequencies. Now we will illustrate that.

If $\Omega_1 = 100$ Hz, $\Omega_2 = 120$ Hz, then the following vibrations of frequencies would occur: 20, 80, 100, 120, 140, 220, 300, 340, and 360 Hz.

(7) The superposition principles are not valid for nonlinear vibration systems

The superposition principles are often used to solve the linear system solutions. However, they are not valid for the nonlinear systems anymore due to the nonlinear and higher order terms.

(8) The frequency capture phenomenon

If there are two harmonic vibrations in a linear system with frequencies ω_0 and Ω, when the two frequencies are close, it will generate a beating phenomenon. The smaller the difference between the two frequencies, the larger the beating periods. When the two frequencies are equal, the beating disappears and the two vibrations become one harmonic vibration.

In the nonlinear vibrations, the results are different. For example, a self-excited system has a self-excitation of frequency ω_0 and another excitation frequency is Ω. If Ω and ω_0 is very close, then the system would have vibration with just one frequency, i.e., the frequencies ω_0 and Ω are entering a synchronous state. This phenomenon is called "frequency capture". The frequency band in which the frequency capture occurs is called "frequency capture band".

For example, for a self-excitation system, when $|\Omega - \omega_0|$ is less than a certain value, the frequency capture will appear, Ω and ω_0 will coincide to a single frequency. $\Delta \omega$ is the capture region.

The frequency capture phenomenon has been utilized in engineering. For the self-synchronous vibration, driven by the exciters with two electric induction motors, its working principle is based on the frequency capture phenomenon. The results of a test show that when the two excitation motors rotate individually, their speed are 962 and 940 rpm respectively, while they rotate simultaneously to drive a vibration machine, its rotations coincide to 950 rpm. This is the frequency capture.

(9) The self-excitation vibrations in some nonlinear vibration systems

In the linear systems, the free vibrations are always decayed due to the existence of the damping. The periodical motions occur only under the periodical excitation forces. While in the nonlinear vibration systems, even the damping exists, the periodical movement could occur. The loss of the energy can be compensated by the energy transferred into the system and the amount and time for the transferred energy are adjusted by the system itself. This is the self-excited vibration.

(10) Chaos

Chaos is a new branch of science. Since its start in the 60's last century, it developed very quickly. It is called another important discovery after the relativity and quantum mechanics. The appearance of study of chaos is associated closely with the modern science and technology, especially the development of the computer technology. Chaos phenomena have practical importance in studying of rotor diagnostics and as a means for the rotor diagnostics.

5.2 Utilization of Smooth Nonlinear Vibration Systems

There are two methods to determine empirically the friction coefficients between an axle and axle bushing: The first one is to calculate friction coefficients by the decay rate in a vibration period of the double pendulum; the second one is to measure and calculate the friction coefficient by the Flode pendulum working principle. The latter method is more accurate.

5.2.1 Measurement of Dry Friction Coefficients Between Axis and Its Bushing Using Double Pendulum

In the swing process of a double pendulum, due to the friction between the axle and axle bushing, the swing amplitude will decay until it stops without outside excitations. Assuming that the friction between the axle and axle bushing is not dependent on velocity and is a constant, the equation of motion for the double pendulum is:

$$J\ddot{\varphi} = \varepsilon M_r - mgl \sin \varphi \qquad (5.25)$$

in which J is the moment of inertial of the pendulum; M_r is the friction moment of force; l is the distance between the center of mass of the pendulum and the suspension point.

The friction moment of force M_r can be expressed as:

$$M_r = \left(mg\cos\varphi + ml\dot\varphi^2\right) r\, f(\dot\varphi) \tag{5.26}$$

Friction coefficient $f(\dot\varphi)$ is in the opposite direction with the relative velocity, and substituting Eq. (5.26) into Eq. (5.25) one gets:

$$J\ddot\varphi = \varepsilon \left(mg\cos\varphi + ml\dot\varphi^2\right) r\, f(\dot\varphi) - mgl\sin\varphi$$

Due to the existence of the friction, when the initial swing angle is φ_0, after one period, the vibration amplitude will reduce to φ_1, and after another period, it further reduces to φ_2. For every vibration period, the vibration amplitude reduces $\Delta\varphi = \varphi_n - \varphi_{n-1}$. This value is associated with friction coefficient directly, and friction coefficient can be calculated from this relationship.

Assume that the sliding friction coefficient $f(\dot\varphi)$ can be expressed as:

$$f(\dot\varphi) = \mathrm{sgn}\dot\varphi = \begin{cases} -f, & \dot\varphi \geq 0 \\ f, & \dot\varphi \leq 0 \end{cases} \tag{5.27}$$

in which f is the sliding friction coefficient.

Now we are going to find out the relationship between $\Delta\varphi$ and f. Assume that the first order of approximate solutions is:

$$\begin{aligned} \varphi &= a\cos\omega_e t \\ \dot\varphi &= -a\omega_e \sin\omega_e t \\ \frac{da}{dt} &= -\delta_e a \end{aligned} \tag{5.28}$$

in which

$$\begin{aligned} \delta_e &= \frac{c_e}{2J}, \quad \omega_e = \sqrt{\frac{k_e}{J}}, \quad \omega_0^2 = \frac{mgl}{J} \\ c_e &= -\frac{1}{\pi\omega_0 a}\int_0^{2\pi} f_0(a\cos\psi, -a\omega_0\sin\psi)\sin\psi\, d\psi \\ k_e &= \frac{1}{\pi a}\int_0^{2\pi} f_0(a\cos\psi, -a\omega_0\sin\psi)\cos\psi\, d\psi \end{aligned} \tag{5.29}$$

$$\begin{aligned} &f_0(a\cos\psi, -a\omega_0\sin\psi) \\ &= -mgla\cos\psi + \left(mg - mg\frac{a^2\cos^2\varphi}{2} + mla^2\omega_0^2\sin^2\psi\right) r\, f(-a\omega_0\sin\psi) \\ &\quad + \frac{1}{6}mg\, la^3\cos^3\psi \end{aligned}$$

5.2 Utilization of Smooth Nonlinear Vibration Systems

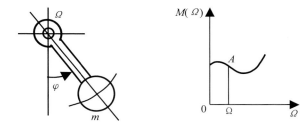

Fig. 5.1 Flode pendulum

Substituting the nonlinear function into above equation, one has:

$$
\begin{aligned}
c_e &= -\frac{4m}{\pi\,\omega_0}\left(\frac{g}{a}+a\omega_0^2 l\right) r\, f \\
k_e &= mgl\cdot\left(1-\frac{1}{8}a^2\right) \\
\delta_e &= \frac{c_e}{2J} = \frac{2}{\pi}\left(\frac{1}{la}+\frac{a^2\omega_0^2}{g}\right) r\, f\omega_0 \\
\omega_e &= \sqrt{\frac{mgl}{J}\left(1-\frac{1}{8}a^2\right)}
\end{aligned}
\qquad (5.30)
$$

Due to the existence of the friction moment of the force, the amplitudes decay and its equation is:

$$
\frac{da}{dt} = -\delta_e a = -\frac{4m}{\pi\,\omega_0}(g+a^2\omega_0^2 l)r\, f \qquad (5.31)
$$

The dry friction coefficient can be computed from the amplitude decay curves by the above equation.

5.2.2 Measurement of Dynamic Friction Coefficients of Rolling Bearing Using Flode Pendulum

(1) Measurement of the dynamic friction coefficient between the axle and axle bushing

The dynamic friction coefficient between the axle and axle bushing can be measured by the friction pendulum test. Figure 5.1 is the mechanic block diagram. The outer bushing of the rotation shaft and the inner bushing is a set of sample. Radius of the rolling surface is r, the distance between the center of gravity of the pendulum and the rotating axle center is l. When the shaft rotates at the angular velocity anticlockwise, the friction force moves the pendulum an angle of φ.

The differential equation of the motion for the pendulum is:

$$J\ddot{\varphi} = M_r - mgl\sin\varphi - ul\dot{\varphi} \tag{5.32}$$

in which J is the inertial of moment of the pendulum; m is the mass of the pendulum; M_r is the friction moment, and u is the air resistance coefficients.

Friction moment M_r can be expresses as:

$$M_r = \left(mg\cos\varphi + ml\dot{\varphi}^2\right) r\, f(\Omega - \dot{\varphi}) \tag{5.33}$$

where r is the radius of the shaft. The friction coefficient $f(\Omega - \dot{\varphi})$ is a function of relative velocity. Substituting Eq. (5.33) into Eq. (5.32) one gets:

$$J\ddot{\varphi} = \left(mg\cos\varphi + ml\dot{\varphi}^2\right) r\, f(\Omega - \dot{\varphi}) - mgl\sin\varphi - ul\dot{\varphi} \tag{5.34}$$

When shaft rotates in the angular velocity Ω while the pendulum is in a static equilibrium state, one has $\ddot{\varphi} = \dot{\varphi} = 0$, at the same time, $\varphi = \varphi_0$, then

$$r\, f(\Omega)\cos\varphi_0 - l\sin\varphi_0 = 0 \tag{5.35}$$

$$f(\Omega) = \frac{l}{r}\tan\varphi_0 \quad \text{or} \quad f(v) = \frac{l}{r}\tan\varphi_0 \tag{5.36}$$

It can be seen that when the pendulum is in still state in the test, the angle φ_0 is very easy to be measured, then the dynamic friction coefficient $f(v)$ can be computed from Eq. (5.37). Then change the rotation velocity for the shaft and measure the dynamic friction coefficient for another specimen.

The dynamic sliding friction coefficient $f(v)$ can be expressed in the following form via test results:

$$f(v) = a - bv + c|v|v + dv^3 \tag{5.37}$$

where a, b, c, d are coefficients, and can be determined by tests.

(2) Determination of the friction coefficient when the pendulum is vibrating in its equilibrium position

Assume:

$$\begin{aligned}F(\Omega - \dot{x}) &= A - B(\Omega - \dot{x}) + C(\Omega - \dot{x})^2 + D(\Omega - \dot{x})^3 \\ &= \left(A - B\Omega + C\Omega^2 + D\Omega^3\right) + \left(B - 2C\Omega - 3D\Omega^2\right)\dot{x} \\ &\quad + (C + 3D\Omega)\dot{x}^2 - D\dot{x}^3\end{aligned} \tag{5.38}$$

5.2 Utilization of Smooth Nonlinear Vibration Systems

Substituting Eq. (5.38) into (5.34) yields:

$$J\ddot{x} + [ul - (B - 2C\Omega - 3D\Omega^2) - (C + 3D\Omega)\dot{x} + D\dot{x}^2]\dot{x} \\ -(A - B\Omega + C\Omega^2 + D\Omega^3) + mgl\sin\varphi_0 + mgl\cos\varphi_0 x = 0 \quad (5.39)$$

Assume:

$$\begin{aligned} A_0 &= -(A - B\Omega + C\Omega^2 + D\Omega^3) + mgl\sin\varphi_0 \\ B_0 &= ul - B - 2C\Omega - 3D\Omega^2 \\ C_0 &= C + 3D\Omega \end{aligned} \quad (5.40)$$

Then

$$J\ddot{x} + (B_0 - C_0\dot{x} + D\dot{x}^2)\dot{x} + A_0 + mgl\cos\varphi_0 \cdot x = 0 \quad (5.41)$$

i.e.

$$\ddot{x} + \frac{1}{J}(B_0 - C_0\dot{x} + D\dot{x}^2)\dot{x} + \frac{A_0}{J} + \omega_0^2 \cdot x = 0 \quad (5.42)$$

where

$$\omega_0^2 = \frac{mgl\cos\varphi_0}{J} \quad (5.43)$$

Assuming $x_1 = \frac{A_0}{\omega_0^2 J} + x$, when rotation velocity is fixed A_0 is a constant, then Eq. (5.42) becomes:

$$\begin{aligned} \ddot{x}_1 + \omega_0^2 x_1 &= \varepsilon F(x_1, \dot{x}_1) \\ \varepsilon F(x_1, \dot{x}_1) &= -\frac{1}{J}(B_0 - C_0\dot{x}_1 + D\dot{x}_1^2)\dot{x}_1 \end{aligned} \quad (5.44)$$

The first order approximate solution for Eq. (5.44) is:

$$\begin{aligned} x_1 &= \varphi_0 \cos\psi \\ \frac{d\varphi_0}{dt} &= \varepsilon \delta_e(\varphi_0)\varphi_0 \\ \frac{d\psi}{dt} &= \omega_0 + \varepsilon \omega_1(\varphi_0) \end{aligned} \quad (5.45)$$

in which

$$\varepsilon\delta_e(\varphi_0)\varphi_0 = -\frac{1}{2\pi\,\omega_0}\int_0^{2\pi} F(\varphi_0\cos\psi, -\varphi_0\omega_0\sin\psi)\sin\psi\,d\psi$$

$$= -\frac{1}{2\pi\,\omega_0}\int_0^{2\pi} -\frac{1}{J}\left(-B_0\varphi_0\omega_0\sin\psi - C_0\varphi_0^2\omega_0^2\sin^2\psi - D\varphi_0^3\omega_0^3\sin^3\psi\right)\sin\psi\,d\psi \quad (5.46)$$

$$= -\frac{B_0}{2J}\varphi_0 - \frac{3D\omega_0^2}{8J}\varphi_0^3$$

It can be obtained:

$$\frac{d\varphi_0}{dt} = -\frac{\varphi_0}{2J}\left(B_0 + \frac{3}{4}D\omega_0^2\varphi_0^2\right) \quad (5.47)$$

Integrating Eq. (5.47), one gets:

$$\varphi = \frac{\varphi_0}{\sqrt{e^{\frac{B_0}{2MJ}t} + \frac{3D\omega_0^2}{4B_0}\varphi_0^2\left(e^{\frac{B_0}{2MJ}t} - 1\right)}} \quad (5.48)$$

It can be seen from (5.48) that when angular velocity is small, $B_0 < 0$; when angular velocity is large, $B_0 > 0$.

When $B_0 > 0$, the denominator of Eq. (5.48) will be approaching to indefinite as time is increasing indefinitely, so $\varphi_0 \to 0$ and φ_0 is asymptotically stable. When $B_0 < 0$, i.e. slow rotating, then when time is increasing indefinitely, one has:

$$\varphi = \sqrt{\frac{4B_0^2}{3D\omega_0^2}} \quad (5.49)$$

When $B_0 = 0$, $\varphi = \varphi_0$, then there is the stable and equal-amplitude vibration. φ_0 can be found from the below equations:

$$\varphi_{\max} = \varphi_0 + \varphi_{01} \qquad \varphi_{\min} = \varphi_0 - \varphi_{01}$$
$$\varphi_0 = \frac{1}{2}(\varphi_{\max} + \varphi_{\min}) \quad (5.50)$$

The $\varphi_{\max}$ and $\varphi_{\min}$ can be read from the instrument meters, Eq. (5.50) can be used to find out the angle φ_0 for the pendulum at the equilibrium position, substituting it into Eq. (5.36) one can get the value of the friction coefficient $f(v)$.

5.2.3 Increase the Stability of Vibrating Machines Using Hard-Smooth Nonlinear Vibrating Systems

For some vibration machines, such as electric–magnetic feeder, near-resonance vibration conveyer and resonance screen, working in the region of the resonance, the vibration amplitudes are, in general, not in a stable condition. The amplitude instability would bring some adverse effects to the machine performance. To overcome the shortcomings, the harden nonlinear restoring forces are adapted in the vibration system.

Other engineers proposed that the two fix ends of the main vibration springs can be made into a curved form. When the vibration amplitude increases, the working length of the leaf springs becomes shorter, therefore the stiffness becomes larger. The phenomenon can be expressed as:

$$\varphi(x) = k\,x + Q(\dot{x}, x) = f\dot{x} + k\,x + bx^3 + dx^5 \qquad (5.51)$$

where k is the stiffness of the linear portion for the springs; f is the damping coefficient; b and d are the constants.

The electric–magnetic vibration feeder is a double-body vibration system, the mechanics diagram is shown in Fig. 5.2. The differential equations of motion for the machines are:

$$\begin{aligned} & m_1\ddot{x}_1 + k_1 x_1 + k(x) = F_0 + F_1 \sin v\, t_1 + F_2 \sin 2v\, t_2 - \varepsilon Q(\dot{x}, x) \\ & m_2\ddot{x}_2 + k_2 x_2 - k(x) = -(F_0 + F_1 \sin v\, t_1 + F_2 \sin 2v\, t_2) + \varepsilon Q(\dot{x}, x) \\ & x = x_1 - x_2, \quad \dot{x} = \dot{x}_1 - \dot{x}_2 \end{aligned} \qquad (5.52)$$

where m_1, m_2 are the masses of body 1 and body 2;

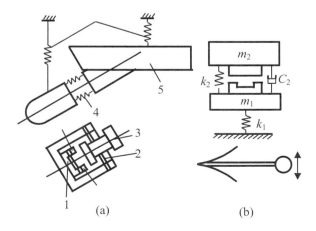

Fig. 5.2 The working mechanism and mechanics block diagram for an electric–magnetic vibration feeder. **a** The working principle figure **b** vibration system and spring structures

k_1, k_1 and k are the springs on body 1, body 2 and between the two bodies respectively;

x_1, x_2 and x are the displacements of body 1, body 2 and the relative displacement between body 1 and body 2; v is the excitation frequency.

The above equations can be written as a matrix form:

$$M\ddot{x} + Kx = F + \varepsilon Q$$

$$M = \begin{bmatrix} m_{11} & 0 \\ 0 & m_{22} \end{bmatrix} = \begin{bmatrix} m_1 & 0 \\ 0 & m_2 \end{bmatrix}, \quad K = \begin{bmatrix} k_{11} & k_{12} \\ k_{21} & k_{22} \end{bmatrix} = \begin{bmatrix} k_1+k & -k \\ -k & k_2+k \end{bmatrix} \quad (5.53)$$

$$F = \left\{ \begin{array}{c} F_0 + F_1 \sin v\, t_1 + F_2 \sin 2v\, t_2 \\ -(F_0 + F_1 \sin v\, t_1 + F_2 \sin 2v\, t_2) \end{array} \right\}, \quad Q = \left\{ \begin{array}{c} Q(\dot{x}, x) \\ -Q(\dot{x}, x) \end{array} \right\}$$

In order to find out the solutions, first transform the equation into a principal coordinate system. Thus, the two natural frequencies and modal function must be found, i.e.:

$$\begin{vmatrix} k_{11} - m_{11}\omega^2 & k_{12} \\ k_{21} & k_{22} - m_{22}\omega^2 \end{vmatrix} = 0 \quad (5.54)$$

$$\omega_{0i} = \sqrt{\frac{-b \pm \sqrt{b^2 - 4ac}}{2a}}, \quad a = m_{11}m_{22}, \quad b = -(m_{11}k_{22} + m_{22}k_{11}),$$

$$c = k_{11}k_{22} - k_{12}^2, \quad k_{12} = k_{21}, \quad i = 1, 2$$

The normal mode function can be found as:

$$\psi_1^{(i)} = 1, \quad \psi_2^{(i)} = -\frac{k_{11} - m_{11}\omega_{0i}^2}{k_{12}} = -\frac{k_{12}}{k_{22} - m_{22}\omega_{0i}^2}, \quad i = 1, 2$$

The electric–magnetic vibration feeder normally works in the region near the second natural frequency. When the natural frequency and normal mode function are given, the solution can be assumed to be:

$$\begin{cases} x_1 = \psi_1^{(2)} x_0 = \psi_1^{(2)}(a_1 \sin \varphi_1 + a_2 \sin 2\varphi_2) \\ x_2 = \psi_2^{(2)} x_0 = \psi_2^{(2)}(a_1 \sin \varphi_1 + a_2 \sin 2\varphi_2) \\ x_0 = a_1 \sin \varphi_1 + a_2 \sin 2\varphi_2, \quad \varphi_i = v\, t_i + \theta_r \end{cases} \quad (5.55)$$

Substituting the above equations into the vibration equations and simplifying them one gets the equation in the second principal coordinate system:

$$m(\ddot{x} - v^2 x_0) = \sum_{r=1}^{2} \varepsilon Q_r \psi_r^{(2)} + \sum_{r=1}^{2} \psi_r^{(2)}(F_{0r} + F_{1r} \sin \varphi_1 + F_{2r} \sin 2\varphi_2) \quad (5.56)$$

$$m = m_1 \psi_1^{(2)^2} + m_2 \psi_2^{(2)^2}, \quad F_{01} = -F_{02}, \quad F_{11} = -F_{12}, \quad F_{21} = F_{22}$$

5.2 Utilization of Smooth Nonlinear Vibration Systems

Terms a and θ can be found from the following equations:

$$\frac{da_i}{dt} = -a_i \delta_e^{(2)} - \sum_{i=1}^{2}\sum_{r=1}^{2} F_i \psi_r^{(2)} \cos\theta_i / m(iv + \omega_2)$$
$$\frac{d\theta_i}{dt} = \omega_e^{(2)} - v + \sum_{i=1}^{2}\sum_{r=1}^{2} F_i \psi_r^{(2)} \sin\theta_i / ma_i(iv + \omega_2)$$
(5.57)

where $\delta_e^{(2)}$ and $\omega_e^{(2)}$ are the equivalent decay rate and natural frequency and can be solved from the following equations:

$$\delta_e^{(2)} = \frac{1}{2\pi ma_i iv} \int_0^{2\pi} \sum_{r=1}^{2} \varepsilon Q_{r0}^{(2)}(a_i, \varphi_i) \psi_r^{(2)} \sin i\varphi_i d\varphi$$
$$\omega_e^{(2)} = \omega_2 - \frac{1}{2\pi miv} \int_0^{2\pi} \sum_{r=1}^{2} \varepsilon Q_{r0}^{(2)}(a_i, \varphi_i) \psi_r^{(2)} \cos i\varphi_i d\varphi_i$$
(5.58)

In the steady constant conditions, $\frac{da_i}{dt} = 0$, $\frac{d\theta_i}{dt} = 0$ hold, thus one has:

$$m^2 a_i^2 [(\omega_e^{(2)^2} - (iv)^2)^2 + 4\delta_e^{(2)^2}(iv)^2] = F_i^{(2)^2}$$
$$\theta_i = \arctan(\omega_e^{(2)^2} - (iv)^2)/2iv\,\delta_e^{(2)}$$
$$F_i^{(2)} = F_{i1}\psi_1^{(2)} + F_{i2}\psi_2^{(2)}$$
(5.59)

As the system is the hardening nonlinear system, its resonant curve has obvious difference as shown in Fig. 5.3. When frequency and damping are changed, the vibration amplitude stability in the sub-resonance region is obviously better than the linear systems.

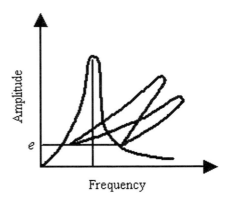

Fig. 5.3 The stable amplitude comparison between a nonlinear and a hardening nonlinear system

5.3 Engineering Utilization of Piece-Wise-Linear Nonlinear Vibration Systems

5.3.1 Hard-Symmetric Piece-Wise-Linear Vibration Systems

For some of the elastic connecting rod type of vibration conveyers, the hardening piece-wise-linear nonlinear system is adapted. The nonlinear system mechanism and mechanics block diagram are shown in Fig. 5.4. The upper body is used to convey the materials, the lower body is used to balance. There are linear springs and gap-piece-wise nonlinear springs between the two bodies. The connecting rod springs are installed on the end of the connecting rod, and the vibration isolation springs are installed beneath the lower body.

The differential equations of the motion for the systems are:

$$\begin{aligned} m_1\ddot{x}_1 + c_1\dot{x}_1 + c\dot{x} + (k_0 + k)x &= k_0 r \sin \nu t - Q(\dot{x}, x) \\ m_2\ddot{x}_2 + c_2\dot{x}_2 - c\dot{x} + k_2 x - (k_0 + k)x &= -k_0 r \sin \nu t + Q(\dot{x}, x) \\ x &= x_1 - x_2 \end{aligned} \qquad (5.60)$$

Where x_1, x_2 and x are the displacements of body 1 and 2 relatived to its static equilibrium position and the relative displacement between body 1 and 2 respectively;

m_1, m_2 are the masses of body 1 and 2;

ν is the rotation angular velocity of the eccentric rotor;

r is the eccentricity;

k_0 is the spring stiffness of the connecting rod;

k is the stiffness of the linear spring;

k_2 is the stiffness of spring between the body 2 and the base;

c_1, c_2 are damping coefficients directly proportional to $\dot{x}_1$ and $\dot{x}_2$;

$Q(\dot{x}, x)$ is the piece-wise force of the gap spring and linear spring:

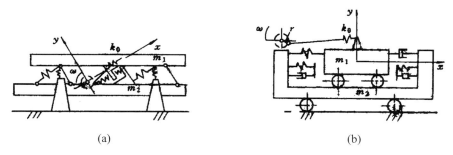

Fig. 5.4 Elastic connecting rod type of vibration conveyer and its block diagram

5.3 Engineering Utilization of Piece-Wise-Linear ...

$$Q(\dot{x}, x) = \begin{cases} 0, & -e < x < e \\ \Delta c\dot{x} + \Delta k(x - e), & x > e \\ \Delta c\dot{x} + \Delta k(x + e), & x < -e \end{cases} \quad (5.61)$$

where Δk is the stiffness of the gap spring on the two sides of m_1;

Δc is the damping coefficient of m_1 which is directly proportional to the stiffness of the gap spring on the body's two sides; e is the gap of m_1 sides.

Due to the symmetry between the gaps and gap springs on the two sides, when the machine is working normally, the vibration center will not shift for case of the first approximation.

For the calculation convenience, use the transformation $y = x + d$, the equations become:

$$\begin{aligned} &m_1\ddot{x}_1 + C\dot{x} + c_1\dot{x}_1 + K(x_1 - x_2) = k_0 r \sin \nu t + \varepsilon Q_1 \\ &m_2\ddot{x}_2 - C\dot{x} + c_2\dot{x}_2 - K(x_1 - x_2) + k_2 x_2 = -k_0 r \sin \nu t + \varepsilon Q_2 \\ &C = c + c_e, \quad K = k_0 + k + k_e \\ &k_e = k_0 + k + \Delta k \left[1 - \frac{2}{\pi}\left(\varphi_e - \frac{1}{2}\sin 2\varphi_e\right)\right] \\ &\varphi_e = \arccos \frac{e}{a} \end{aligned} \quad (5.62)$$

Where c_e and k_e are the equivalent linearized damping coefficient and stiffness. Then the nonlinear force can be expresses as:

$$\varepsilon Q_1 = -\varepsilon Q_2 = c_e\dot{x} + k_e x + \begin{cases} 0, & -e < x < e \\ -c\dot{x} - \Delta k(x - e), & x > e \\ -c\dot{x} - \Delta k(x + e), & x < -e \end{cases} \quad (5.63)$$

The above equation can be rewritten in a form of matrix:

$$M\ddot{x} + C\dot{x} + Kx = F + \varepsilon Q \quad (5.64)$$

$$M = \begin{bmatrix} m_1 & 0 \\ 0 & m_2 \end{bmatrix}, \quad C = \begin{bmatrix} c_1 + C & -C \\ -C & c_2 + C \end{bmatrix}, \quad K = \begin{bmatrix} K & -K \\ -K & k_2 + K \end{bmatrix}$$

$$F = \begin{Bmatrix} k_0 r \sin \nu t \\ -k_0 r \sin \nu t \end{Bmatrix}, \quad Q = \begin{Bmatrix} Q_1 \\ Q_2 \end{Bmatrix}$$

Now we will find the solutions.

As this machine works near the region of the second natural frequency, we may resolve its first order approximate solution: First find the second order natural frequency and normal mode function then transform it into its second principal coordinate system.

$$\psi_1^{(2)} = 1$$
$$\psi_2^{(2)} = (K - m_1\omega_2^2)/K = K/(K + k_2 - m_2\omega_2^2) \quad (5.65)$$
$$m_1 m_2 \omega_j^4 - [(K + k_2)m_1 + K m_2]\omega_j^2 + K k_2 = 0, \quad j = 1, 2$$

When the normal mode function and natural frequency are known, assume:

$$\begin{cases} x_1 = \psi_1^{(2)} x_0 = \psi_1^{(2)} a \cos\varphi \\ x_2 = \psi_2^{(2)} x_0 = \psi_2^{(2)} a \cos\varphi \quad \varphi = \omega_2 t + \theta \\ x_0 = a \cos\varphi \end{cases} \quad (5.66)$$

Substituting it into Eq. (5.64) one gets the equations on the second principal coordinates:

$$m(\ddot{x}_0 - v^2 x_0) = \sum_{r=1}^{2} \varepsilon Q_r \psi_r^{(2)} + \sum_{r=1}^{2} \varepsilon E_r \psi_r^{(2)} \cos vt \quad (5.67)$$
$$m = m_1 \psi_1^{(2)^2} + m_2 \psi_2^{(2)^2}, \quad E_1 = m_0 v^2 r, \quad E_2 = 0$$

a and θ in Eq. (5.66) can be found by the following equations:

$$\frac{da_i}{dt} = -a\delta_e^{(2)} - \sum_{r=1}^{2} \varepsilon E_r \psi_r^{(2)} \cos\theta / m(v + \omega_2)$$
$$\frac{d\theta_i}{dt} = \omega_e^{(2)} - v + \sum_{r=1}^{2} \varepsilon E_r \psi_r^{(2)} \sin\theta / ma(v + \omega_2) \quad (5.68)$$

where $\delta_e^{(2)}$, $\omega_e^{(2)}$ are the equivalent decay rate and equivalent natural frequency which can be found:

$$\delta_e^{(2)} = \frac{1}{2\pi m a v} \int_0^{2\pi} \sum_{r=1}^{2} \varepsilon Q_{r0}^{(2)}(a, \varphi) \psi_r^{(2)} \sin\varphi d\varphi$$
$$\omega_e^{(2)} = \omega_2 - \frac{1}{2\pi m v} \int_0^{2\pi} \sum_{r=1}^{2} \varepsilon Q_{r0}^{(2)}(a, \varphi) \psi_r^{(2)} \cos\varphi d\varphi \quad (5.69)$$

In the steady constant case, $\frac{da}{dt} = 0$, $\frac{d\theta}{dt} = 0$ hold, thus one has:

$$m^2 a^2 [(\omega_e^{(2)^2} - v^2)^2 + 4\delta_e^{(2)^2} v^2] = E^{(2)^2}$$
$$\theta = \arctan(\omega_e^{(2)^2} - v^2)/2v\, \delta_e^{(2)} \quad (5.70)$$
$$E^{(2)} = E_1 \psi_1^{(2)} + E_2 \psi_2^{(2)}$$

5.3 Engineering Utilization of Piece-Wise-Linear ...

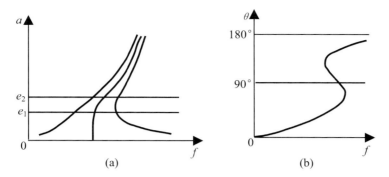

Fig. 5.5 F.R.F and phase angle response function

The frequency response function and phase angle response function for the system can be obtained from the above equations as shown in Fig. 5.5.

Figure 5.5 exhibits the features of the hard-nonlinear system. Since the real working frequency of the system is lower than the second natural frequency and in the range of $(0.85$–$0.95)$ $\omega_e^{(2)}$, in order to increase the calculation accuracy, we may improve the solution accuracy by finding the further improved approximate solutions:

$$x_s = \psi_s^{(2)} a \cos\varphi + \varepsilon u_s^{(2)}(a, \varphi), s = 1, 2$$

$$u_s^{(2)} = \frac{1}{2\pi} \sum_{\substack{n=-\infty \\ n \neq 1, -1}}^{n=\infty} \psi_r^{(2)} \frac{\int_0^{2\pi} \sum \psi_r^{(2)} Q_{r0}^{(2)}(a, \varphi) e^{-\mathrm{inv}\,t} d\varphi}{m(\omega_r^{(2)^2} - n^2 v^2)} e^{\mathrm{inv}\,t} \quad (5.71)$$

5.3.2 Soft-Asymmetric Piece-Wise-Linear Vibration Systems

In the industries, the technological processes need to accomplish some of the usually requirement that the working body has certain moving trace. The spring table used in mineral separation process is a special nonlinear vibration mechanism and system with a required movement trajectory. Figure 5.6 shows the working mechanism and mechanics diagram. It can be seen from the figure that the machine is composed of two vibration bodies. There are soft-linear springs and gap-hard-springs installed on the two sides of the bodies and the single-axle inertial exciter is installed on the body 2.

The differential equations of the motion for the system are:

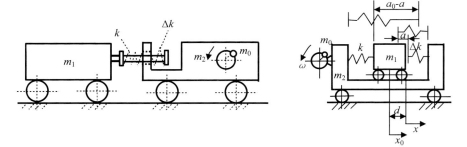

(a) Table working mechanism   (b) Table mechanics block diagram

Fig. 5.6 The spring table

$$m_1\ddot{x}_1 + k_1 x_1 + k_e x = \varepsilon \Delta Q(\dot{x}, x)$$
$$m_2\ddot{x}_2 + k_2 x_2 - k_e x = m_0 r v^2 \cos v t - \varepsilon \Delta Q(\dot{x}, x) \qquad (5.72)$$
$$x = x_1 - x_2, \quad \dot{x} = \dot{x}_1 - \dot{x}_2$$

$$\Delta Q(\dot{x}, x) = \begin{cases} -k(b_0 - b - x) - \Delta k(b + x) - (f + \Delta f)\dot{x} + k_c x, & x \geq -b \\ -k(b_0 - b - x) - f\dot{x} + k_c x, & x \leq -b \end{cases}$$
(5.73)

Where x_1, x_2 and x are the displacements of body 1 and 2 relative to its static equilibrium position and the relative displacement between body 1 and 2 respectively; m_1, m_2 are the masses of body 1 and 2 respectively; m_0 is the mass of the eccentric; v is the rotation angular velocity of the eccentric rotor; r is the eccentricity; k_c is the spring stiffness of the connecting rod; k_1 and k_2 is the stiffness of springs of the body 1 and body 2 respectively; $Q(\dot{x}, x)$ and $\Delta Q(\dot{x}, x)$ are the nonlinear force and residual nonlinear force respectively; b is the compression of the hard springs in the static state; $b_0 - b$ is the compression of the soft springs in the static state; k and Δk are the stiffness of the soft and hard springs; f and Δf are the damping coefficients of the soft and hard springs in the working regions.

We will use a simple method to find their solutions. The vibration isolation spring stiffness is small and can be ignored in the approximate calculation. Then the equations can be written as:

$$m_1\ddot{x}_1 + k_e x = \varepsilon \Delta Q(\dot{x}, x)$$
$$m_2\ddot{x}_2 - k_e x = m_0 r v^2 \cos v t - \varepsilon \Delta Q(\dot{x}, x) \qquad (5.74)$$
$$x = x_1 - x_2, \quad \dot{x} = \dot{x}_1 - \dot{x}_2$$

Multiplying the first equation in Eq. (5.74) by $\frac{m_2}{m_1+m_2}$ and subtracting it from the second one multiplied by $\frac{m_1}{m_1+m_2}$ one has:

5.3 Engineering Utilization of Piece-Wise-Linear ...

$$m\ddot{x} + k_c x = \varepsilon \Delta Q(\dot{x}, x) + E_1 \cos \nu t$$
$$m_2 \ddot{x}_2 - k_e x = E_2 \cos \nu t - \varepsilon \Delta Q(\dot{x}, x) \qquad (5.75)$$
$$m = \frac{m_1 m_2}{m_1 + m_2}, \quad x = x_1 - x_2, \quad E_1 = -\frac{m}{m_1} m_0 r \nu^2, \quad E_2 = m_0 r \nu^2$$

The equivalent stiffness is introduced in the above expressions, the purpose of doing that is that the first order approximate solutions are closer to those of the real systems. We can also express the nonlinear forces as a sum of some equivalent linear restoring forces and a residual nonlinear restoring force:

$$Q(\dot{x}, x) = k_e x + \Delta Q(\dot{x}, x)$$
$$k_c = \frac{-1}{2\pi \nu m} \int_0^{2\pi} Q(a \cos \varphi, -a\nu \sin \varphi) \cos \varphi \, d\varphi \qquad (5.76)$$
$$\Delta Q(\dot{x}, x) = Q(\dot{x}, x) - k_e x$$

Since the spring forces are asymmetric, the distance shifting from the initial static position from the vibration center when the machines are working in normal conditions can be found from the potential balance conditions:

$$\int_{a-d}^{a+d} Q(x) \, dx = 0 \qquad (5.77)$$

i.e.:

$$\frac{1}{2} \Delta k [a - (d - b)]^2 = k(b_0 - b + d) 2a$$

where b_0 is the pre-compression sum of the soft and hard springs, the pre-compression for the soft-spring is $b_0 - b$; b is the pre-compression of the hard-spring.

From the balance of the forces one can obtain $\Delta k b = k(b_0 - b)$, i.e.:

$$b = \frac{k b_0}{k + \Delta k} \qquad (5.78)$$

d can be found from the following equation:

$$Ad^2 + Bd + C = 0$$
$$A = 1, \quad B = -2\left[\left(\frac{2b}{b_0 - b} + 1\right)a + b\right], \quad C = (a + b)^2 - 4ab \qquad (5.79)$$

The machine works near the resonance state. Now we will find the resonant solutions. The solutions for the differential equation of the relative moment can be assumed to be:

$$x = -d + a\cos(vt + \theta) = d + a\cos\varphi \tag{5.80}$$

where a is the relative vibration amplitude.

$$\begin{aligned}\frac{da}{dt} &= -\varepsilon\delta_e(a) + \cdots \\ \frac{d\theta}{dt} &= \omega_0 - v + \varepsilon\omega_1(a) + \cdots\end{aligned} \tag{5.81}$$

where δ_e, ω_e are the equivalent decay rate and natural frequency and can be found as follows:

$$\begin{aligned}\delta_e &= \frac{1}{2\pi m a v}\int_0^{2\pi}\Delta Q_0(a,\varphi)\sin\varphi d\varphi \\ &= \frac{1}{2m}\left\{f + \frac{1}{2}\Delta f\left(1 - \frac{2}{\pi}\right)\left(\varphi_e + \frac{1}{2}\sin 2\varphi_e\right)\right\} \\ \omega_e &= \omega_0 - \frac{1}{2\pi m a v}\int_0^{2\pi}\Delta Q_0(a,\varphi)\cos\varphi d\varphi \quad \cdots \\ &= \sqrt{\frac{k + \frac{1}{2}\Delta k\left[1 - \frac{2}{\pi}\left(\varphi_e + \frac{1}{2}\sin 2\varphi_e\right)\right]}{m}}\end{aligned} \tag{5.82}$$

In steady constant case, $\frac{da}{dt} = 0$, $\frac{d\theta}{dt} = 0$ hold we can get the following equations:

$$\begin{aligned}m^2 a^2[(\omega_e^2 - v^2)^2 + 4\delta_e^2 v^2)] &= E^2 \\ \theta &= \arctan(\omega_e^2 - v^2)^2/2v\delta_e\end{aligned} \tag{5.83}$$

The Frequency Response Function from Eq. (5.83) is shown in Fig. 5.7 which reflects the characteristics of the soft-nonlinear system.

The working frequency of the system is a little bit lower than the second order natural frequency of the system, $(0.85$–$0.95)$ $\omega_e^{(2)}$. The reason why the machine uses the asymmetric nonlinear system is that it acquires the asymmetric trajectory.

Fig. 5.7 Frequency response function

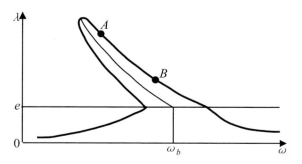

However the first order approximate solution contains only the first order harmonics, i.e. its vibration is in harmonic form. The asymmetric vibration systems contain many second harmonic components. Thus the higher order harmonics, especially the second order harmonics, must be solved. As the curves of the resultant harmonics reveal asymmetric shapes, the curves could make the materials on the table move in a direction.

The improved approximate solution can be found by the following assumption:

$$x = a\cos\varphi + \varepsilon u_1(a, \varphi)$$

$$\varepsilon u = \frac{1}{2\pi} \sum_{\substack{n=-\infty \\ n \neq 1, -1}}^{n=\infty} \frac{\int_0^{2\pi} Q_0(a, \varphi)e^{-inv\,t}d\varphi}{m(\omega_r^{(2)2} - n^2v^2)} e^{inv\,t} \quad (5.84)$$

Substituting $Q_{r0}^{(2)}(a, \varphi)$ into Eq. (5.84) and ignoring $F_m(\ddot{x}_1)$, one gets:

$$\begin{aligned}
x_s &= \psi_s^{(2)} a \cos\varphi + \sum_{n=2}^{\infty} \psi_s^{(2)}[\psi_1^{(2)} - \psi_2^{(2)}]^2 \cdot \frac{a}{\pi m v_2^2(1-n^2)} \\
&\quad \cdot [(k_{01}c_{n0} - k_{02}c_{n2})\cos n\varphi + v_2(c_{01}d_{n1} - c_{02}d_{n2})\sin n\varphi] \\
c_{n1} &= \frac{2}{n}\cos\varphi_{01}\sin n\varphi_{01} - \frac{\sin(n-1)\varphi_{01}}{n-1} - \frac{\sin(n+1)\varphi_{01}}{n+1} \\
c_{n2} &= \frac{2}{n}\cos\varphi_{02}\sin n\varphi_{02} - \frac{\sin(n-1)\varphi_{02}}{n-1} - \frac{\sin(n+1)\varphi_{02}}{n+1} \\
d_{n2} &= \frac{\sin(n-1)\varphi_{02}}{n-1} - \frac{\sin(n+1)\varphi_{02}}{n+1} \\
d_{n1} &= \frac{\sin(n-1)\varphi_{01}}{n-1} - \frac{\sin(n+1)\varphi_{01}}{n+1}
\end{aligned} \quad (5.85)$$

Adding the second order harmonic solution to the first order harmonics solution one can get the curve shown in Fig. 5.8. This curve correlates to the test results well.

Fig. 5.8 Comparison between the calculated and tested results **a** theoretical curve **b** measured curve

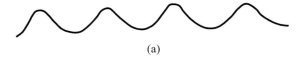
(a)

(b)

5.3.3 Nonlinear Vibration Systems with Complex Piece-Wise Linearity

(1) Establishment of the vibration system equations

The nonlinear resonance vibration screens excited by the single-axle inertial exciter, as shown in Fig. 5.9, belong to this kind of machines. The screen box, balancing mass, eccentric mass, spring of guidance plate, vibration isolation spring, rubber spring with gap are shown in Fig. 5.9. When the screen box is approaching to the mass center of the balancing mass, the first order natural frequency by the supporting spring is far less than the working frequency, the obvious swing vibration would not appear in the screen box. In this case the equation of motions for the balancing body and screen box in the x and y directions can be expresses as:

$$
\begin{aligned}
&m_1 \ddot{x}_1 + c_{12}\dot{x} + k_{12}x = F_0 \sin \nu t \\
&m_2 \ddot{x}_2 - c_{12}\dot{x} - k_{12}x + k_{3x}x_2 = 0 \\
&(m_1 + m_2)\ddot{y}_1 + k_3 y = F_0 \cos \nu t \\
&x = x_1 - x_2, \quad \dot{x} = \dot{x}_1 - \dot{x}_2, \quad y_1 = y, \quad F_0 = m_0 \nu^2 r
\end{aligned} \quad (5.86)
$$

$$
f_c(\dot{x}) = \begin{cases} c'\dot{x} & -\infty \leq x \leq e \\ 0 & e_1 \leq x \leq e_2 \\ c''\dot{x} & e_2 \leq x \leq \infty \end{cases}
$$

$$
f_k(x) = \begin{cases} k'x & -\infty \leq x \leq e \\ m_1 g \sin \alpha & e_1 \leq x \leq e_2 \\ k''(x - e_2) + m_1 g \sin \alpha & e_2 \leq x \leq \infty \end{cases} \quad (5.87)
$$

where m_1 is the balancing mass, m_2 is the mass of the screen box, m_0 is the mass of the eccentric; $f_x(s)$ is the nonlinear damping force; $f_k(x)$ is the nonlinear spring

Fig. 5.9 Inertial resonance screen diagram

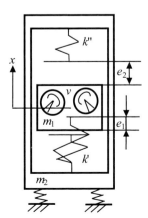

force; c' and c'' are the damping coefficients of the upper and lower springs between the eccentric and screen box; k' and k'' are the stiffnesses of the upper and lower springs between the eccentric and screen box; e_1 and e_2 are the gaps between the eccentric and upper and lower springs; k_{3x} and k_{3y} are the stiffnesses of the vibration isolation springs in x and y directions; x_1, x_2 and x are the displacements of the balancing body and the screen box and its relative displacement; y_1 and y_2 are the displacements of the balancing body and the screen box in y direction; F_0 is the excitation amplitude; v is the excitation frequency, r is the eccentricity and t is the time.

When the screen box swing vibration is ignored, the system has three degrees of freedom. If the nonlinearity is considered to be weak, the linear methods can be used to this equation to transform the coordinates into the principal coordinates and then use the approximate method to solve the single degree of freedom vibration problems.

(2) Determination of the vibration center

Since the system is asymmetric the vibration center in the vibration process will shift away from its original position. The vibration center can be found by the potential energy balance condition:

$$\int_{-a+d}^{a+d} f_k(x) \mathrm{d}x = 0 \tag{5.88}$$

where a is the vibration amplitude of the screen box relatived to the balancing body; d is the coordinate of the vibration center.

When the machine is working normally, $d + a > e_2$, integrating Eq. (5.88) piece-wisely, one gets:

$$(k'' - k')d^2 + 2[(k'' + k')a + (k'e_1 - k''e_2)]d + \left[(k'' + k')a^2 + 2a(k'e_1 - k''e_2) - \left(k'e_1^2 - k''e_2^2\right)\right] = 0 \tag{5.89}$$

Then d can be found from the above equation:

$$d = \frac{-\left[(k'' + k')a + (k'e_1 - k''e_2) + \sqrt{4k''k'a^2 + 4k'(k'e_1 - k''e_2)a + k''k'(e_2 - e_1)^2}\right]}{k'' - k'} \tag{5.90}$$

When $k'' = k'$, d can be found as follows:

$$d = \frac{e_2 - e_1}{2} \cdot \frac{2a - (e_2 + e_1)}{2a - (e_2 - e_1)} \tag{5.91}$$

For a general situation, it can be solved by three cases:

$$d = \begin{cases} 0 & 0 \leq a_x \leq e_1 \\ (\sqrt{a_x} - \sqrt{e_1})^2 & e_1 \leq a_x \leq a^* \\ -\{[(k''+k')a_x + (k''e_1 - k''e_2)] + \sqrt{4k''k'a_x^2 + 4k'(k''e_1 - k''e_2)a_x + k''k'(e_2 - e_1)^2}\} & a^* \leq a_x \leq \infty \\ \cdot (k'' - k')^{-1} \end{cases}$$

(5.92)

when $k' = k''$, then

$$d = \begin{cases} 0 & 0 \leq a_x \leq e_1 \\ (\sqrt{a_x} - \sqrt{e_1})^2 & e_1 \leq a_x \leq a^* \\ \frac{e_2 - e_1}{2} \cdot \frac{2a_x - (e_2 + e_1)}{2a_x - (e_2 - e_1)} & a^* \leq a_x \leq \infty \end{cases}$$ (5.93)

(3) Approximate method for solutions

Since the vibration isolation spring stiffness is very small and has small effect on the forced vibration of the screen box, it can be ignored. Then one has:

$$y_1 = y_2 = a_y \cos \nu t = -\frac{m_0 r}{m_1 + m_2} \cos \nu t$$
$$a_y = -\frac{m_0 r}{m_1 + m_2}$$ (5.94)

where a_y is the vibration amplitude in y direction.

Simplifying the first and the second equations in Eq. (5.87) one gets:

$$m\ddot{x} + f_c(\dot{x}) + f_k(x) = \frac{m}{m_1} F_0 \sin \nu t$$
$$m = \frac{m_1 m_2}{m_1 + m_2}$$ (5.95)

where m is the induced mass and x is the relative displacement.

$$x = d + a_x \sin(\nu t + \beta)$$
$$\frac{da_x}{dt} = \delta_e(a_x) + \frac{F_0}{m(\omega_0 + \nu)} \sin \beta$$
$$\frac{d\beta}{dt} = \omega_e - \nu - \frac{F_0}{m a_{x0}}$$ (5.96)

in which the equivalent damping ratio δ_e and the equivalent natural frequency ω_e can be found in the following equations:

5.3 Engineering Utilization of Piece-Wise-Linear ...

$$\delta_e = \frac{1}{2\pi m \omega_0 a_x} \int_0^{2\pi} f_c(a_x \omega_0 \cos \psi) \cos \psi d\psi$$

$$\omega_e^2 = \frac{1}{\pi m a_x} \int_0^{2\pi} f_c(a_x \sin \psi) \sin \psi d\psi \quad (5.97)$$

Substituting the nonlinear forces into above equation, one has:

$$\delta_e = \begin{cases} \delta' & 0 \leq a_x \leq e_1 \\ \delta'\left[\frac{1}{2} + \frac{1}{\pi}\left(\arcsin\frac{1}{z_1} + \frac{1}{z_1}\sqrt{1 - \frac{1}{z_1^2}}\right)\right] & e_1 \leq a_x \leq a^* \\ \delta'\left\{\frac{1 + \frac{\delta''}{\delta'}}{2} + \frac{1}{\pi}\left[\left(\arcsin\frac{1}{z_1} + \frac{1}{z_1}\sqrt{1 - \frac{1}{z_1^2}}\right) \right. \right. \\ \left. \left. -\frac{\delta''}{\delta'}\left(\arcsin\frac{1}{z_2} + \frac{1}{z_2}\sqrt{1 - \frac{1}{z_2^2}}\right)\right]\right\} & a^* \leq a_x \leq \infty \end{cases} \quad (5.98)$$

$$\omega_e^2 = \begin{cases} \omega_0'^2 & 0 \leq a_x \leq e_1 \\ \omega_0'^2\left[\frac{1}{2} + \frac{1}{\pi}\left(\arcsin\frac{1}{z_1} + \frac{1}{z_1}\sqrt{1 - \frac{1}{z_1^2}}\right)\right] & e_1 \leq a_x \leq a^* \\ \omega_0'^2\left\{\frac{1 + \frac{k''}{k'}}{2} + \frac{1}{\pi}\left[\left(\arcsin\frac{1}{z_1} + \frac{1}{z_1}\sqrt{1 - \frac{1}{z_1^2}}\right) \right. \right. \\ \left. \left. -\frac{k''}{k'}\left(\arcsin\frac{1}{z_2} + \frac{1}{z_2}\sqrt{1 - \frac{1}{z_2^2}}\right)\right]\right\} & a^* \leq a_x \leq \infty \end{cases} \quad (5.99)$$

The stable vibration is in the steady constant situation, and the relative amplitude in stable condition can be found as:

$$a_x = \frac{F_0}{m}\sqrt{[\omega_e^2 - \omega_0^2]^2 + 4\delta_e^2\omega_0^2}$$

$$\beta = \arctan\left[-\frac{2\omega_0\delta_e}{\omega_e^2 - \omega_0^2}\right] \quad (5.100)$$

The resonance curve can be drawn from Eq. (5.100) as shown in Fig. 5.10. We can see from Fig. 5.10 that as the amplitudes increase the resonance curve can be partitioned into three parts: linear, soft-nonlinear and hard-linear.

The relative displacement and displacements of body 1 and 2 can be found as follows:

$$x = x_1 - x_2 = d + a_x \sin(\nu t + \beta)$$

$$x_1 = \frac{m}{m_1}d + \frac{1}{m_1 + m_2}\sqrt{(m_1 a_x)^2 + (m_0 r)^2 - 2m_1 a_x m_0 r \cos\beta}\sin(\nu t + \beta)$$

$$x_2 = -\frac{m}{m_2}d - \frac{1}{m_1 + m_2}\sqrt{(m_1 a_x)^2 + (m_0 r)^2 + 2m_1 a_x m_0 r \cos\beta}\sin(\nu t + \beta)$$

(5.101)

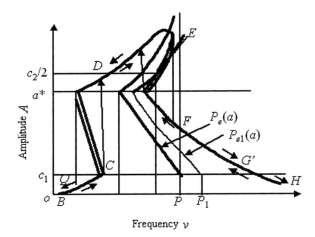

Fig. 5.10 The resonance curves of the inertial type of resonance screen

when $\beta = 0$, one has:

$$x_1 = \frac{m_2}{m_1 + m_2}d + \frac{m_2 a_x - m_0 r}{m_1 + m_2}\sin \nu t$$
$$x_2 = -\frac{m_1}{m_1 + m_2}d - \frac{m_1 a_x + m_0 r}{m_1 + m_2}\sin \nu t \quad (5.102)$$

Hence the approximate values for the amplitudes are:

$$a_{x1} = \frac{m_2 a_x - m_0 r}{m_1 + m_2}$$
$$a_{x2} = -\frac{m_1 a_x + m_0 r}{m_1 + m_2} \quad (5.103)$$

5.4 Utilization of Vibration Systems with Hysteresis Nonlinear Force

The vibration systems with hysteresis nonlinear force have been used in the vibration shaping machines, road rammers, etc. They all utilize the vibrations to generate the plastic deformation of practical significance in ramming the material of elastic transformation, or make the sliding among the particles in the materials thus to ram the materials. In the vibration old rolling or hot rolling of steal or non-ferrous metal, the rolling process can be finished unless the material itself deforms. Accompanying the occurrence of the elastic–plastic deformation, the hysteresis restoring force model would appear in the coordinate systems composed of displacement and restoring force. Hence only utilizating the vibration systems with the hysteresis nonlinear

5.4 Utilization of Vibration Systems with Hysteresis Nonlinear Force

restoring forces could accomplish the ramming of the road or technological process of deforming the metal material.

5.4.1 Simplest Hysteresis Systems

The hysteresis nonlinear forces have a variety of forms. The simple parallelogram form is shown in Fig. 5.11. The differential equation of movement for a single degree of freedom vibration system with a hysteresis nonlinear force is as follows:

$$m\ddot{y} + F(y, \dot{y}) = F_0 \sin \nu t \qquad (5.104)$$

$$F(y, \dot{y}) = \begin{cases} k(a \sin \nu t - b) & 0 \leq \nu t \leq \pi/2 \\ k(a - b) & \pi/2 \leq \nu t \leq (\pi - \psi_1) \\ k(a \sin \nu t + b) & (\pi - \psi_1) \leq \nu t \leq \pi \end{cases}$$

$$\psi_1 = \arcsin\left(\frac{a - 2b}{a}\right)$$

where a is the first order approximate vibration amplitude; k is the slope of the line, i.e. the stiffness of the rammed material; b is the coordinates of the pivotal point between two segments of lines.

To speed up to reach the real work state, it can take the nonlinear function to be the equivalent linear force plus the additional nonlinear force:

$$F(y, \dot{y}) = c_e \dot{y} + k_e y - \varepsilon f(y, \dot{y}) \qquad (5.105)$$

where k_e is the equivalent stiffness and can be computed by the methods mentioned in Chap. 3; ε is a small parameter; $f(y, \dot{y})$ is a nonlinear function.

Equation (5.104) can be expressed as:

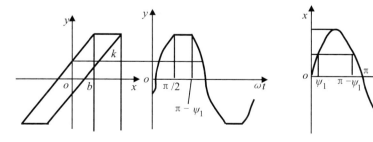

Fig. 5.11 Parallelogram hysteresis model

$$m\ddot{y} + c_e \dot{y} + k_e y = \varepsilon f(y, \dot{y}) + \varepsilon F_0 \sin \nu t$$

$$f(y, \dot{y}) = c_e \nu \cos \nu t + k_e \sin \nu t - \begin{cases} k(a \sin \nu t - b) & 0 \leq \nu t \leq \pi/2 \\ k(a - b) & \pi/2 \leq \nu t \leq (\pi - \psi_1) \\ k(a \sin \nu t + b) & (\pi - \psi_1) \leq \nu t \leq \pi \end{cases}$$
(5.106)

Now we will illustrate how to find the non-resonance solution and resonance solutions.

(1) The non-resonance solution of the equations

Now we have transformation:

$$y = x + A \sin(\nu t - \alpha) \tag{5.107}$$

Substituting it into Eq. (5.106), one has:

$$m\ddot{x} + k_e x = \varepsilon f[x + A \sin(\nu t - \alpha)] - \varepsilon c_e \dot{x} \\ + \varepsilon \left[F_0 \sin \nu t - (k_e - m\nu^2) A \sin(\nu t - \alpha) - c_e \nu A \cos(\nu t - \alpha) \right] \tag{5.108}$$

Assume that the solution is:

$$x = a \cos(\omega_0 t + \beta) = a \cos \psi$$

$$\frac{da}{dt} = -\delta_e(a) a = -\frac{1}{4\pi^2 \omega_0} \int_0^{2\pi} \int_0^{2\pi} f_0(a, \psi, \theta) \sin \psi \, d\psi \, d\theta \tag{5.109}$$

$$\frac{d\psi}{dt} = \omega_0 + \varepsilon \omega_1 = \omega_0 - \frac{1}{4\pi^2 \omega_0 a} \int_0^{2\pi} \int_0^{2\pi} f_0(a, \psi, \theta) \cos \psi \, d\psi \, d\theta$$

where $\theta = \nu t$, $\omega_0 = \sqrt{\frac{k_e}{m}}$

$$f_0(a, \psi, \theta) = f(a \cos \psi, -a\omega_0 \sin \psi, \nu t) \tag{5.110}$$

The improved first order approximate solution is:

$$x = a \cos \psi + \varepsilon u_1(a, \psi, \theta)$$

$$u_1(a, \psi, \theta) = \frac{1}{4\pi^2} \sum_n \sum_m \frac{e^{i(n\theta + m\psi)}}{\omega_0^2 - (n\theta + m\omega_0)^2} \int_0^{2\pi} \int_0^{2\pi} f_0(a, \psi, \theta) e^{-i(n\theta + m\psi)} d\theta \, d\psi \tag{5.111}$$

$$\left[n^2 + (m^2 - 1)^2 \neq 0 \right]$$

Due to the existence of the damping, the free vibration would decay to zero, i.e. $x \to 0$. In fact, the significant solution is the solution for the forced vibration. Its vibration amplitude and phase angle are:

5.4 Utilization of Vibration Systems with Hysteresis Nonlinear Force

$$A = \frac{F_0 \cos \alpha}{k_e - mv^2}, \quad \alpha = \arctan \frac{c_e v}{k_e - mv^2} \quad (5.112)$$

The equivalent linear stiffness and damping are:

$$c_e = \frac{4kb}{\pi}\left(\frac{b-a}{a}\right)$$

$$k_e = \frac{ka}{\pi}\left[\frac{\pi}{2} + \arcsin\left(\frac{a-2b}{a}\right) + \left(\frac{a-2b}{a}\right)\sqrt{1-\left(\frac{a-2b}{a}\right)^2}\right] \quad (5.113)$$

(2) The resonance solutions

Assume the solutions of the equations are:

$$y = a\cos(vt + \beta) = a\cos\varphi$$

$$\frac{da}{dt} = -\delta_e(a)\,a - \frac{\varepsilon F_0}{m(\omega_0 + v)}\cos\beta \quad (5.114)$$

$$\frac{d\beta}{dt} = \omega_e(a) - v + \frac{\varepsilon F_0}{ma(\omega_0 + v)}\sin\beta$$

$$\delta_e(a) = \frac{\varepsilon}{2\pi\omega_0 a m}\int_0^{2\pi} f(a\cos\varphi, -av\sin\varphi)\sin\varphi\,d\varphi$$

$$\omega_e(a) = \omega_0 - \frac{\varepsilon}{2\pi\omega_0 a m}\int_0^{2\pi} f(a\cos\varphi, -av\sin\varphi)\cos s\varphi\,d\varphi \quad (5.115)$$

where $\omega_0 = \sqrt{\frac{k_e}{m}}$.

Substituting Eq. (5.106) into above equation one gets:

$$\delta_e = \frac{2kb}{\pi m}\left(\frac{b-a}{a}\right)$$

$$\omega_e = \sqrt{\frac{ka}{\pi m}\left[\frac{\pi}{2} + \arcsin\left(\frac{a-2b}{a}\right) + \left(\frac{a-2b}{a}\right)\sqrt{1-\left(\frac{a-2b}{a}\right)^2}\right]} \quad (5.116)$$

In the steady and constant situation, $\frac{da}{dt} = 0$, $\frac{d\beta}{dt} = 0$, thus:

$$a = \frac{F_0}{\sqrt{\left(\omega_e^2 - v^2\right)^2 + \delta_e^2}} \quad (5.117)$$

In fact the hysteresis process is the plastic deformation process of the materials. The larger the plastic deformation, the speedier the materials being rammed. Thus, from improving the technological process effect, the measures must be taken to speed up the technological process.

5.4.2 Hysteresis Systems with Gaps

The vibration utility nonlinear systems sometimes acquire better technological indices than linear systems, such as the vibration centrifugal for hydro-extracting and vibration conveyers and vibration coolers. The nonlinear forces for the vibration machines come from the effects of the materials. Based on the study on inertial vibration cone crashers, one proposed a bi-directional symmetric hysteresis model with a gap as shown in Fig. 5.12. This is a typical piece-wise linear type of hysteresis model.

The feature of the nonlinear hysteresis is the description of characteristics of deformation stiffness energy loss of the structure and systems. The hysteresis nonlinear forces have big effects on the machine technological processes, it is imperative that its study is valuable for theoretical and practical applications.

(1) Model of bi-directional symmetric hysteresis with a gap

Assuming that the body of an inertial cone vibration crasher is still, the moving cone is moving in the x-direction reciprocally under the action of the eccentric mass. The moving cone and the body are interacted by the materials in between. Taking the hysteresis into account, then the moving cone is considered as a single degree of freedom vibration systems as shown in Fig. 5.13 and its differential equation of motion is:

$$m\ddot{x} + c\dot{x} + kx + F(x) = P\sin\theta t \tag{5.118}$$

where m is the machine mass, c is the damping, k is the linear stiffness of the system; the bi-directional hysteresis can be expressed as:

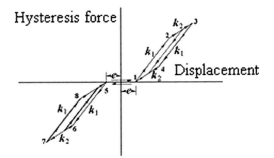

Fig. 5.12 Bi-directional symmetric hysteresis

5.4 Utilization of Vibration Systems with Hysteresis Nonlinear Force

Fig. 5.13 Vibration crasher mechanics model

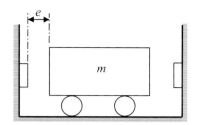

$$F(x) = \begin{cases} 0 & |x| \leq e \\ k_1 x + \alpha_1 \mathrm{sgn}(x) & e \leq |x| \leq x_2 \quad \dot{x} > 0 \\ k_2 x + \alpha_2 \mathrm{sgn}(x) & |x| \geq x_2 \quad \dot{x} > 0 \\ k_1 x + \alpha_3 \mathrm{sgn}(x) & |x| \geq x_4 \quad \dot{x} < 0 \\ k_2 x + \alpha_4 \mathrm{sgn}(x) & e \leq |x| \leq x_4 \quad \dot{x} < 0 \end{cases} \tag{5.119}$$

where k_1 is the elastic load stiffness; k_2 is the plastic load stiffness; e is the gap,

$$a_1 = -k_1 e, \ a_2 = -k_2 x_2 + F_2, \ a_3 = -k_1 x_3 + F_2, \ a_4 = -k_2 e$$

(2) Numerical analysis and results

The nonlinear system with a bi-directional symmetric hysteresis and with a gap is a typical energy loss system. Due to the special characteristics of the piece-wise linear hysteresis system, it lacks of studies on its chaotic problems. Under the condition of a set of special parameters the bifurcation and singularity can generate a very complex nonlinear vibration patterns for the system. The nonlinear system with a bi-directional symmetric hysteresis with a gap can be considered to be a simple piece-wise hysteresis system. As a special case of the piece-wise hysteresis system the chaotic behavior of the systems attracts wide range of attention.

In the specific numerical solution process, the whole time period can be divided into many small time steps Δt_k, and assigned a constant restoring force $f(x)$ in each time step, that is to say that using a piece-wise linearization method to solve the approximate solutions. After each calculations, the $f(x)$ values are re-determined by the displacement changes. There are many step-integral methods, such as Wilson-θ method, Newmark-β method, Center-difference method, Houbolt method and classical Runge–Kutta method to treat a variety of stiffness degeneration piece-wise hysteresis line segments. The authors use the improved 4th order Runge–Kutta method.

(1) System damping

In order to observe if this nonlinear hysteresis force could generate the chaos phenomenon, for the time-being we assume that the system restoring force is all subjected by the nonlinear hysteresis force. The equation can be written as:

$$\ddot{x} + 2\varsigma\omega\dot{x} + f(x) = p\sin\theta t \qquad (5.120)$$

where $p = P/m$. The nonlinear hysteresis force is

$$f(x) = \begin{cases} 0 & |x| \le e \\ \omega_1^2 x + a_1 \mathrm{sgn}(x) & e \le |x| \le x_2 \quad \dot{x} > 0 \\ \omega_2^2 x + a_2 \mathrm{sgn}(x) & |x| \ge x_2 \quad \dot{x} > 0 \\ \omega_1^2 x + a_3 \mathrm{sgn}(x) & |x| \ge x_4 \quad \dot{x} < 0 \\ \omega_2^2 x + a_4 \mathrm{sgn}(x) & e \le |x| \le x_4 \quad \dot{x} < 0 \end{cases} \qquad (5.121)$$

where $\omega_1^2 = k_1/m$; $\omega_2^2 = k_2/m$; $a_1 = -\omega_1^2 e$, $a_2 = -\omega_2^2 e$, $a_3 = -\omega_2^2 x_3 + f_2$, $a_4 = -\omega_2^2 e$.

When the damping of the systems changes from -0.3 to 1.7 the bifurcation and maximum Lyapunov index corresponding to $p = 2.0$, $\theta = 2.0$, $\omega_1^2 = 40.0$, $\omega_2^2 = 20.0$, $e = 0.03$, $x_2 = x_4 = 0.1$ are shown in Fig.5.14.

When the system vibrates, the hysteresis restoring forces dissipate the system energy. When the system damping is negative, for example the self-excitation system, the system could generate the chaotic behavior easily. The Poincare surface and maximum Lyapunov index for $2\xi\omega = -0.3$ are shown in Fig. 5.14a. It is a typical strange chaos attractor; after that a periodical movement appears in the system. However when $2\xi\omega = 0.7$, i.e. the damping is positive, the chaotic behavior still appears in the system. Its chaos attractor shape and maximum Lyapunov index are shown in Fig. 5.14b. The chaos movement is a globally stable and locally unstable behavior, while the damping plays a role in maintaining system stability. In some of the nonlinear systems with hysteresis restoring force and large energy losses, the chaos appears even the damping is positive. It is possibly due to the non-smooth characteristics of the systems.

(2) Outside excitation frequency. The bifurcation is shown in Fig. 5.15 with $2\xi\omega = -0.3$, $p = 2.0$, $\theta = 2.0$, $\omega_1^2 = 40.0$, $\omega_2^2 = 20.0$, $e = 0.03$, $x_2 = x_4 = 0.1$, θ is in the range of 0.6–2.0.

It can be seen from Fig. 5.16 that when the system excitation frequency changes from 0.6 to 2.0, there appear several chaos regions in the initial stage and the system constant changes from chaotic to periodical and from periodical to chaotic. When frequency is near $\theta = 2.8$, there appears a double period inverse bifurcation. Figure 5.17 shows the Poincare section and its corresponding maximum Lyapunov index when frequency $\theta = 0.6$ and $\theta = 2.88$. Comparing with the shapes of other nonlinear chaos attractors, as the hysteresis storing forces are piece-wise linear, its chaos attractors appear mottled and discontinuous in its geometric structures. However it contains infinitive levels of structures. This phenomenon exists perhaps uniquely in the piece-wise hysteresis system.

5.4 Utilization of Vibration Systems with Hysteresis Nonlinear Force

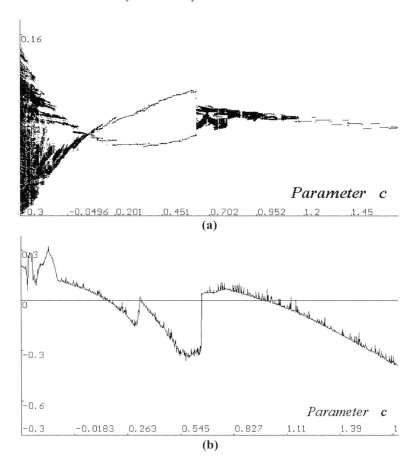

Fig. 5.14 The bifurcation and maximum Lyapunov index

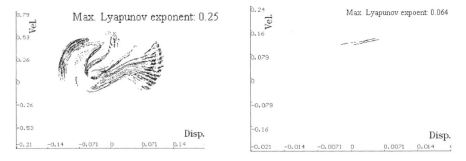

Fig. 5.15 Poincare sections with different parameters

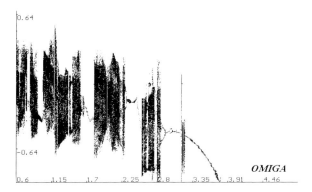

Fig. 5.16 Bifurcation with different parameters

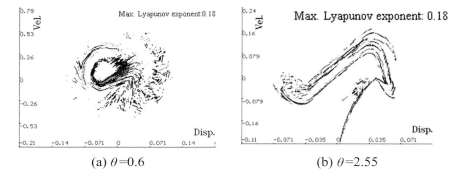

(a) $\theta=0.6$ (b) $\theta=2.55$

Fig. 5.17 The Poincare sections under different system parameters for system 3

(3) Conclusions

The study shows that when the system damping is negative, such as in the self-excitation system, the system shows a variety of nonlinear vibration patterns, the chaotic behaviors appear easily; when the system damping is positive, the system vibration patterns are mainly the periodical vibration, however the chaotic behavior could occur locally.

The vibration systems are subjected to hysteresis nonlinear forces, such as vibration shaping machines, vibration rammers, vibration rollers and inertial vibration cone crushers, their working efficiency could be greatly reduced when the disorderly chaotic movement under some special conditions. We should pay attention to it in system designs, synthesis and system controls.

5.5 Utilization of Self-excited Vibration Systems

The self-excited vibration has been utilized in many aspects of the industry applications, such as pneumatic type and hydraulic type of rock drills and crushers in the mining industry, pneumatic pick in the coal mining; the air shovel used in removing the sands on cast-parts, steam hammer used in the forging shop, etc. The working process of the steam locomotive is a self-excitation vibration; the reciprocal oil cylinder is controlled by hydraulic valves or the system driven by pistons. In the radio communication and instrument and meter of the industry branches, the oscillators and wave shape generators, switch thermal-adjustors in the thermostat; the watch and wall clock in everyday life; string music instruments. The heart beats of human is also a self-excited vibration.

Now take two examples: the first one is the self-excitation vibration of the pneumatic impactor and its working principle is shown in Fig. 5.18.

It can be seen from Fig. 5.18 that the differential equation of the moment for the piston is:

$$m \frac{d^2 x}{dt^2} = F_2 p_2(t) - F_1 p_1(t) - G + f\left(\frac{|\dot{x}|}{\dot{x}}\right) \tag{5.122}$$

where m is the mass of the piston; F_1 and F_2 are the effective areas of the front and rear chambers of the cylinder; $p_1(t)$ and $p_2(t)$ are the pressures in rear and front chambers of the cylinder; G is the component in the impact direction of the piston gravity; $f(|\dot{x}|/\dot{x})$ is the nonlinear force:

$$f\left(\frac{|\dot{x}|}{\dot{x}}\right) = \begin{cases} R_1 & \dot{x} \leq 0 \\ -R_2 & \dot{x} \geq 0 \end{cases} \tag{5.123}$$

where R_1 and R_2 are the dry friction forces.

According to the above equations the vibration curves can be drawn in the phase plane as shown in Fig. 5.19. It shows that the system acquires a stable periodical vibration.

Fig. 5.18 The working principle diagram of pneumatic impactor

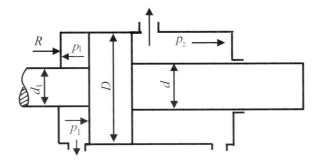

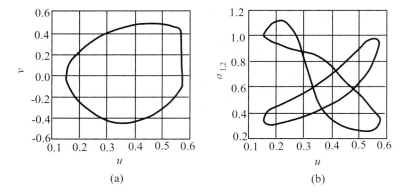

Fig. 5.19 The vibration curves in the phase plane for the piston of the pneumatic impactor

Fig. 5.20 The working principle diagram of the electronic oscillator

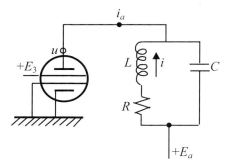

The second example is the self-excitation vibration for an electronic oscillator as shown in Fig. 5.20.

It can be seen from Fig. 5.20 that the differential equation of the oscillator currents is:

$$CL\ddot{x} - \alpha \dot{x} + \frac{\beta}{3}\dot{x}^3 + x = 0 \quad (5.124)$$

where C is the capacity in the oscillation circuit; L is the induction in the circuit; α and β are the positive constants.

Assume:

$$\omega_0^2 = \frac{1}{CL}, \quad \varepsilon = \frac{\alpha}{CL}, \quad \mu^2 = \frac{3\alpha}{\beta} \quad (5.125)$$

Equation (5.124) becomes a standard Rayleigh Equation:

Fig. 5.21 Current wave diagram of the electronic oscillator

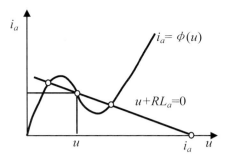

$$\ddot{x} - \varepsilon\left(\dot{x} - \frac{1}{\mu^2}\dot{x}^3\right) + \omega_0^2 x = 0 \qquad (5.126)$$

It can be transformed as the first order equations as follows:

$$\frac{dx}{dt} = y, \quad \frac{dy}{dt} = \varepsilon\left(y - \frac{1}{\mu^2}y^3\right) - \omega_0^2 x \qquad (5.127)$$

The slope for the trace in the phase plane is:

$$\frac{dy}{dx} = \left(1 - \frac{1}{\mu^2}y^2\right) - \omega_0^2\frac{x}{y} \qquad (5.128)$$

Figure 5.21 is drawn by this equation and it shows that the system acquires the stable periodical vibration.

5.6 Utilization of Nonlinear Vibration Systems with Impact

The vibration machines using impact to accomplish technological process include the frog rammer, vibration hammering, vibration shakeout machines with impact and vibration drilling machine.

The impact type of vibration machines is a special example for the nonlinear vibration machines. The theoretical and experimental results show that the acceleration generated in short period by the impact is several times, 10 times and even hundreds times larger than the maximum acceleration a general linear vibration machine could generate. The huge impact forces are important for ramming soil, making plastic deformation, rock cracking and crushing and sand falling from the cast parts.

The impact type of vibration machines can be driven by an elastic connecting rod type of exciter, inertial type of exciter or electromagnetic exciter, pneumatic type or hydraulic type of exciter. It can work near the resonance region or far away from it.

The frog type of rammers is used to illustrate the mechanical characteristics of the impact vibration machines.

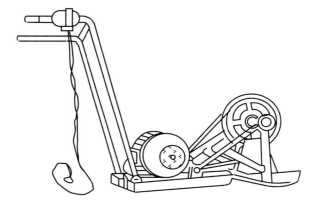

Fig. 5.22 Schematic of a frog rammer mechanism

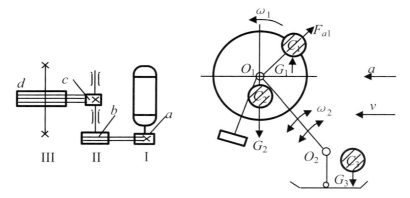

Fig. 5.23 Working structure and model of a frog rammer mechanism

Figure 5.22 shows the working mechanism of a frog type of rammer. The machine moves a distance after each impact in a vibration period.

Referring to Fig. 5.23 one can get the differential equations of the rammer head swing around point O_1 and the shift of the transmission machine seat:

$$J\ddot{\psi} = m_1 r \omega^2 l_1 \sin(\varphi + \alpha_0 + \psi) - G_1 l_1 \cos(\alpha_0 + \psi) - G_2 l_2 \cos(\alpha_0 + \psi - \delta)$$

$$\frac{G}{g}\ddot{x} \approx m_1 r \omega^2 (-\cos\varphi + f \sin\varphi) + (G_1 + G_2)\omega_2^2 l_c [\cos(\alpha_0 + \psi) + f \sin(\alpha_0 + \psi)] - fG \quad (5.129)$$

$$J = \frac{G_1}{g} l_1^2 + \frac{G_2}{g} l_2^2, \quad G = G_1 + G_2 + G_3$$

Where J is the total rotational moment of inertia around axle O_1 of the rammer head; m_1 and r are the mass and eccentricity of the eccentric lump respectively; ω is the angular velocity of the eccentric; G_1, G_2 and G_3 are the gravity of the eccentric, rammer head, and the tray of the motor support frame respectively; ψ and $\dot{\psi}$ are the

5.6 Utilization of Nonlinear Vibration Systems with Impact

swing angle and swing angular acceleration; x is the shift acceleration; l_1, l_2 and l_c are the swing rod's length, distance from the rammer head mass center to the swing center and the distance from the resultant mass center to swing center; α_0 is the angle between the swing rod and the horizontal plane; f is the friction coefficient between the bottom frame and the ground surface; φ is the angle $\varphi = \omega t$.

The phase angle φ_d of the eccentric at the time when the rammer head is about to rise can be determined from the first equation:

$$\sin(\varphi_d + \alpha_0 + \psi) = \frac{G_1 l_1 \cos(\alpha_0 + \psi) + G_2 l_2 \cos(\alpha_0 + \psi - \delta)}{m_1 r \omega^2 l_1} \quad (5.130)$$

Furthermore the phase angle of the eccentric lump during the rammer head rising can be calculated—the rising angle and the rammer head falling phase angle—the falling angle $\varphi_z = \varphi_d + \theta$. Then the shift distance of the tray can be calculated from Eq. (5.129). In the simplifying case; the above equation can be simplified as:

$$\sin \varphi_d = \frac{\sum m_i g \, l_i}{m_1 \omega^2 r \, l_1}, \quad \varphi_d = \arcsin \frac{1}{D}, \quad D = \frac{\sum m_i g \, l_i}{m_1 \omega^2 r \, l_1} \quad (5.131)$$

When $D \leq 1$, φ_d has no solutions; when $D > 1$, φ_d has solutions. At this time the rammer head is rising and its equation is:

$$\sum m_i l_i^2 \ddot{\psi} = m_1 r \omega^2 l \sin(\varphi_d + \alpha_0 + \psi) - G_1 l_1 \cos(\alpha_0 + \psi) + G_2 l_2 \cos(\alpha_0 + \psi - \delta) \quad (5.132)$$

As ψ is smaller than α_0 the approximate calculation can be replaced with the average value, or it can be ignored. The equation can be rewritten as:

$$\psi = \frac{m_1 r l_1}{\sum m_i l_i^2} \left[(\sin \varphi_d - \sin \varphi) + \cos \varphi_d (\varphi - \varphi_d) - \frac{1}{2} g \frac{(\varphi - \varphi_d)^2}{\omega^2} \right] \quad (5.133)$$

When $\psi = 0$ the rammer head returns to its original position and accomplishs the ramming. If $\varphi = \varphi_z$, while the rammer head rise angle is $\theta = \varphi_z - \varphi_d$, the relation between the φ_d and θ can be found:

$$\tan \varphi_d = \frac{\theta - \sin \theta}{-1 + \cos \theta + \frac{1}{2} \theta^2} \quad (5.134)$$

Integrating Eq. (5.129) one gets the moment of momentum when the rammer head begins to fall:

$$J \dot{\psi} = G_1 l_0 \cos \beta \left[\frac{1}{\tan \varphi_d} (1 - \cos \theta) + \sin \theta - \theta \right] \quad (5.135)$$

While the impact force can be calculated as follows:

$$F = \frac{J\dot{\psi}}{\Delta t \cdot l_0 \cos\beta} = \frac{JG_1}{\Delta t}\left[\frac{1}{\tan\psi_d}(1-\cos\theta) + \sin\theta - \theta\right] \quad (5.136)$$

in which Δt is the impact time.

It is obvious that the rise angle of the rammer head should not be larger than 360°, while too small θ could reduce the impact force and lower the working efficiency. The rise angle should be selected in the range of 180°–270° and the corresponding rise leave angle φ_d should be 45°–28°. The rotation number of the machine can be calculated as follows:

$$n = \frac{30}{\pi}\sqrt{\frac{G_1\cos\beta}{m_0 r \sin\varphi_d}} \quad (5.137)$$

According to the second equation, the initial sliding angle for the forward movement φ_k is:

$$\cos\varphi_k = -\frac{1}{m_0 r \omega^2}\left[G_2 \frac{\sin\mu_0}{\cos(\mu_0 - \beta)} + G_1 \sin\beta\right] \quad (5.138)$$

where $\mu_0 = \arctan f_0$.

We can further find the sliding stop angle φ_m (from Fig. 5.24) and the distance of the machine forward movement for each impact.

5.7 Utilization of Frequency-Entrainment Principles

The frequency entrainment principle has been widely used in the self-synchronous vibration machines. In the past in order to keep two machines realizing the synchronous operations a pair of gears with gear ratio 1:1 (reverse rotation) or chain wheel, was used. In 60's of the twentieth century the self-synchronous machines were invented and the gear sets were not needed anymore. According to frequency entrainment principles, the two exciters driven by two electric-induction motors realize the self-synchronous operations and make the machines move in line, circle trajectory as required. The self-synchronous machines can be planer movement and spatial movement, single body or double body, reverse rotation or same rotation direction etc. There are many types and forms used in the industry branches: the self-synchronous vibration feeder; self-synchronous vibration conveyers, self-synchronous vibration screens, self-synchronous driers; self-synchronous vibration sanding machines; self-synchronous line vibration screens; self-synchronous cold mine vibration screen and hot mine vibration screen etc. Their advantages are:

(1) Simple structure: the gear sets are deleted and the transmission device is rather simple;
(2) As the gear sets are deleted, the maintenance and lubrication are simplified;

5.7 Utilization of Frequency-Entrainment Principles

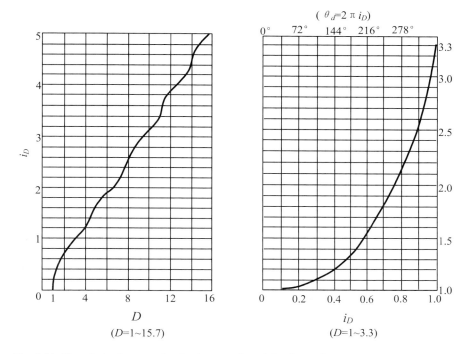

Fig. 5.24 The relation between the rise-leave angle and the rise angle

(3) The vibration amplitudes in starting and stopping through the resonance region can be reduced for some self-synchronous vibration machines.
(4) Most of the self-synchronous vibration machines are directly driven by excitation motors and make them simpler and cost lower and easy to install.
(5) The two main axles of the exciters in the self-synchronous vibration machines can be installed under the condition of large distance.
(6) The self-synchronous vibration machines can be standardized, generalized and seriated.

There are a lot of references on the study of the topic and there are some significant results on it. However the above problems have not been thoroughly studied and the self-synchronous mechanisms have not yet been entirely revealed. The authors will discuss on the problems on the self-synchronous line vibration screen and self-synchronous vibration feeder by the designs and experiment results the authors conducted.

5.7.1 Synchronous Theory of Self-synchronous Vibrating Machine with Eccentric Exciter

The self-synchronous vibration machines with the exciters installed away from the mass center of the body (or called exciter deflection type of self-synchronous vibration machine) have been used widely in industry. The advantage of this type of machines is that the height of machine is low which is good for enhancing the strength and stiffness of the body. However the laws of movement of the machines have not been studied in details yet in the current references on it, such as the lack of practical and feasible formulations on calculating the vibration angles and movement trajectories on body components therefore there is some blindness in design of the machines. It is imperative that the effects of the exciter installation position on the body vibration angle and movement trajectory are theoretically expounded. The authors, based on the theoretical analysis, have proven that the satisfactory results can be achieved by applying the theoretical formulas proposed in this book through the self-synchronous theory lab tests on the self-synchronous line vibration screens, self-synchronous vibration conveyers, self-synchronous probability screen, self-synchronous cold mine screen and self-synchronous theory test table.

(1) The vibration equation and solutions of the self-synchronous vibration machine body

Figure 5.25 shows the schematic of the exciter deflection type of the self-synchronous vibration machine. Figure 5.26 is the installation diagram for the machine. Point o_1' is the body center, o_1''' is the center point in the line segment between the two axis centers of the exciters, i.e. the mass center of the exciters. o_1' is the resultant mass center; while o_{xy} is the fixed coordinates. $o'_{x'y'}$ is the moving coordinates. By using Lagrange Equations to obtain the equation of moments we need to list the kinetic and potential energy of the system:

$$T = \frac{1}{2}m\left\{[\dot{x} - l_0\dot{\psi}\sin(\beta_0 + \pi + \psi + \psi_0)]^2 + [\dot{y} + l_0\dot{\psi}\cos(\beta_0 + \pi + \psi + \psi_0)]^2\right\}$$
$$+ \frac{1}{2}\sum_{i=1}^{2}m_{0i}\left\{[\dot{x} - l_i\dot{\psi}\sin(\beta_i + \psi + \psi_0) - r_i\dot{\varphi}_i\sin\varphi_i]^2\right.$$
$$\left. + [\dot{y} + l_i\dot{\psi}\cos(\beta_i + \psi + \psi_0) + r_i\dot{\varphi}_i\cos\varphi_i]^2\right\} + \frac{1}{2}J_p\dot{\psi}^2 + \frac{1}{2}\sum_{i=1}^{2}J_i\dot{\varphi}_i^2$$

(5.139)

5.7 Utilization of Frequency-Entrainment Principles

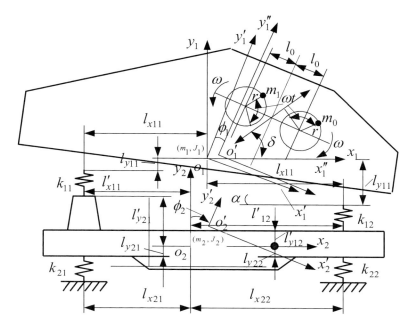

Fig. 5.25 Schematic of the exciter deflection type of self-synchronous vibration machines

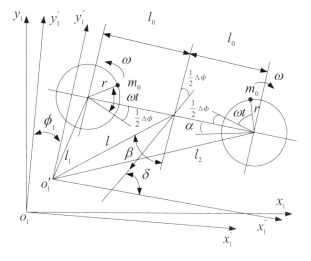

Fig. 5.26 The installation of the exciter deflection type of self-synchronous vibration machines

$$V = \frac{1}{2}\sum_{i=1}^{2}\{k_{xi}[x_0 + x - \rho_i(\psi + \psi_0)\sin(\theta_i + \psi + \psi_0)]^2$$
$$+ k_{yi}[y_0 - y - \rho_i(\psi + \psi_0)\cos(\theta_i + \psi + \psi_0)]^2\} \quad (5.140)$$
$$+ \left(m + \sum_{i=1}^{2} m_{0i}\right)g[-y_0 + y + \rho_i(\psi + \psi_0)\cos(\theta_i + \psi + \psi_0)]$$

When the system contains the damping force directly proportional to elastic deformation speed, the energy dissipation function can be written as:

$$D_0 = \frac{1}{2}\sum_{i=1}^{2}\{f_{xi}[\dot{x} - \rho_i\dot{\psi}\sin(\theta_i + \psi + \psi_0)]^2 + f_{yi}[-\dot{y} - \rho_i\dot{\psi}\cos(\theta_i + \psi + \psi_0)]^2 + f_i\dot{\varphi}_i^2\} \quad (5.141)$$

In the above equations, ψ and ψ_0 are much smaller compared with β_i and θ_i. They can be neglected, then ignoring the gravity potential and spring static deformation energy, Eqs. (5.139)–(5.141) can be simplified as:

$$T = \frac{1}{2}m\left\{[\dot{x} - l_0\dot{\psi}\sin(\beta_0 + \pi)]^2 + [\dot{y} + l_0\dot{\psi}\cos(\beta_0 + \pi)]^2\right\}$$
$$+ \frac{1}{2}\sum_{i=1}^{2}m_{0i}\left\{[\dot{x} - l_i\dot{\psi}\sin\beta_i - r_i\dot{\varphi}_i\sin\varphi_i]^2 + [\dot{y} + l_i\dot{\psi}\cos\beta_i + r_i\dot{\varphi}_i\cos\varphi_i]^2\right\}$$
$$+ \frac{1}{2}J_p\dot{\psi}^2 + \frac{1}{2}\sum_{i=1}^{2}J_i\dot{\varphi}_i^2$$
$$V = \frac{1}{2}\sum_{i=1}^{2}\left[k_{xi}(x - \rho_i\psi\sin\theta_i)^2 + k_{yi}(-y - \rho_i\psi\cos\theta_i)^2\right] \quad (5.142)$$
$$D_0 = \frac{1}{2}\sum_{i=1}^{2}\left\{f_{xi}[\dot{x} - \rho_i\dot{\psi}\sin\theta_i]^2 + f_{yi}[-\dot{y} - \rho_i\dot{\psi}\cos\theta_i]^2 + f_i\dot{\varphi}_i^2\right\}$$

Substituting the kinetic and potential energy and energy dissipation function into Lagrange Equation:

$$\frac{d}{dt}\frac{\partial T}{\partial \dot{q}_i} - \frac{\partial T}{\partial g_i} + \frac{\partial V}{\partial q_i} + \frac{\partial D_0}{\partial \dot{q}_i} = Q_i$$

and taking x, y, ψ, φ_i as the generalized coordinates q_i, we can obtain the five differential equations of moment of the body. After simplified they are:

$$m_0 l_0 \sin(\beta_0 + \pi) + \sum_{i=1}^{2} m_{0i} l_i \sin\beta_i = 0$$

$$m_0 l_0 \cos(\beta_0 + \pi) + \sum_{i=1}^{2} m_{0i} l_i \cos\beta_i = 0$$

5.7 Utilization of Frequency-Entrainment Principles

Canceling the terms with the same value with opposite signs, one can get four differential equations of the movement for the system:

$$\left(m + \sum_{i=1}^{2} m_{0i}\right)\ddot{x} + f_x\dot{x} + f_{x\psi}\dot{\psi} + k_x x + k_{x\psi}\psi = \sum_{i=1}^{2} m_{0i}\left(r_i\dot{\varphi}_i^2 \cos\varphi_i + r_i\ddot{\varphi}_i \sin\varphi_i\right)$$

$$\left(m + \sum_{i=1}^{2} m_{0i}\right)\ddot{y} + f_y\dot{y} + f_{y\psi}\dot{\psi} + k_y y + k_{y\psi}\psi = \sum_{i=1}^{2} m_{0i}\left(r_i\dot{\varphi}_i^2 \sin\varphi_i - r_i\ddot{\varphi}_i \cos\varphi_i\right)$$

$$J\ddot{\psi} + f_\psi\dot{\psi} + f_{\psi x}\dot{x} + f_{\psi y}\dot{y} + k_\psi\psi + k_{\psi x}x + k_{\psi y}y$$
$$= \sum_{i=1}^{2} \{m_{0i}r_i l_i[\dot{\varphi}_i^2 \sin(\varphi_i - \beta_i) - \ddot{\varphi}_i \cos(\varphi_i - \beta_i)] - L_i\}$$

$$J_{0j}\ddot{\varphi}_j + f_j\varphi_j - \sum_{j=1}^{2} m_{0j}\left\{\left[\ddot{x} - l_j\ddot{\psi}\sin\beta_j - l_j\dot{\psi}^2\cos\beta_j\right]l_j\sin\varphi\right.$$
$$\left. - \left[\ddot{y} + l_j\ddot{\psi}\cos\beta_j - l_j\dot{\psi}^2\sin\beta_j\right]r_j\cos\varphi_j\right\} = L_j, \quad i, j = 1, 2$$
(5.143)

where

$$f_x = f_{x1} + f_{x2}$$
$$f_{x\psi} = f_{\psi x} = f_{x1}\rho_1 \sin\theta_1 + f_{x2}\rho_2 \sin\theta_2$$
$$f_y = f_{y1} + f_{y2}$$
$$f_{y\psi} = f_{\psi y} = f_{y1}\rho_1 \cos\theta_1 + f_{y2}\rho_2 \cos\theta_2$$
$$f_\psi = f_{x1}\rho_1^2 \sin^2\theta_1 + f_{y1}\rho_1^2 \cos^2\theta_1 + f_{x2}\rho_2^2 \sin^2\theta_2 + f_{y2}\rho_2^2 \cos^2\theta_2$$
$$k_x = k_{x1} + k_{x2}$$
$$k_{x\psi} = k_{\psi x} = k_{x1}\rho_1 \sin\theta_1 + k_{x2}\rho_2 \sin\theta_2 \quad (5.144)$$
$$k_y = k_{y1} + k_{y2}$$
$$k_{y\psi} = k_{\psi y} = k_{y1}\rho_1 \cos\theta_1 + k_{y2}\rho_2 \cos\theta_2$$
$$J = J_\rho + ml_0^2 + \sum_{i=1}^{2} m_{0i}l_i^2$$
$$J_{0i} = J_i + m_{0i}r_i^2$$

To find the stable solutions for the first three equations in (5.143), the solutions are assumed to have the following forms:

$$x = \sum_{i=1}^{2} (A_{xi} \sin\varphi_i + B_{xi} \cos\varphi_i)$$

$$y = \sum_{i=1}^{2} (A_{yi} \sin\varphi_i + B_{yi} \cos\varphi_i) \quad (5.145)$$

$$\psi = \sum_{i=1}^{2} (A_{\psi i} \sin\varphi_i + B_{\psi i} \cos\varphi_i)$$

where the coefficients $A_{xi}, B_{xi}, A_{yi}, B_{yi}, A_{\psi i}, B_{\psi i}$ could be found by a general analytical method.

Most of the self-synchronous vibration machines work in a frequency domain away from the resonance frequency, which can simplify the problem and at the same time meet the engineering accuracy. Now we consider only the system in which $f_{x\psi} = f_{\psi x} = 0$, $f_{y\psi} = f_{\psi y} = 0$, $k_{x\psi} = k_{\psi x} = 0$, $k_{y\psi} = k_{\psi y} = 0$. The first three equations in Eq. (5.139) are independent. We consider the reverse rotation of the two exciter motors:

$$\dot{\varphi}_1 = -\dot{\varphi}_2 = \dot{\varphi}, \quad \varphi_1 = \varphi - \left(\nu + \frac{1}{2}\Delta\varphi\right),$$

$$\varphi_2 = 180° - \left(\nu + \varphi + \frac{1}{2}\Delta\varphi\right)$$

Taking $m_{01} = m_{02} = m_0$, $r_1 = r_2 = r$, Eq. (5.143) can be rewritten as:

$$M\ddot{x} + f_x\dot{x} + k_x x = 2m_0\dot{\varphi}^2 r \sin\left(\nu + \frac{1}{2}\Delta\varphi\right) \sin\varphi$$
$$M\ddot{y} + f_y\dot{y} + k_y y = 2m_0\dot{\varphi}^2 r \cos\left(\nu + \frac{1}{2}\Delta\varphi\right) \sin\varphi \quad (5.146)$$

$$J\ddot{\psi} + f_\psi\dot{\psi} + k_\psi\psi = 2m_0\dot{\varphi}^2 r \left[l' \sin\left(\gamma - \tfrac{1}{2}\Delta\varphi\right) \sin\varphi - l_a \sin\tfrac{1}{2}\Delta\varphi \cos\varphi\right]$$

The steady solutions for the first three equations are:

$$x = -\frac{2m_0 r \sin\left(\nu + \frac{1}{2}\Delta\varphi\right)}{m'_x} \cos\alpha_x \sin(\varphi - \alpha_x)$$

$$y = -\frac{2m_0 r \cos\left(\nu + \frac{1}{2}\Delta\varphi\right)}{m'_y} \cos\alpha_y \sin(\varphi - \alpha_y)$$

$$\psi = -\frac{2m_0 r}{J'_\psi} \cos\alpha_\psi \left[l \sin\left(\gamma - \frac{1}{2}\Delta\varphi\right) \sin(\varphi - \alpha_{\psi 1}) - l_a \sin\frac{1}{2}\Delta\varphi \cos(\varphi - \alpha_{\psi 2})\right]$$

(5.147)

where $m'_x = M - \frac{k_x}{\dot{\varphi}^2}$, $m'_y = M - \frac{k_y}{\dot{\varphi}^2}$, $J'_\psi = J - \frac{k_\psi}{\dot{\varphi}^2}$.

Find the synchronization conditions using Hamilton's principle.

Using Eq. (5.147) to find the velocities x, y and the angular velocity ψ and substituting them into Eq. (5.146) one obtains the expressions for kinetic and potential energy:

5.7 Utilization of Frequency-Entrainment Principles

$$T = \frac{1}{2}m\left\{\frac{-2m_0\dot{\varphi}r\sin\left(\nu + \frac{1}{2}\Delta\varphi\right)}{m'_x}\cos\alpha_x\cos(\varphi-\alpha_x) + l_0\sin(\beta+\pi)\frac{2m_0\dot{\varphi}r}{J'_\psi}\cos\alpha_\psi\right.$$

$$\left. \times \left[l\sin\left(\gamma-\frac{1}{2}\Delta\varphi\right)\cos(\varphi-\alpha_\psi) + l_a\sin\frac{1}{2}\Delta\varphi\sin(\varphi-\alpha_\psi)\right]\right\}^2$$

$$+ \frac{1}{2}m\left\{\frac{-2m_0\dot{\varphi}r\sin\left(\nu + \frac{1}{2}\Delta\varphi\right)}{m'_y}\cos\alpha_y\cos(\varphi-\alpha_y) - l_0\cos(\beta+\pi)\frac{2m_0\dot{\varphi}r}{J'_\psi}\cos\alpha_\psi\right.$$

$$\left. \times\left[l\sin\left(\gamma-\frac{1}{2}\Delta\varphi\right)\cos(\varphi-\alpha_\psi) + l_a\sin\frac{1}{2}\Delta\varphi\sin(\varphi-\alpha_\psi)\right]\right\}^2$$

$$+ \frac{1}{2}\sum_{i=1}^{2}m_{0i}\left\{\frac{-2m_0\dot{\varphi}r\sin\left(\nu+\frac{1}{2}\Delta\varphi\right)}{m'_x}\cos\alpha_x\cos(\varphi-\alpha_x) + l_i\sin\beta_i\frac{2m_0\dot{\varphi}r}{J'_\psi}\cos\alpha_\psi\right. \quad (5.148)$$

$$\left. \times\left[l\sin\left(\gamma-\frac{1}{2}\Delta\varphi\right)\cos(\varphi-\alpha_\psi) + l_a\sin\frac{1}{2}\Delta\varphi\sin(\varphi-\alpha_\psi)\right] - r_i\dot{\varphi}_i\sin\varphi_i\right\}^2$$

$$+ \frac{1}{2}\sum_{i=1}^{2}m_{0i}\left\{\frac{-2m_0\dot{\varphi}r\sin\left(\nu+\frac{1}{2}\Delta\varphi\right)}{m'_y}\cos\alpha_y\cos(\varphi-\alpha_y) + l_i\sin\beta_i\frac{2m_0\dot{\varphi}r}{J'_\psi}\cos\alpha_\psi\right.$$

$$\left. \times\left[l\sin\left(\gamma-\frac{1}{2}\Delta\varphi\right)\cos(\varphi-\alpha_\psi) + l_a\sin\frac{1}{2}\Delta\varphi\sin(\varphi-\alpha_\psi)\right] - r_i\dot{\varphi}_i\sin\varphi_i\right\}^2$$

$$+ \frac{1}{2}J_p\left\{\frac{-2m_0\dot{\varphi}r}{J'_\psi}\cos\alpha_\psi\left[l\sin\left(\gamma-\frac{1}{2}\Delta\varphi\right)\cos(\varphi-\alpha_\psi) + l_a\sin\frac{1}{2}\Delta\varphi\sin(\varphi-\alpha_\psi)\right]\right\}^2$$

$$+ \frac{1}{2}\sum_{i=1}^{2}J_i\dot{\varphi}_i^2$$

$$V = \frac{1}{2}\sum_{i=1}^{2}k_{xi}\left\{\left[\frac{-2m_0r}{m'_x}\sin\left(\nu+k_{xi}\left\{\left[\frac{-2m_0r}{m'_x}\sin\left(\nu+\frac{1}{2}\Delta\varphi\right)\cos\alpha_x\sin(\varphi-\alpha_x)\right)\right.\right.\right.$$

$$\times\cos\alpha_x\sin(\varphi-\alpha_x)$$

$$+ \rho_i\sin\theta_i\frac{2m_0r}{J'_\psi}\cos\alpha_\psi\left[l\sin\left(\gamma-\frac{1}{2}\Delta\varphi\right)\sin(\varphi-\alpha_x) - l_a\sin\frac{1}{2}\Delta\varphi\cos(\varphi-\alpha_x)\right]\right\}^2 \quad (5.149)$$

$$+ \frac{1}{2}\sum_{i=1}^{2}k_{yi}\left\{\frac{2m_0r}{m'_y}\cos\left(\nu+\frac{1}{2}\Delta\varphi\right)\cos\alpha_y\sin(\varphi-\alpha_y)\right.$$

$$+ \rho_i\cos\theta_i\frac{2m_0r}{J'_\psi}\cos\alpha_\psi\left[l\sin\left(\gamma-\frac{1}{2}\Delta\varphi\right)\sin(\varphi-\alpha_y) - l_a\sin\frac{1}{2}\Delta\varphi\cos(\varphi-\alpha_y)\right]\right\}^2$$

The Lagrange Function is:

$$L = T - V \quad (5.150)$$

In one vibration period, the Hamilton Action is

$$I = \int_0^{T_1} L\,\mathrm{d}t = \int_0^{2\pi} L\,\mathrm{d}\varphi \quad (5.151)$$

where T_1 is the vibration period.

Substituting Eq. (5.148) into Eq. (5.149) one gets:

$$I \approx 2\pi \left\{ \left(m + \sum_{i=1}^{2} m_{0i} \right) \times \right.$$

$$\left[\frac{m_0^2 \dot{\varphi}^2 r^2 \sin^2\left(\nu + \frac{1}{2}\Delta\varphi\right) \cos^2\alpha_x}{m_x'^2} + \frac{m_0^2 \dot{\varphi}^2 r^2 \cos^2\left(\nu + \frac{1}{2}\Delta\varphi\right) \cos^2\alpha_y}{m_y'^2} \right]$$

$$\left. + J \frac{m_0^2 \dot{\varphi}^2 r^2 \cos^2\alpha_\psi}{J_\psi'^2} \left[l^2 \sin^2\left(\gamma - \frac{1}{2}\Delta\varphi\right) + l_\alpha^2 \sin^2 \frac{1}{2}\Delta\varphi \right] \right\}$$

$$-2\pi m_0^2 r^2 \left\{ \frac{k_x}{m_x'^2} \sin^2\left(\nu + \frac{1}{2}\Delta\varphi\right) \cos^2\alpha_x + \frac{k_y}{m_y'^2} \cos^2\left(\nu + \frac{1}{2}\Delta\varphi\right) \cos^2\alpha_y \right.$$

$$\left. + \frac{k_\psi}{J_\psi'^2} \cos^2\alpha_\psi \left[l^2 \sin^2\left(\gamma - \frac{1}{2}\Delta\varphi\right) + l_a^2 \sin^2 \frac{1}{2}\Delta\varphi \right] \right.$$

$$= 2\pi m_C^2 \dot{\varphi}^2 r^2 \left\{ \frac{\sin^2\left(\nu + \frac{1}{2}\Delta\varphi\right) \cos^2\alpha_x}{m_x'^2} + \frac{\cos^2\left(\nu + \frac{1}{2}\Delta\varphi\right) \cos^2\alpha_y}{m_y'^2} \right.$$

$$\left. + \frac{\cos^2\alpha_\psi}{J_\psi'^2} \left[l^2 \sin^2\left(\gamma - \frac{1}{2}\Delta\varphi\right) + l_\alpha^2 \sin^2 \frac{1}{2}\Delta\varphi \right] \right\}$$

(5.152)

For the vibration systems of the self-synchronous vibration machines, except the above mentioned potential forces, there are non-potential actions. For example, the outputs of the friction moments M_{g1} and M_{g2} from the two exciters and M_{f1} and M_{f2} from the bearings are subject to the rotation shafts.

According to Hamilton's principle, when the sum of the variation on the definite integral Eq. (5.152) and the integration over the virtual work by the non-potential force in a vibration period is equal to zero the system total action renders a minimum:

$$\pi m_0^2 \dot{\varphi}^2 r^2 \left[\left(\frac{\cos^2\alpha_x}{m_x'} - \frac{\cos^2\alpha_y}{m_y'} \right) \sin(2\nu + \Delta\varphi) + \frac{l_a^2 \cos^2\alpha_\psi}{J_\psi'} \sin\Delta\varphi \right.$$

$$\left. - \frac{l^2 \cos^2\alpha_\psi}{J_\psi'} \sin(2\gamma - \Delta\varphi) - 2\pi \left[(M_{g1} - M_{g2}) - (M_{f1} - M_{f2}) \right] = 0 \right.$$

$$\frac{1}{2} m_0^2 \dot{\varphi}^2 r^2 \left\{ \left[\left(\frac{\cos^2\alpha_x}{m_x'} - \frac{\cos^2\alpha_y}{m_y'} \right) \cos 2\nu + \frac{l_a^2 \cos^2\alpha_\psi}{J_\psi'} + \frac{l^2 \cos^2\alpha_\psi}{J_\psi'} \cos 2\nu \right]^2 \right.$$

$$\left. + \left[\left(\frac{\cos^2\alpha_x}{m_x'} - \frac{\cos^2\alpha_y}{m_y'} \right) \sin 2\nu - \frac{l^2 \cos^2\alpha_\psi}{J_\psi'} \sin 2\nu \right]^2 \right\}^{1/2} \sin(\Delta\varphi + \varepsilon)$$

$$- \left[(M_{g1} - M_{g2}) - (M_{f1} - M_{f2}) \right] = 0$$

(5.153)

5.7 Utilization of Frequency-Entrainment Principles

where

$$\varepsilon = \arctan \frac{\left(\frac{\cos^2 \alpha_x}{m'_x} - \frac{\cos^2 \alpha_y}{m'_y}\right) \sin 2\nu - \frac{l_a^2}{J'_\psi} \cos^2 \alpha_\psi \sin 2\gamma}{\left(\frac{\cos^2 \alpha_x}{m'_x} - \frac{\cos^2 \alpha_y}{m'_y}\right) \cos 2\nu + \frac{l_a^2}{J'_\psi} \cos^2 \alpha_\psi + \frac{l^2}{J'_\psi} \cos^2 \alpha_\psi \cos 2\gamma} \quad (5.154)$$

Equation (5.152) can be rewritten as:

$$\sin(\Delta\varphi + \varepsilon) = \frac{1}{D} \text{ or } \Delta\varphi = \left(\arcsin \frac{1}{D}\right) - \varepsilon \quad (5.155)$$

in which D is the synchronization index and its value is:

$$D = \frac{m_0^2 \dot{\varphi}^2 r^2 W}{2\left[(M_{g1} - M_{g2}) - (M_{f1} - M_{f2})\right]} \quad (5.156)$$

in which W is the stability index and its value is:

$$W = \left\{ \left[\left(\frac{\cos^2 \alpha_x}{m'_x} - \frac{\cos^2 \alpha_y}{m'_y}\right) \cos 2\nu + \frac{l_a^2 \cos^2 \alpha_\psi}{J'_\psi} + \frac{l^2 \cos^2 \alpha_\psi}{J'_\psi} \cos 2\nu \right]^2 \right.$$
$$\left. + \left[\left(\frac{\cos^2 \alpha_x}{m'_x} - \frac{\cos^2 \alpha_y}{m'_y}\right) \sin 2\nu - \frac{l^2 \cos^2 \alpha_\psi}{J'_\psi} \sin 2\nu \right]^2 \right\}^{1/2}$$

(5.157)

It can be seen from Eq. (5.155) that the necessary condition realizing the synchronization operation is the absolute values of the synchronization index D is larger or equal to 1:

$$|D| \geq 1 \quad (5.158)$$

If $|D| < 1$, then $\Delta\varphi$ in Eq. (5.155) has no solution, and the balance of the equation can be met, i.e. the synchronization operation can not be realized., that is to say that "self-synchronous" vibration machines can operate only in the non-synchronization conditions. If one excitation motor stops operating, i.e. $M_{g2} = 0$ (Taking $M_{g2} = 0$ in Eq. (5.156)), replacing D with D_1, if $|D_1| \geq 1$, then when one excitation motor's power supply is cut off, the excitation motor without power would acquire a certain energy to overcome the dissipated energy from the friction moment due to the system vibrations. That is the so-called "vibration synchronous transmission" working state.

(2) Stability condition for the synchronous operation state

When D and D_1 are determined, and when $|D| \geq 1$ and $|D_1| > 1$ we can find out the corresponding $\Delta\varphi$. $\Delta\varphi$ has two D values (one is to D, another to D_1). For example, when $D = 2$, $\Delta\varphi + \varepsilon = 30°$ or $150°$. When $D \approx \infty$, $\Delta\varphi + \varepsilon = 0°$ or $180°$. According to the synchronization conditions, two corresponding synchronization states can be determined. For their stability analysis, when the second order derivatives of the definite integral with respect to the $\Delta\varphi$ is larger than zero, the stability conditions are found to be:

$$W \cos(\Delta\varphi + \varepsilon) > 0 \tag{5.159}$$

It can be seen from the equation that when $\Delta\varphi + \varepsilon = -90° - 90°$ the synchronization is stable while $90° < \Delta\varphi + \varepsilon < 270°$ the synchronization is not stable (as W is positive). Therefore the stable value of phase angle difference $\Delta\varphi$ has something to do with ε and the stable value for $\Delta\varphi$ is:

$$\Delta\varphi = -90° - \varepsilon \sim 90° - \varepsilon \tag{5.160}$$

ε can be calculated from Eq. (5.154).

When the central line between the two centers of the exciters for the self-synchronous vibration machines is parallel to the x axis, i.e. the angle ν is equal to zero, ε in Eq. (5.154) can be calculated by the following formulas:

$$\varepsilon = \arctan \frac{-\frac{l^2 \cos^2 \alpha_\psi}{J'_\psi} \sin 2\gamma}{\frac{\cos^2 \alpha_x}{m'_x} - \frac{\cos^2 \alpha_y}{m'_y} + \frac{l_a^2 + l^2 \cos 2\gamma}{J'_\psi} \cos^2 \alpha_\psi} \tag{5.161}$$

While the synchronization index is:

$$D = \frac{1}{2} \frac{m_0^2 \dot{\varphi}^2 r^2}{(M_{g1} - M_{g2}) - (M_{f1} - M_{f2})} \left[\left(\frac{\cos^2 \alpha_x}{m'_x} - \frac{\cos^2 \alpha_y}{m'_y} + \frac{l_a^2 + l^2 \cos 2\gamma}{J'_\psi} \cos^2 \alpha_\psi \right)^2 + \left(\frac{l^2 \sin 2\gamma}{J'_\psi} \cos^2 \alpha_\psi \right)^2 \right]^{1/2} \tag{5.162}$$

The stability index is:

$$W = \left[\left(\frac{\cos^2 \alpha_x}{m'_x} - \frac{\cos^2 \alpha_y}{m'_y} + \frac{l_a^2 + l^2 \cos 2\gamma}{J'_\psi} \cos^2 \alpha_\psi \right)^2 + \left(\frac{l^2 \sin 2\gamma}{J'_\psi} \cos^2 \alpha_\psi \right)^2 \right]^{1/2} \tag{5.163}$$

From Eq. (5.155) when $|D| \gg 1$ the lead phase angle of the exciter 1 over the exciter 2 is:

$$\Delta\varphi = -\varepsilon = \arctan \frac{\frac{l^2 \cos^2 \alpha_\psi}{J'_\psi} \sin 2\gamma}{\frac{\cos^2 \alpha_x}{m'_x} - \frac{\cos^2 \alpha_y}{m'_y} + \frac{l_a^2 + l^2 \cos 2\gamma}{J'_\psi} \cos^2 \alpha_\psi} \tag{5.164}$$

5.7 Utilization of Frequency-Entrainment Principles

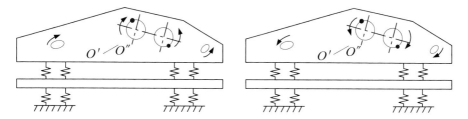

Fig. 5.27 Relation between the exciter rotation direction and rotation direction of the trajectory eclipses on the two ends of the body

(3) Vibration angle β of the body mass center and movement trajectories of the body componente

We know from Fig. 5.27 that when $|D| \gg 1$, the vibration angle of the body mass center is:

$$\beta = 90° - \nu - \frac{1}{2}\Delta\varphi = 90° - \nu - \varepsilon \tag{5.165}$$

For the working situation of Eq. (5.165) and as $\Delta\varphi = -\varepsilon$ we have:

$$\beta = 90° - \nu + \frac{1}{2}\arctan \frac{\left(\frac{\cos^2 \alpha_x}{m'_x} - \frac{\cos^2 \alpha_y}{m'_y}\right) \sin 2\nu - \frac{l_a^2}{J'_y} \cos^2 \alpha_\psi \sin 2\gamma}{\left(\frac{\cos^2 \alpha_x}{m'_x} - \frac{\cos^2 \alpha_y}{m'_y}\right) \cos 2\nu + \frac{l_a^2}{J'_\psi} \cos^2 \alpha_\psi \frac{l_a^2}{J'_y} \cos^2 \alpha_\psi \cos 2\gamma} \tag{5.166}$$

Now we will study some special cases:

i. When the machine works far beyond the resonance region, and $l^2 \gg l_a^2$. In this case, $m'_x \approx m'_y \approx m$, $\cos^2 \alpha_x \approx \cos^2 \alpha_y \approx \cos^2 \alpha_\psi \approx 1$, l_a^2 can be neglected, $J_\psi^2 \approx J^2$. Then the vibration angle is

$$\beta = 90° - \nu - \gamma \approx \beta_0 \tag{5.167}$$

As shown in Fig. 5.27 the vibration direction line of the body is line $O'O'''$. The movement trajectory at an arbitrary point on the body is:

$$\begin{aligned} x_e &= x - l_{ey}\psi \\ y_e &= y + l_{ey}\psi \end{aligned} \tag{5.168}$$

The movement trajectory of the body mass center is a straight line while the long eclipse at the ends of the body. The rotation direction of the eclipse and the value of ψ have something to do with the rotation direction of the exciters. Figure 5.27 shows the relationship.

ii. When the machine works far beyond the resonance region, and $l_a^2 \gg l^2$. From Eq. (5.166) one can find that $\frac{1}{2}\Delta\varphi = -\frac{1}{2}\varepsilon \approx 0$, the vibration angle of the body is:

$$\beta = 90° - \nu \qquad (5.169)$$

The vibration of the body mass center is perpendicular to the connecting line between the centers of the two exciters. According to Eq. (5.167), if the center of the connecting line is away from the body mass center, the swing vibration will occur. In this case the movement trajectory of the ends of the body is an oval.

iii. The machine works beyond the resonance region, and $l_a^2 \approx l^2$. The vibration direction of the body is an inclined line between the vibration directions as mentioned above.

The vibration angle of the mass center of the body can be calculated from Eq. (5.167) and the trajectory is a long oval approximating line.

Example 5.1 Given a self-synchronous vibration machine, $l_a^2 = 255$ mm, $l^2 = 270$ mm, $\gamma = 16°$, $\gamma = 35°$. Find the vibration angle of the mass center of the body.

Solution:

$$\begin{aligned}
\beta &= 90° - \gamma - \frac{1}{2}\arctan\left(\frac{l^2 \sin 2\gamma}{l_a^2 + l^2 \cos 2\gamma}\right) \\
&= 90° - 35° - \frac{1}{2}\arctan\left[\frac{0.27^2 \sin(6 \times 16°)}{0.255^2 + 0.27^2 \cos(2 \times 16°)}\right] \\
&= 45°45'
\end{aligned}$$

(4) Experiment results and their brief analysis

From the aforementioned theoretical analysis, the authors designed and manufactured multi exciter deflection type of self-synchronous vibration machines and also conducted dynamics experiments on the machines, measured the dynamic parameters and analyzed the dynamic characteristics.

i. The synchronous tests. The simplest method to measure if the two eccentric rotors of the two exciters are operating synchronously is the direct observation of the movement trajectory of the vibration body. In the synchronous operation state, the movement trajectories on different locations are shown in Fig. 5.27: straight line for mass center and ovals for the body ends. Other methods for measuring the synchronization are light-electronics transducer method and high-speed photographing method. A white paper strip is attached to each eccentric on rotors. When two rotors rotate in a period, the light-electronics transducer detects the strip and a pulse signal is displayed in an oscilloscope. The high speed camera directly photographs the phase to judge if the synchronization is realized or not. The two eccentric rotors on the large cold mineral screens never realize the synchronous rotations. However for other synchronous vibration machines, sometimes non-synchronization may occur. In this case the vibration curves recorded are the vibration curve with quasi-periods.

ii. The stability test of the synchronous operation. As mentioned before that there are two phase angle differences corresponding to the synchronous operation and one is stable and another one is not. For a synchronous vibration machine with the two rotors rotate in opposite directions, $\Delta\varphi = -90° - \varepsilon \sim 90° - \varepsilon$ is corresponding to the stable state. When $|D| \gg 1, \Delta\varphi = -\varepsilon$. For a large cold mineral screen, theoretical phase angle difference $\Delta\varphi = 45°, \nu = 20°$, the vibration angle $\beta = 90° - \nu - \Delta\varphi/2 = 90° - 20° - 45° = 25°$. The measured phase angle between light-electronic transducer and high-speed camera is $80°$ which reflects some error with theoretical results. However the error is still within the range of Eq. (5.166). Therefore it is proven that the exciter deflection type of self-synchronous vibration machines cold mine screen is working in the stable synchronous state as proposed in this chapter. In near two years operation no unstable synchronous operation in $90° - \varepsilon < \Delta\varphi < 270° - \varepsilon$ ever occurred.

iii. Measurements of the movement trajectory, vibration amplitudes, vibration angle and vibration curves of the vibration body. The movement trajectories on the different locations on the large cold mine synchronous screen are very similar to what shown in Fig. 5.27, i.e. the trajectory near the mass center of the body is a straight line and a long oval on the ends of the body. The vibration amplitudes and angles on different locations of the body are shown in Table 5.1.

When the machines start and then transit to normal working state, the vibration curve on some point of the body is shown in Fig. 5.27. It can be seen from the Figure that the body vibrates close as a harmonic vibration. Therefore the tow eccentric rotors rotate in stable synchronous operations.

We reached some conclusions based on the analysis and test results we conducted:

(1) Derived the synchronization condition and synchronization stability conditions for the exciter deflection type of self-synchronous vibration machines.
(2) When the exciters are installed away from the mass center of the body, the vibration of the body mass center could vibrate in a certain direction as long as the parameters l_a, l, γ and ν are chosen properly.
(3) The vibration direction line of the body mass center could be:
   i. when $l \gg l_a$ the vibration direction is parallel to line $O''O'''$;
   ii. when $l_a \gg l$ the vibration direction is perpendicular to the connecting line between the two center of the exciter axles;

Table 5.1 Vibration amplitudes and angles on the body

Along direction of feed materiel	Parameter	Feed materiel end	Middle end	Discharged material end
Left side	Amplitude/mm	4.75	3.95	3.6
	Angle of vibration/(°)	25	15	4
Right side	Amplitude/mm	4.7	3.95	3.55
	Angle of vibration/(°)	25	15	4

iii. when $l \approx l_a$ the vibration angle can be calculated by Eq. (5.166).

(4) The movement trajectory on the two ends of the body has something to do with the rotation direction of the main axle and its pattern is shown in Fig. 5.27.

5.7.2 Double Frequency Synchronization of Nonlinear Self-synchronous Vibration Machines

The self-synchronous vibration machines can realize the synchronization in basic frequency and double frequency as well. We will discuss the synchronization theory of the nonlinear self-synchronous vibration machine—discuss the basic and double frequency synchronization of the self-synchronous vibration machine under excitations of two or more than two excitation motors, i.e. to discuss the frequency entrainment of basic harmonics, the higher order harmonics and sub-harmonics.

(1) Equations of motions and their solutions

Figure 5.28 shows the working mechanism of a nonlinear self-synchronous vibration machines with 2- degree of freedom, and two excitation motors or two exciters (the two eccentric rotors in each exciter are synchronized by a gear-synchronizer). The equations of motions for body 1 and body 2 are:

$$m_{11}\ddot{y}_1 + f_{11}\dot{y}_1 + f_{12}\dot{y}_2 + k_{11}y_1 + k_{12}y_2$$
$$= \sum m_{0g}\dot{\varphi}_j^2 r_j \sin \varphi_j + \varepsilon Q_1(y_1, \dot{y}_1, y_2, \dot{y}_2, \varphi_1, \dot{\varphi}_1, \varphi_2, \dot{\varphi}_2) \quad (5.170)$$

$$m_{22}\ddot{y}_2 + f_{21}\dot{y}_1 + f_{22}\dot{y}_2 + k_{21}y_1 + k_{22}y_2 = \varepsilon Q_2(y_1, \dot{y}_1, y_2, \dot{y}_2, \varphi_1, \dot{\varphi}_1, \varphi_2, \dot{\varphi}_2) \quad (5.171)$$

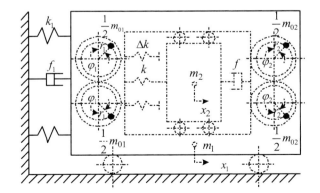

Fig. 5.28 Working principle diagram for the 2- degree of freedom self-synchronous vibration machine

5.7 Utilization of Frequency-Entrainment Principles

where

$$m_{11} = m_1, \; m_{22} = m_2, \; f_{11} = f + f_1, \; f_{12} = -f$$
$$f_{21} = -f, \; f_{22} = f, \; k_{11} = k + k_1, \; k_{12} = -k$$
$$k_{21} = -k, \; k_{22} = k \tag{5.172}$$
$$\varepsilon Q_1 = F_1(y_1, \dot{y}_1, y_2, \dot{y}_2, \varphi_1, \dot{\varphi}_1, \varphi_2, \dot{\varphi}_2) + \sum m_{j0}\ddot{\varphi}_j r_j \cos \varphi_j$$
$$\varepsilon Q_2 = -F_1(y_1, \dot{y}_1, y_2, \dot{y}_2, \varphi_1, \dot{\varphi}_1, \varphi_2, \dot{\varphi}_2)$$

where m_1 and m_2 are the masses of body 1 and body 2; f_1 and f are the damping coefficients of body 1 and relative movement between two bodies; k_1 and k the spring stiffnesses of body 1 and between the two bodies; $y_1, y_2, \dot{y}_1, \dot{y}_2, \ddot{y}_1, \ddot{y}_2$ are the displacements, velocities and accelerations of body 1 and body 2 respectively; Δk is the gap spring stiffness; εQ_1 and εQ_2 are the nonlinear forces acting upon body 1 and body 2.

The rotation equations of the two shafts are:

$$\varepsilon J_j \ddot{\varphi}_j + (L_j + c_j)\dot{\varphi}_j = c_j \dot{\varphi}_{sj} - M_j \dot{\varphi}_j^2 + m_{0j} r_j (\ddot{y} + \varepsilon g) \cos \varphi_j \quad j = 1,2 \tag{5.173}$$

When the system is weak nonlinear, the solution can be found by approximate methods. Now find solution of Eq. (5.170). Transform the coordinates into principal coordinates. From Eq. (5.170) it is not difficult to determine the normal mode functions $\gamma_1^{(1)}, \gamma_2^{(1)}, \gamma_1^{(2)}, \gamma_2^{(2)}$ and the corresponding natural frequencies ω_1 and ω_2. The vibration equations in the principal coordinates are:

$$m^{(1)}\ddot{\varphi}^{(1)} + f^{(1)}\dot{\varphi}^{(1)} + k^{(1)}\varphi^{(1)} = \sum P_j^{(1)} \sin \varphi_j + \varepsilon(Q_1 \gamma_1^{(1)} + Q_2 \gamma_2^{(1)})$$
$$m^{(2)}\ddot{\varphi}^{(2)} + f^{(2)}\dot{\varphi}^{(2)} + k^{(2)}\varphi^{(2)} = \sum P_j^{(2)} \sin \varphi_j + \varepsilon(Q_1 \gamma_1^{(2)} + Q_2 \gamma_2^{(2)}) \tag{5.174}$$

where

$$m^{(1)} = m_{11}\gamma_1^{(1)2} + m_{22}\gamma_2^{(1)2}$$
$$m^{(2)} = m_{11}\gamma_1^{(2)2} + m_{22}\gamma_2^{(2)2}$$
$$f^{(1)} = f_{11}\gamma_1^{(1)2} + f_{22}\gamma_2^{(1)2} + 2f_{12}\gamma_1^{(1)}\gamma_2^{(1)}$$
$$f^{(2)} = f_{11}\gamma_1^{(2)2} + f_{22}\gamma_2^{(2)2} + 2f_{12}\gamma_1^{(2)}\gamma_2^{(2)} \tag{5.175}$$
$$k^{(1)} = k_{11}\gamma_1^{(1)2} + k_{22}\gamma_2^{(1)2} + 2k_{12}\gamma_1^{(1)}\gamma_2^{(1)}$$
$$k^{(2)} = k_{11}\gamma_1^{(2)2} + k_{22}\gamma_2^{(2)2} + 2k_{12}\gamma_1^{(2)}\gamma_2^{(2)}$$

The vibration equations in the principal coordinates are:

$$\ddot{\varphi}_N^{(1)} + 2n^{(1)}\dot{\varphi}_N^{(1)} + \omega^{(1)2}\varphi_N^{(1)} = \frac{1}{m^{(1)}}\left(\sum p_j^{(1)} \sin\varphi_j + \varepsilon(Q_1\gamma_1^{(1)} + Q_2 r_2^{(1)})\right)$$
$$\ddot{\varphi}_N^{(2)} + 2n^{(2)}\dot{\varphi}_N^{(2)} + \omega^{(2)2}\varphi_N^{(2)} = \frac{1}{m^{(2)}}\left(\sum p_j^{(2)} \sin\varphi_j + \varepsilon(Q_1\gamma_1^{(2)} + Q_2 r_2^{(2)})\right)$$
(5.176)

As two equations are independent, it is easy to find out the approximate solutions for $\varphi_N^{(1)}$ and $\varphi_N^{(2)}$. Then transform them back to the original coordinates:

$$y_2 = \frac{\gamma_2^{(1)}}{\sqrt{m^{(1)}}}\varphi_N^{(1)} + \frac{\gamma_2^{(2)}}{\sqrt{m^{(2)}}}\varphi_N^{(2)} \tag{5.177}$$

It can be seen that method of finding the solutions for multi-degree of freedom vibration systems is based on the method of finding solutions for single degree of freedom systems. The difference is that to transform the original coordinates into the principal coordinates, to find the solutions in the principal coordinates and then to transform the solutions back to the original coordinates.

(2) The necessary conditions for double-frequency synchronizations

The so-called frequency entrainment is when two or multi excitation motors are operating, through the vibration of the synchronous vibration machines the systems acquire one frequency ratio in a certain frequency region (for example: $\varphi_{21} \pm \Delta\varphi_{21}$ and $p\varphi_{11}/q$ or φ_{21} and $(\varphi_{11} \pm \Delta\varphi_{11})p/q$):

$$\dot{\varphi}_{21} = \frac{p}{q}\dot{\varphi}_{11}q \tag{5.178}$$

in which p and q are integers and they are a pair of mutual prime numbers.

This phenomenon is called frequency entrainment or synchronization.

When $\dot{\varphi}_{21} = \dot{\varphi}_{11}$ it is called base frequency synchronization; when $\dot{\varphi}_{21} = p\dot{\varphi}_{11}$ p is called p-fold frequency synchronization; when $\dot{\varphi}_{21} = \dot{\varphi}_{11}/q$ it is called $1/q$ decreasing frequency synchronization; when $\dot{\varphi}_{21} = \dot{\varphi}_{11}p/q$ it is called fraction frequency synchronization. The base frequency entrainment is also called basic harmonic frequency entrainment; when $\dot{\varphi}_{21} > \dot{\varphi}_{11}$, the synchronization is also called high order harmonic frequency entrainment; when $\dot{\varphi}_{21} < \dot{\varphi}_{11}$, the synchronization is called sub-harmonic frequency entrainment.

Whether the frequency entrainment is realized or not is dependent upon that some conditions are satisfied. First let's discuss the synchronization conditions, i.e. analyze the rotation equations of the self-synchronous vibration machines driven by two motors. In the first order approximations, from Eq. (5.173) we have:

$$\varepsilon J_1 \ddot{\varphi}_1 + (L_1 + C_1)\dot{\varphi}_1 = C_1\dot{\varphi}_{s1} - M_1\dot{\varphi}_1^2 + m_{01}r_1(\ddot{y} + \varepsilon g)\cos\varphi_1$$
$$\varepsilon J_2 \ddot{\varphi}_2 + (L_2 + C_2)\dot{\varphi}_2 = C_2\dot{\varphi}_{s2} - M_2\dot{\varphi}_2^2 + m_{02}r_2(\ddot{y} + \varepsilon g)\cos\varphi_2$$
(5.179)

where

5.7 Utilization of Frequency-Entrainment Principles

$$y_1 = a\cos\psi + \sum C_{j1}\sin(\varphi_j - \alpha_j) + \varepsilon u_1(a, \psi, \theta_1, \theta_2) \tag{5.180}$$

The stable acceleration solutions can be found from Eq. (5.180):

$$\ddot{y}_1 = -a\dot{\psi}^2 \cos\psi - \sum C_{j1}\dot{\varphi}_{j1}^2 \sin(\varphi_{j1} - \alpha_{j1}) + \varepsilon \ddot{u}_1(a, \psi, \theta_1, \theta_2) \tag{5.181}$$

Substituting it into Eq. (5.179) one has:

$$\begin{aligned}
&J_1\ddot{\varphi}_1 + (L_1 + C_1)\dot{\varphi}_1 = C_1\dot{\varphi}_{s1} - M_1\dot{\varphi}_1^2 + m_{01}r_1[-a\dot{\psi}^2 \cos\psi \\
&- \sum C_{j1}\dot{\varphi}_{j1}^2 \sin(\varphi_{j1} - \alpha_{j1}) + \varepsilon(\ddot{u}_1 + g)]\cos\varphi_1 \\
&J_2\ddot{\varphi}_2 + (L_2 + C_2)\dot{\varphi}_2 = C_2\dot{\varphi}_{s2} - M_2\dot{\varphi}_2^2 + m_{02}r_2[-a\dot{\psi}^2 \cos\psi \\
&- \sum C_{j1}\dot{\varphi}_{j1}^2 \sin(\varphi_{j1} - \alpha_{j1}) + \varepsilon(\ddot{u}_1 + g)]\cos\varphi_2
\end{aligned} \tag{5.182}$$

Now we are interested in realizing the higher order harmonic synchronization and sub-harmonic synchronization. First express the u_1 as follows:

$$u_1(a, \psi, \theta_{11}, \theta_{21}) = \frac{1}{2\pi^2}\sum_{n,m,l}[A_c^{(n,m,l)}\cos(n\theta_{11} + m\theta_{21} + l\varphi) + A_s^{(n,m,l)}\sin(n\theta_{11} + m\theta_{21} + l\varphi)] \tag{5.183}$$

$n, m, l = \pm 1, \pm 2, \pm 3, \pm 4$

where

$$\begin{aligned}
A_c^{(n,m,l)} &= \frac{-(n\dot{\theta}_{11} + m\dot{\theta}_{11} + l\omega)^2}{\omega^2 - (n\dot{\theta}_{11} + m\dot{\theta}_{21} + l\omega)^2}\int_0^{2\pi}\int_0^{2\pi}\int_0^{2\pi} F(a, \psi, \theta_{11}, \theta_{21})\cos(n\theta_{11} + m\theta_{21} + l\omega)\mathrm{d}\theta_{11}\mathrm{d}\theta_{21}\mathrm{d}\psi \\
A_s^{(n,m,l)} &= \frac{-(n\dot{\theta}_{11} + m\dot{\theta}_{21} + l\omega)^2}{\omega^2 - (n\dot{\theta}_{11} + m\dot{\theta}_{21} + l\omega)^2}\int_0^{2\pi}\int_0^{2\pi}\int_0^{2\pi} F(a, \psi, \theta_{11}, \theta_{21})\sin(n\theta_{11} + m\theta_{21} + l\omega)\mathrm{d}\theta_{11}\mathrm{d}\theta_{21}\mathrm{d}\psi
\end{aligned} \tag{5.184}$$

Substituting $\ddot{u}_1$ into Eq. (5.181) and assuming $\dot{\varphi}_{21} = \frac{p}{q}\dot{\varphi}_{11}$ i.e. $q\dot{\varphi}_{21} = p\dot{\varphi}_{11}$, $\varphi_{21} \neq \varphi_{11}$, also assuming $\theta_{11} = \dot{\varphi}_{11}t + \delta_{11} - \alpha_{11}$, $\theta_{21} = \dot{\varphi}_{21}t + \delta_{21} - \alpha_{21}$, taking average on Eq. (5.181), one gets the general expressions on the average moments:

$$\begin{aligned}
&(L_1 + C_1)\dot{\varphi}_{11} = C_1\dot{\varphi}_{s1} - M_1\dot{\varphi}_{11}^2 - \frac{1}{2}m_{01}r_1\dot{\varphi}_{11}^2 b_{11}\sin\alpha_{11} + \frac{\varepsilon}{2}m_{01}r_1\frac{1}{2\pi^2}\varphi_1 \\
&\varphi_1 = \varphi_1(m, n, l, \delta_{11}, \delta_{21}, \alpha_{11}, \alpha_{21}) \\
&(L_2 + C_2)\dot{\varphi}_{21} = C_2\dot{\varphi}_{s2} - M_2\dot{\varphi}_{21}^2 - \frac{1}{2}m_{02}r_2\dot{\varphi}_{21}^2 b_{21}\sin\alpha_{21} + \frac{\varepsilon}{2}m_{02}r_2\frac{1}{2\pi^2}\varphi_2 \\
&\varphi_2 = \varphi_2(m, n, l, \delta_{11}, \delta_{21}, \alpha_{11}, \alpha_{21})
\end{aligned} \tag{5.185}$$

It can be seen from Eq. (5.183) that in the case in which u_1 can be expressed a complex function of θ_{11}, θ_{21}, ψ, whether it is the basic harmonic synchronization or higher order harmonic synchronization, or sub-harmonic synchronization,they all could occur.

Now we discuss a simple case. For example, for a nonlinear vibration machine with one impact within q vibration periods, i.e. the material impact period is $1/q$, the frequency is $\frac{\dot{\varphi}_{11}}{q}$, the material impact forces and piece-wise mass inertial force can be expanded in Fourier series. The stable solution of the vibration machines (improved the first approximation solution) can be expressed as:

$$y = \sum_i C_{ji}\sin(\varphi_{ji} - \alpha_{ji}) + \varepsilon u_1(\theta_{ji}) \qquad (5.186)$$

We will express $\dot{\varphi}_{ji}$, $\dot{\varphi}_{ji}$, $\ddot{\varphi}_{ji}$ in the following forms:

$$\varphi_{ji} = i\frac{\sigma_i}{q}(\dot{\varphi}_{10}t + \delta_{1i}), \quad \dot{\varphi}_{ji} = i\frac{\sigma_i}{q}(\dot{\varphi}_{10} + \dot{\delta}_{1i}) \quad \ddot{\varphi}_{ji} = i\sigma_i\ddot{\delta}_{ji}$$
$$\varphi_{1q} = \sigma_1(\dot{\varphi}_{10}t + \delta_{1q}), \quad \varphi_{2p} = \frac{p}{q}\sigma_2(\dot{\varphi}_{10}t + \delta_{2q}) \qquad (5.187)$$

where σ takes positive sign when axle 2 rotates in the same direction as axle 1, and takes negative sign in the opposite case; p/q is the frequency ratio of axle 2 and axle 1.

(3) Stability condition for the double frequency synchronization operation state.

Substituting Eqs. (5.186) and (5.187) into Eq. (5.182) one has:

$$J_1\ddot{\delta}_{11} = C_1\dot{\varphi}_{s1} - (L_1 + C_1)(\dot{\varphi}_{10} + \dot{\delta}_{11}) - M_1(\dot{\varphi}_{10} + \dot{\delta}_{11})^2$$
$$-\frac{1}{2}m_{01}r_1\sum_i C_{ji}\dot{\varphi}_{ji}{}^2\sin(\varphi_{ji} - \alpha_{ji} \pm \varphi_1) \qquad (5.188)$$

where $\pm$ denotes the positive and negative sign exist.

Averaging Eq. (5.188), integrations of $\ddot{\delta}_{11}, \dot{\delta}_{11}, \ddot{\delta}_{21}, \dot{\delta}_{21}$ are zero. Equation (5.188) can be expressed as the balancing equations for motor rotation moments M_{gj}, rotation friction moments M_{fj}, and vibration moments M_{zj}:

$$\begin{aligned} M_{g1} - M_{f1} - M_{z1} = 0 \\ M_{g2} - M_{f2} - M_{z2} = 0 \end{aligned} \qquad (5.189)$$

where

$$M_{gj} = C_j\dot{\varphi}_{sj} - C_j\dot{\varphi}_{j0}, \quad M_{fj} = L_j\dot{\varphi}_{j0} - M_j\dot{\varphi}_{j0}^2, \quad j = 1, 2$$

5.7 Utilization of Frequency-Entrainment Principles

$$M_{z1} = \frac{1}{2}m_{01}r_1\left[-C_{1q}\dot{\varphi}_{1q}^2 \sin\alpha_{1q} + C_{2p}\dot{\varphi}_{2p}^2 \sin\left(\sigma_2\frac{p}{q}\delta_{2p} - \alpha_{2p} \pm \sigma_1\delta_{1q}\right)\right]$$

$$M_{z2} = \frac{1}{2}m_{02}r_2\left[-C_{2q}\dot{\varphi}_{2q}^2 \sin\alpha_{2q} + C_{1p}\dot{\varphi}_{1p}^2 \sin\left(\sigma_1\delta_p - \alpha_{1p} \pm \sigma_2\frac{p}{q}\delta_{2q}\right)\right]$$

(5.190)

where " $\pm$ " takes "$-$" when σ_1 and σ_2 have the same sign and "$+$" when it is opposite.

In the general situation, φ_{2p} and φ_{2p} are small and they can be neglected. Then Eq. (5.189) can be written as (taking $\sigma_1 = 1$, $\sigma_2 = 1$).

$$M_{g1} - M_{f1} = \frac{1}{2}m_{01}r_1\left\{C_{1p}\dot{\varphi}_{1q}^2 \sin\alpha_{1q} + C_{2p}\dot{\varphi}_{2p}^2 \sin\left(\alpha_{2p} - \sigma_2\frac{p}{q}\delta_{2p} + \sigma_1\delta_{1q}\right)\right\}$$

(5.191)

Assuming the rotating frequencies of the vibration machine main transmission $\dot{\varphi}_1$ and $\dot{\varphi}_2$ are much larger than the system natural frequency ω_0, $\alpha_{2p} \approx \alpha_{1p} \to 180°$, and subtracting the first equation from the second equation in Eq. (5.191) one gets:

$$\sin\left(\delta_{1q} - \frac{p}{q}\delta_{2p} + \gamma\right) = $$

$$\frac{M_{g1} - M_{g2} - (M_{f1} - M_{f2}) - \frac{1}{2}\sin\alpha_{1q}\left(m_{01}\gamma_1 C_{1q}\dot{\varphi}_{1q}^2 - m_{02}\gamma_2 C_{2p}\dot{\varphi}_{2p}^2\right)}{\frac{1}{2}\sqrt{\left(C_{2p}\dot{\varphi}_{2p}^2 \cos\alpha_{2p} + C_{1q}\dot{\varphi}_{1q} \cos\alpha_{1q}\right)^2 + \left(C_{2p}\dot{\varphi}_{2p}^2 \sin\alpha_{2p} - C_{1q}\dot{\varphi}_{1q}^2 \sin\alpha_{1q}\right)^2}}$$

(5.192)

$$\gamma = \arctan\frac{C_{2p}\dot{\varphi}_{2p}^2 \sin\alpha_{2p} - C_{1q}\dot{\varphi}_{1q}^2 \sin\alpha_{1q}}{C_{2p}\dot{\varphi}_{2p}^2 \cos\alpha_{2p} + C_{1q}\dot{\varphi}_{1q}^2 \cos\alpha_{1q}}$$

(5.193)

We assume:

$$\frac{1}{D} = \frac{M_{g1} - M_{g2} - (M_{f1} - M_{f2}) - \frac{1}{2}\sin\alpha_{1q}\left(m_{01}\gamma_1 C_{1q}\dot{\varphi}_{1q}^2 - m_{02}\gamma_2 C_{2p}\dot{\varphi}_{2p}^2\right)}{\frac{1}{2}\sqrt{\left(C_{2p}\dot{\varphi}_{2p}^2 \cos\alpha_{2p} + C_{1q}\dot{\varphi}_{1q} \cos\alpha_{1q}\right)^2 + \left(C_{2p}\dot{\varphi}_{2p}^2 \sin\alpha_{2p} - C_{1q}\dot{\varphi}_{1q}^2 \sin\alpha_{1q}\right)^2}}$$

(5.194)

and D is the synchronization index.

It can be seen from Eq. (5.192) that when synchronization index $|D| < 1$, Eq. (5.192) has no solutions. It indicates that the balance can not be acquired in the synchronization for the vibration systems, while it can operate only in the non-synchronization conditions. Therefore, the necessary condition for the synchronization operation or the synchronization criterion is:

$$|D| \geq 1$$

The D value can be computed from Eqs. (5.192), (5.193), then $\delta_{1q} - \frac{p}{q}\delta_{2p}$ can be determined, i.e.

$$\Delta\delta = \delta_{1q} - \frac{p}{q}\delta_{2p} = \arcsin D - \gamma \qquad (5.195)$$

We can see from Eq. (5.189) that when $|D| \geq 1$ we have the basic harmonic synchronization; when $q = 1$, $p = 2, 3, 4, \ldots$, we have the super harmonic synchronization; when $p = 1$, $q = 2, 3, 4, \ldots$, we have the sub-harmonic synchronization; when p and q are a pair of mutual prime numbers and not equal to 1 then we have the fraction harmonic synchronization.

From Eq. (5.189) $|D| \geq 1$ we can find two phase angles corresponding to arcsin D. For example $D = 0.5$, then $\Delta\delta = -150° - \gamma$; if $D = -0.5$ then $\Delta\delta = -30° - \gamma$ or $\Delta\delta = -150° - \gamma$. Now we will determine, via the stability analysis, the two balanced states or the synchronization movement states which sate is stable and which state is unstable. We must establish the perturbation equation. Assume the perturbations are: $\Delta\delta_{11}$, $\Delta\dot\delta_{11}$, $\Delta\ddot\delta_{11}$, $\Delta\delta_{21}$, $\Delta\dot\delta_{21}$, $\Delta\ddot\delta_{21}$; $\Delta\delta_{ji}$, $\Delta\dot\delta_{ji}$, $\Delta\ddot\delta_{ji}$.

Referring to Eqs. (5.188) and (5.191) we can write the following perturbation equations:

$$J_1\Delta\ddot\delta_{11} = -(L_1 + c_1)\Delta\dot\delta_{11} - M_1[2\dot\varphi_{11}\Delta\dot\delta_{11} + \Delta\dot\delta_{11}^2] - \frac{1}{2}m_{01}r_1\{C_{1q}(2\dot\varphi_{1q}\Delta\dot\delta_{1q}$$

$$+\Delta\dot\delta_{1q}^2)\sin\alpha_{2q} + C_{2p}(2\dot\varphi_{1p}\Delta\dot\delta_{2p} + \Delta\dot\delta_{2p}^2)\sin(\alpha_{2p} - \frac{p}{q}\delta_{2p} + \delta_{1q})$$

$$+C_{2p}\dot\varphi_{2p}^2(\Delta\delta_{1q} - \frac{p}{q}\Delta\delta_{2p})\cos(\alpha_{2p} - \frac{p}{q}\delta_{1p} + \delta_{1q})\}$$

$$J_2\Delta\ddot\delta_{21} = -(L_2 + c_2)\Delta\dot\delta_{21} - M_2[2\dot\varphi_{21}\Delta\dot\delta_{21} + \Delta\dot\delta_{21}^2] - \frac{1}{2}m_{02}r_2\{C_{1q}(2\dot\varphi_{2q}\Delta\dot\delta_{2q}$$

$$+\Delta\dot\delta_{2p}^2)\sin\alpha_{2p} + C_{1p}(2\dot\varphi_{1q}\Delta\dot\delta_{1q} + \Delta\dot\delta_{1q}^2)\sin(\alpha_{1q} - \frac{p}{q}\delta_{2p} + \delta_{1q})$$

$$+C_{1q}\dot\varphi_{1q}^2(\Delta\delta_{2p} - \frac{p}{q}\Delta\delta_{1q})\cos(\alpha_{1q} - \frac{p}{q}\delta_{2p} + \delta_{1q})\}$$

$$(5.196)$$

According to stability theorems we can determine the stability for different operation states, and will not give details here.

(4)   Determination of the frequency entrainment regions

We can see from Eq. (5.194) that after the dynamic parameters of the self-synchronization (C_{1q}, C_{2p}, $\dot\varphi_{1q}$, $\dot\varphi_{2p}$, M_{f1}, M_{f2}, α_{1q}, α_{1p}, $m_{01}r_1$, $m_{02}r_2$), the motor rotation moment difference ΔM_g and friction moment difference ΔM_i between the two axles can be determined by the synchronization operation limit condition

5.7 Utilization of Frequency-Entrainment Principles

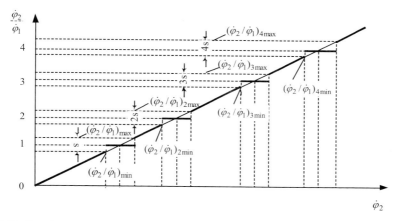

Fig. 5.29 Entrainment region of basic and double frequency entrainment of vibration machines driven by two motors

$D = 1$ or $D = -1$. When ΔM_i is known the motor characteristics index C_{1m} and C_{2m} can be found as follows:

$$\frac{\Delta M_g - \Delta M_f - \frac{1}{2}\sin\alpha_{1q}\left(m_{01}r_1 C_q\dot{\varphi}^2_{1q} - m_{02}r_2 C_{2p}\dot{\varphi}^2_{2p}\right)}{\frac{1}{2}\sqrt{\left(C_{2p}\dot{\varphi}^2_{2p}\cos\alpha_{2p} + C_{1q}\dot{\varphi}^2_{1q}\cos\alpha_{1q}\right)^2 + \left(C_{2p}\dot{\varphi}^2_{2p}\sin\alpha_{2p} - C_{1q}\dot{\varphi}^2_{1q}\sin\alpha_{1q}\right)^2}} = \pm 1$$

or

$$\frac{C_{1m}(\dot{\varphi}_{s1} - \dot{\varphi}_{1q}) - C_{2m}(\dot{\varphi}_{s2} - \dot{\varphi}_{2p}) - M_{f1} + M_{f2} - \frac{1}{2}\sin\alpha_{1q}\left(m_{01}r_1 C_{1q}\dot{\varphi}^2_{1q} - m_{02}r_2 C_{2p}\dot{\varphi}^2_{2p}\right)}{\frac{1}{2}\sqrt{\left(C_{2p}\dot{\varphi}^2_{2p}\cos\alpha_{2p} + C_{1q}\dot{\varphi}^2_{1q}\cos\alpha_{1q}\right)^2 + \left(C_{2p}\dot{\varphi}^2_{2p}\sin\alpha_{2p} - C_{1q}\dot{\varphi}^2_{1q}\sin\alpha_{1q}\right)^2}} = \pm 1$$

We can find C_{1m} and C_{2m} and we can also find the maximum or rotation speed ratio $\left(\frac{p}{q}\right)_{max}$ or minimum ratio $\left(\frac{p}{q}\right)_{min}$ under different operation situations. Thus we can determine the boundary of realizing frequency entrainment and its condition can be found from Eq. (5.195):

$$C_{1m}\left[\dot{\varphi}_{s1} - \left(\frac{p}{q}\right)_m \dot{\varphi}_{2p}\right] - C_{2m}(\dot{\varphi}_{s2} - \dot{\varphi}_{2p}) - M_{f1} + M_{f2} \\ -\frac{1}{2}\sin\alpha_{1q}\left(m_{01}r_1 c_{1q}\dot{\varphi}^2_{1q} - m_{02}r_2 c_{2p}\dot{\varphi}^2_{2p}\right) = 0 \quad (5.197)$$

We can find the upper and lower boundaries of the frequency entrainment from Fig. 5.29. Due to the fact that the nonlinear synchronous vibration machines have basic frequency and double frequency synchronization, i.e. there exist the basic and sub-harmonic frequency entrainment, hence there are many frequency entrainment regions. The size of the frequency entrainment regions is associated closely with the strength of the self-synchronous vibration machines. The larger the entrainment regions the stronger the self-synchronization. We can determine the synchronization is strengthened or weakened by changing which parameters in the formulations derived above. Hence the size of the entrainment regions has significance in the realization of the synchronization of the self-synchronous vibration machines.

It is worth noticing that generally the frequency entrainment for high order harmonics is much more difficult than the basic harmonics and the entrainment regions for higher order harmonics are much narrower than that of the basic harmonics. For sub-harmonic frequency entrainment it may occur only when system has sub-harmonic vibration.

(5) Conclusions

  i. The dynamic characteristics of the self-synchronous vibration machines have been studied in many planer movements, the synchronization conditions and synchronization stability conditions are found, and they can be used to calculate the synchronization state and dynamics parameters of the machines.
  ii. The necessary condition for realizing the synchronization for a variety of self-synchronous vibration machines is that the synchronization index D is larger or equal to 1, i.e.:

$$|D| \geq 1$$

When this condition can not be satisfied, the synchronization can not be achieved.

  iii. After the synchronization conditions for the self-synchronous vibration machines are met there may be two synchronization states, one of them is stable. The stability needs to be determined by the stability conditions. The stability condition for synchronization operation states or stability criterion is:

$$W \cos \Delta\alpha_0 > 0$$

The stability index W is different for different self-synchronous vibration machines and the specific expression for each individual case can be found by dynamic analysis. When stability index W is positive, the phase angle difference $\Delta\alpha_0$ is near $0°$; when stability index W is negative, the phase angle difference $\Delta\alpha_0$ is near $180°$. You may further calculate the movement trajectory of the body.

  iv For nonlinear vibration systems, in addition to synchronization for the basic harmonic frequency entrainment, it may acquire 2 and 3 times frequency synchronization, and possible sub-harmonic decreasing frequency synchronization as well.

5.8 Utilization of Nonlinear Vibration Systems with Nonlinear Inertial Force

The vibration centrifuge hydro-extractor is mainly used for hydro-extraction from the fine particle materials. This machine has excellent technological index by strengthening the hydro-extraction effect using vibration. Figure 5.30 shows the working mechanism of it with symmetric and asymmetric elastic forces. The eccentric shaft rotates reversely synchronously and generates the axial excitation force. The main vibration system is composed of body shell, nonlinear rubber spring and screen basket and vibrate axially under the excitation. The materials under the excitation of centrifugal force and axial vibration force slide forward without jumps until they fall into the discharge trough. The symmetric or asymmetric gag springs, and elastic force are linear and have the hardening characteristics which is favorable to hydro-extracting. It is obvious that the nonlinear springs and materials involved in vibration make the machine display a complex nonlinear dynamic characteristics. It is important significance to study and analyze the characteristics.

5.8.1 Movement Equations for Vibration Centrifugal Hydro-Extractor with Nonlinear Inertial Force

According to the technological requirements, the materials move intermittently forward on the inner wall of the cone of the vibration centrifugal hydro-extractor, realize hydro-extracting and automatically discharged without inverse movement and/or jumps. The material movement state in a period is shown in Fig. 5.31. In the sliding region, the force subjected to the cone wall by the materials is the sum of the positive sliding friction force, inertial force component, while in the intermittent region materials are relative still to the cone body and there is only inertial force component.

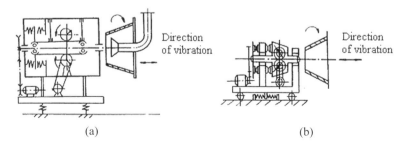

Fig. 5.30 Mechanism diagram for vibration centrifugal hydro-extractor. **a** Symmetric gap spring hydro-extractor **b** Asymmetric gap spring hydro-extractor

Fig. 5.31 One-direction sliding motion in a period

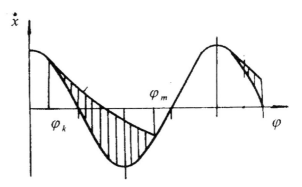

Fig. 5.32 Mechanics diagram for asymmetric vibration centrifuge

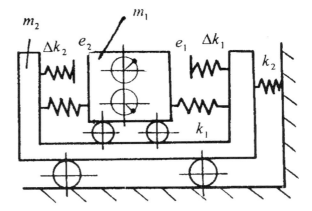

Assume φ_k to be the initial sliding angle, φ_m to be the sliding stop angle. It can be derived that the nonlinear inertial force of the materials in one period is the piece-wise linear equation:

$$F_m(\ddot{x}_1) = \begin{cases} -m_m \ddot{x}_1 \cos^2 \beta & \varphi_m < \varphi < \varphi_k + 2\pi \\ -m_m \mu \cos \beta (\Omega^2 R_c \cos \beta + \ddot{x}_1 \sin \beta) & \varphi_k < \varphi < \varphi_m \end{cases} \quad (5.198)$$

where m_m is the material mass in the cone; β is the half of the cone angle; R_c is the equivalent radius of materials in the cone body; Ω is the cone body rotation angular velocity; μ is the sliding friction coefficient.

The simplified mechanics diagram for the asymmetric piece-wise linear stiffness of the vibration centrifuge is shown in Fig. 5.32.

The equations of motions for the systems can be written in matrix as follows:

5.8 Utilization of Nonlinear Vibration Systems ...

$$\begin{bmatrix} m_1 & 0 \\ 0 & m_2 \end{bmatrix} \begin{Bmatrix} \ddot{x}_1 \\ \ddot{x}_2 \end{Bmatrix} + \begin{bmatrix} k_1 & -k_1 \\ -k_1 & k_1 + k_2 \end{bmatrix} \begin{Bmatrix} x_1 \\ x_2 \end{Bmatrix}$$
$$= -\varepsilon \begin{bmatrix} c_1 & -c_1 \\ -c_1 & c_1 + c_2 \end{bmatrix} \begin{Bmatrix} \dot{x}_1 \\ \dot{x}_2 \end{Bmatrix} - \varepsilon \begin{Bmatrix} f_m(\ddot{x}_1) + f_k(x, \dot{x}) \\ -f_k(x, \dot{x}) \end{Bmatrix} + \varepsilon \begin{Bmatrix} E_0 \sin \nu t \\ 0 \end{Bmatrix}$$
(5.199)

Where x_1, x_2 and x are the displacements for bodies 1 and 2 and the relative displacement between the two bodies respectively; $x = x_1 - x_2$; m_1, m_2 are the masses of bodies 1 and 2, mass 1 contains the material equivalent mass: $m_1 = m'_1 + m_m \sin^2 \beta$, m'_1 is the mass of body 1; k_1, k_2, c_1, c_2 are the spring stiffnesses and damping coefficients in the horizontal direction between body 1 and body 2, body 2 and base; E_0 is the inertial force by eccentric: $E_0 = m_0 \nu^2 r$, m_0 is the mass of the eccentric, r is the eccentricity, ν is the angular velocity of the eccentric $f_k(x, \dot{x})$ is the asymmetric elastic restoring force and damping between body 1 and body 2:

$$f_k(x, \dot{x}) = \begin{cases} c\dot{x} & -e_2 < x < e_1 \\ c\dot{x} + \Delta c_1 \dot{x} + \Delta k_1 (x - e_1) & x \geq e_1 \\ c\dot{x} + \Delta c_2 \dot{x} + \Delta k_2 (x + e_2) & x \leq -e_2 \end{cases}$$
(5.200)

where c is the damping coefficient associated with linear spring k_1; Δk_1, Δk_2 are the gap spring stiffnesses; Δc_1, Δc_2 are the damping coefficients associated with gap spring; e_1, e_2 are the left and right side gap distances.

5.8.2 Nonlinear Vibration Responses of Vibration Centrifugal Hydro-Extractor

The vibration centrifugal hydro-extractors generally work near the second natural frequency region. Assume that there is no inner resonance. We only analyze the vibration of basic frequency ω_2 and the higher order components are small terms which meet the real engineering requirements.

The normal mode matrix corresponding to the natural frequencies ω_1, ω_2 can be easily found:

$$\Phi_M = [\Phi_{m1}, \Phi_{m2}] = \begin{bmatrix} \varphi_{11} & \varphi_{12} \\ \varphi_{21} & \varphi_{22} \end{bmatrix}$$
(5.201)

We can find:

$$\varphi_{12} = 1 \quad \varphi_{22} = (k_1 - \omega_2^2 m_1)/k_1$$
(5.202)

The second principal mass and stiffness are:

$$M_2 = \boldsymbol{\Phi}_{m2}^{\mathrm{T}} M \boldsymbol{\Phi}_{m2} \quad K_2 = \boldsymbol{\Phi}_{m2}^{\mathrm{T}} K \boldsymbol{\Phi}_{m2} \tag{5.203}$$

Now we concentrate on the second natural frequency vibration. Assume:

$$\begin{Bmatrix} x_1 \\ x_2 \end{Bmatrix}_2 = \boldsymbol{\Phi}_{m2} z = \begin{Bmatrix} \varphi_{12} \\ \varphi_{22} \end{Bmatrix} z \tag{5.204}$$

The equation of motion for the second principal coordinate is:

$$M_2 \ddot{z} + K_2 z = \varepsilon \varphi_{m2}^{\mathrm{T}} \{f\}_2 + \varepsilon \varphi_{m2}^{\mathrm{T}} E(t) \equiv \varepsilon f + \varepsilon E_0 \sin \theta \tag{5.205}$$

in which the nonlinear term can be written as:

$$f = a_0 \dot{z} - \begin{cases} a_1 \ddot{z} \\ a_2 + a_3 \ddot{z} \end{cases} + \begin{cases} 0 \\ b_1 \dot{z} + b_2 z + b_3 \\ b_4 \dot{z} + b_5 z + b_6 \end{cases} \triangleq a_0 + f'_m + f'_k \tag{5.206}$$

where $a_0 = -(c_1 + \varphi_{22}^2 c_2 - \varphi_{22}^2 c + 2\varphi_{22} c - c)$, $a_1 = m_m \cos^2 \beta$, $a_2 = -m_m \mu \cos^2 \beta \Omega^2 R_c$,
$a_3 = -m_m \mu \cos \beta \sin \beta$, $b_1 = -(1 - \varphi_{22})^2 \Delta c_1$, $b_2 = -(1 - \varphi_{22})^2 \Delta k_1$, $b_3 = (1 - \varphi_{22}) \Delta k_1 e_1$.
$b_4 = -(1 - \varphi_{22})^2 \Delta c_2$, $b_5 = -(1 - \varphi_{22})^2 \Delta k_2$, $b_6 = (\varphi_{22} - 1) \Delta k_2 e_2$.

Assume the first order approximate solution to be:

$$z = a \cos \varphi = a \cos(\nu t + \theta) + \varepsilon u_1 + \ldots \tag{5.207}$$

In the first approximation case, assume $\varphi_1 = \arccos \frac{e_1}{a(1-\varphi_{22})}$, in the first quadrant and $\varphi_1 = \arccos \frac{-e_2}{a(1-\varphi_{22})}$, in the second quadrant. The equivalent damping coefficients and stiffness are

$$\delta_e(a) = \frac{\varepsilon}{2\pi \omega_2 M_2 a} \{-\pi \omega_2 a_0 a - \frac{1}{2}(a_1 - a_3)\omega_2^2 a(\sin^2 \varphi_m - \sin^2 \varphi_k) + a_2(\cos \varphi_m - \cos \varphi_k) \\ -b_1 \omega_2 a(\varphi_1 - \frac{1}{2}\sin 2\varphi_1) - b_2 \omega_2 a(\pi - \varphi_2 + \frac{1}{2}\sin 2\varphi_2)\} \tag{5.208}$$

$$k_e(a) = K_2 - \frac{\varepsilon}{\pi a} \{\pi \omega_2^2 a_1 a + \frac{1}{4}(a_1 - a_3)\omega_2^2 a(2\varphi_k + \sin 2\varphi_k - 2\varphi_m - \sin 2\varphi_m) - a_2(\sin \varphi_m - \sin \varphi_k) \\ +b_1 a(\varphi_1 + \frac{1}{2}\sin 2\varphi_1) + 2b_3 \sin \varphi_1 + b_5 a(\pi - \varphi_2 - \frac{1}{2}\sin 2\varphi_2) - 2b_6 \sin \varphi_2\} \tag{5.209}$$

The equivalent natural frequency is:

$$\omega_e(a) = \sqrt{k_e(a)/M_2} \tag{5.210}$$

The amplitude a and phase angle θ can be determined by the integration over the following two expressions:

5.8 Utilization of Nonlinear Vibration Systems …

$$\dot{a} = -\delta_e(a)a - \frac{\varepsilon E_0}{M_2(\omega_2 + \nu)} \cos\theta, \quad \dot{\theta} = \omega_e(a) - \theta + \frac{\varepsilon E_0}{M_2(\omega_2 + \nu)a} \sin\theta \tag{5.211}$$

for the constant state concerned in industry application, $a = 0$, θ_0, we obtain the frequency response equations:

$$a^2 M_2^2 \{[\omega_e^2(a) - \nu^2]^2 + 4\delta_e^2(a)\nu^2\} = \varepsilon^2 E_0^2 \tag{5.212}$$

$$\theta = \arctan \frac{\omega_e^2(a) - \nu^2}{2\delta_e(a)\nu} \tag{5.213}$$

Further deriving, we can obtain the improved terms with 2, 3 order harmonic components:

$$\begin{aligned}
u_1 = \frac{1}{\pi \omega_2^2 M_2} &\{ -\frac{1}{3}[\frac{1}{6}\omega_2^2 a(a_1 - a_3)(\sin 3\varphi_k + 3\sin\varphi_k - \sin 3\varphi_m - 3\sin\varphi_m) \\
&+ \frac{1}{2}a_2(\sin 2\varphi_k - \sin 2\varphi_m) + \frac{1}{3}b_2 a(\sin 3\varphi_1 + 3\sin\varphi_1) - \frac{1}{3}b_5 a(\sin 3\varphi_2 + 3\sin\varphi_2) \\
&+ b_3 \sin 2\varphi_1 - b_6 \sin 2\varphi_2] \times \cos 2\varphi \\
&- \frac{1}{3}[\frac{1}{6}\omega_2^2 a(a_1 - a_3)(\cos 3\varphi_m + 3\cos\varphi_m - \cos 3\varphi_k - 3\cos\varphi_k) + \frac{1}{2}a_2(\cos 2\varphi_m - \cos 2\varphi_k) \\
&+ \frac{1}{3}b_1 \omega_2 a(\sin 3\varphi_1 - 3\sin\varphi_1) - \frac{1}{3}b_4 \omega_2 a(\sin 3\varphi_2 - 3\sin\varphi_2)] \times \sin 2\varphi \\
&- \frac{1}{8}[\frac{1}{8}(a_1 - a_3)\omega_2^2 a(\sin 4\varphi_k + 2\sin 2\varphi_k - \sin\varphi_m - 2\sin 2\varphi_m) \\
&- \frac{1}{3}a_2(\sin 3\varphi_m - \sin 3\varphi_k) + \frac{1}{4}b_2 a(\sin 4\varphi_1 + 2\sin 2\varphi_1) + \frac{2}{3}b_3 \sin 3\varphi_1 \\
&+ \frac{1}{4}b_5 a(\sin 4\varphi_2 + 2\sin 2\varphi_2) - \frac{2}{3}b_6 \sin 3\varphi_2] \times \cos 3\varphi \\
&- \frac{1}{8}[\frac{1}{8}(a_1 - a_3)\omega_2^2 a(\cos 4\varphi_m + 2\cos 2\varphi_m - \cos\varphi_k - 2\cos 2\varphi_k) \\
&+ \frac{1}{3}a_2(\cos 3\varphi_m - \cos 3\varphi_k) + \frac{1}{4}b_1 \omega_2 a(\sin 4\varphi_1 - 2\sin 2\varphi_1) \\
&- \frac{1}{4}b_4 \omega_2 a(\sin 4\varphi_2 - 2\sin 2\varphi_2)] \times \sin 3\varphi\} + \ldots
\end{aligned} \tag{5.214}$$

5.8.3 Frequency-Magnitude Characteristics of Vibration Centrifugal Hydro-Extractor

We can see from the above results that the nonlinear springs and materials have direct effects on the system equivalent stiffness and damping coefficient and thus affect the amplitudes and phase angle difference of the system. Figure 5.33 shows the system frequency response function and higher order harmonic frequency response function for a vibration centrifugal hydro-extractor. Figure 5.33a is the principal resonance curve (a_0 is nominal amplitude) for different material masses (m_m, 2 mm); Fig. 5.33b

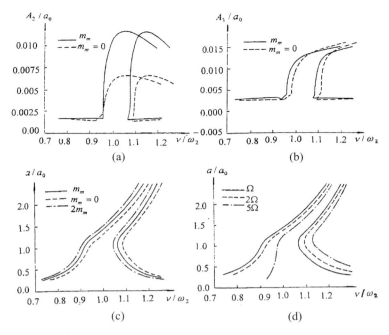

Fig. 5.33 Frequency response functions for vibration centrifugal vibration hydro-extractor. **a** Basic harmonic for different masses **b** basic harmonic for different cone rotations **c** the second harmonic with nonlinear inertial force **d** the third harmonic with nonlinear inertial force

shows the principal resonance curve for different angular velocities of the cone body (Ω, 2Ω, 5Ω); Fig. 5.33c, d display the second and third order harmonic resonance curves for different material masses and different cone angular velocities, in which the corresponding second and third harmonic terms are composed of the sine and cosine terms.

5.8.4 Experiment Vibration Responses of Vibration Centrifugal Hydro-Extractor

The acceleration curves along the screen basket of the hydro-extractor are shown in Fig. 5.34, measured directly from the acceleration transducers. It can be seen from the Figure that the acceleration curves for the asymmetric vibration centrifugal hydro-extractor show strong asymmetric characteristics, while the accelerations for the symmetric hydro-extractor are symmetric. The two curves all show the nonlinearity.

We can draw some conclusions from above analysis:

The material nonlinear inertial force has effects on the principal resonance curves, the increase of the material mass will make the curve shift toward left and display

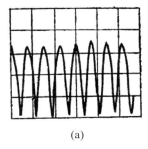
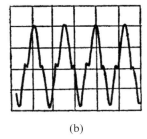

(a) (b)

Fig. 5.34 Experimental acceleration curves for the vibration centrifugal hydro-extractor. **a** Asymmetric hydro-extractor vibration curves **b** symmetric hydro-extractor vibration curves

the harden characteristics. The cone body rotation angular velocity has some effect on the principal resonance curve as well. As Ω increases, the curve shifts toward the right.

The material nonlinear inertial force affects the components of the second and third order harmonics of the system, as shown in Fig. 5.33c, d. The existence of the higher order harmonics results in large vibration acceleration and we should pay enough attention to it. For example when the natural frequency approaches to some of the higher order harmonic frequency the resonance may occur and increase the dynamic stresses. The best practice indicates that their natural frequency must be 5 times of the working frequency.

The study results in this section provide the theoretical basis and analytical methods for the designs of the vibration machines with similar nonlinear elastic force and nonlinear inertial forces.

5.9 Utilization of Slowly-Changing Parameter Nonlinear Systems

In the engineering it often encounters such rotor systems whose parameters change with time. The reasons that cause the change are many: some changes are due to the physical and geometric change of the rotor itself. For example the rotor has cracks, the stiffness of the whole system will change overtime and an additional unbalanced excitation force varying with time is subject to the rotor. When the rotor is in the operation process the change of the outside conditions may also result in its parameter changes. For example the rotor system supported by the oil film may cause the oil film turbulence due to large outside disturbance, result in the damping change with time and even result in the instability of the oil film. During the startup and stop processes of the rotor system if there exists an eccentricity in the rotor the eccentricity may cause the unbalanced excitation change with time in the process. In discussing the time-changing rotor systems, what we should emphasize is the introduction of the active control into the change of the rotor system parameters. In recent years, as with the development of the active controls people apply the modern control theory and intelligent control methods to the rotor active controls. There are many control methods. Their physical significance is to achieve the vibration amplitude control by adjusting the system parameters to reduce the unbalanced excitation. It is obvious that during the control implementation the rotor system is the time-changing system and pace of the parameter changes is used to control rotor's vibration amplitudes.

The time-changing system is a kind of nonlinear systems and can be expressed as follows:

$$\frac{d}{dt}\left[m(t)\frac{dx}{dt}\right] + c(t)\frac{dx}{dt} + k(t)x = f\left(t, x, \frac{dx}{dt}\right) \quad (5.215)$$

If the mass $m(t)$, damping $c(t)$ and stiffness $k(t)$ and excitation $f(t, x, \frac{dx}{dt})$ can be expressed analytically then the numerical or analytical (in the case of degree of freedom is small) methods can be used to find their solutions, thus obtain its responses and analyze its dynamic characteristics. In the nonlinear time-changing systems if the parameter changes are very small compared with the vibration period, the systems are called the slow-changing parameter systems. The slow-changing parameter system is an important special case for the changing parameter systems. It can be expressed as:

$$\frac{d}{dt}\left[m(\tau)\frac{dx}{dt}\right] + c(\tau)\frac{dx}{dt} + k(\tau)x = f\left(\tau, x, \frac{dx}{dt}\right) \quad (5.216)$$

where $\tau = \varepsilon t$, ε is a small parameter.

5.9.1 Slowly-Changing Systems Formed in Processes of Starting and Stopping

For some machines during the start-up and stop process we can consider the excitation changes slowly with time. Taking a single degree of freedom system, when its stiffness is symmetric we may just study the vibration in x direction. The equation of motion for the system is:

$$m\frac{d^2x}{dt^2} + kx = \varepsilon f\left(x, \frac{dx}{dt}\right) + \varepsilon E(\tau)\sin\theta \quad (5.217)$$

in which $\frac{d\theta}{dt} = \nu(\tau)$, $\tau = \varepsilon t$, m and k are constants.

Now we use the asymptotic method to find the solutions. If we only study the principal vibration case, then the first approximation solution is assumed to be:

5.9 Utilization of Slowly-Changing Parameter Nonlinear Systems

$$x = a\cos(\theta + \vartheta) \tag{5.218}$$

in which $\frac{da}{dt} = \varepsilon A_1(\tau, a, \vartheta)$, $\omega = \sqrt{\frac{k}{m}}$ then

$$\frac{d\vartheta}{dt} = w - \vartheta(\tau) + \varepsilon B_1(\tau, a, \vartheta) \tag{5.219}$$

We obtain:

$$\begin{aligned}\frac{dx}{dt} &= \frac{da}{dt}\cos(\theta + \vartheta) - a\sin(\theta + \vartheta)\frac{d(\theta + \vartheta)}{dt} \\ &= \varepsilon A_1 \cos(\theta + \vartheta) - a\omega\sin(\theta + \vartheta) - \varepsilon a B_1 \sin(\theta + \vartheta)\end{aligned} \tag{5.220}$$

$$\begin{aligned}\frac{d^2 x}{dt^2} &= \varepsilon\left[(w - \vartheta(\tau))\frac{\partial A_1}{\partial \vartheta} - 2a\omega B_1\right]\cos(\theta + \vartheta) - a\omega^2 \cos(\theta + \vartheta) \\ &\quad - \varepsilon[(w - \vartheta(\tau))a\frac{\partial B_1}{\partial \vartheta} 1) - 2\omega A_1]\sin(\theta + \vartheta)\end{aligned} \tag{5.221}$$

Letting the coefficients equal for the same order of terms we obtain:

$$\frac{da}{dt} = -\frac{\varepsilon}{2\pi m \omega}\int_0^{2\pi} f_0(a, \varphi)\sin\varphi d\varphi - \frac{\varepsilon E(\tau)}{m(\omega + \vartheta(\tau))}\cos\vartheta$$

$$\frac{d\varphi}{dt} = w - \vartheta(\tau) - \frac{\varepsilon}{2\pi m\omega\, a}\int_0^{2\pi} f_0(a, \varphi)\cos\varphi d\varphi + \frac{\varepsilon E(\tau)}{ma(\omega + \vartheta(\tau))}\sin\vartheta \tag{5.222}$$

Integrating numerically over the above equations one can get the rotor system responses for the excitation amplitude changing slow with time during the start-up and stop processes.

5.9.2 Slowly-Changing Rotor Systems Formed in Active Control Processes

Due to the introduction of the active controls some of the parameters in the rotor systems are changing and thus the systems amplitude is reduced. Now we take the slow-changing stiffness as an example to illustrate and study the vibration characteristics of the rotor systems.

The so-called slowly-changing stiffness is the supporting stiffness of the rotor and is changing with time slowly, and the change of the stiffness is completed in the time period which is equivalent to sum of large numbers of the vibration periods. If the supporting stiffness remains symmetric during the slow-changing process, then the vibration under the controls can be expressed as:

$$m\ddot{x} + c\dot{x} + k(\tau)x = me\,\omega^2 \cos(\omega t) \quad (5.223)$$

where $\tau = \varepsilon t$, ε is a small parameter.

In the real control processes, the change of the supporting stiffness may be a complex nonlinear process. However it doesn't matter how the supporting stiffness is changed as long as it changes slowly with time we may consider it as a piece-wise-linear nonlinear process, i.e.:

$$k(\tau) = \begin{cases} k_0 + k_1\tau & 0 < \tau \leq \tau_1 \\ k_1 + k_2\tau & \tau_1 < \tau \leq \tau_2 \\ k_2 + k_3\tau & \tau_2 < \tau \leq \tau_3 \end{cases} \quad (5.224)$$

We assume that the damping is linear. Then the changing process described by Eq. (5.210) can be expressed as follows:

$$m\ddot{x} + k_0 x = -\varepsilon c\dot{x} - k\tau x + me\,\omega^2 \cos(\omega t) \quad (5.225)$$

The above equation can be simplified as:

$$\ddot{x} + \omega_0^2 x = -\varepsilon c_d \dot{x} - k_d \tau x + e\omega^2 \cos(\omega t) \quad (5.226)$$

where $\omega_0^2 = \frac{k_0}{m}, c_d = \frac{c}{m}, k_d = \frac{k}{m}$.

Use multi-scale method to find its solutions. Assume:

$$T_n = \varepsilon^n t, \ n = 0, 1, 2, \ldots,$$

then

$$\begin{cases} \dfrac{d}{dt} = \dfrac{dT_0}{dt}\dfrac{\partial}{\partial T_0} + \dfrac{dT_1}{dt}\dfrac{\partial}{\partial T_1} + \cdots = D_0 + \varepsilon D_1 + \cdots \\ \dfrac{d^2}{dt^2} = D_0^2 + 2\varepsilon D_0 D_1 + \varepsilon^2(D_1^2 + 2D_0 D_2) + \cdots \end{cases} \quad (5.227)$$

where $D_0, D_1, D_2, \ldots$ denote the partial derivatives with respect to $T_0, T_1, T_2, \ldots$

Substituting Eq. (5.226) into Eq. (5.225) one gets:

$$D_0^2 x + 2\varepsilon D_0 D_1 x + \omega_0^2 x = -\varepsilon c_d D_0 \dot{x} - \varepsilon^2 c_d D_1 x - k_d \tau x + e\omega^2 \cos(\omega t) \quad (5.228)$$

Assume the solution of Eq. (5.227) takes the form of:

$$x = x_0(T_0, T_1) + \varepsilon x_1(T_0, T_1) + \cdots \quad (5.229)$$

5.9 Utilization of Slowly-Changing Parameter Nonlinear Systems

Substituting Eq. (5.228) into Eq. (5.227) one has:

$$D_0^2 x_0 + 2\varepsilon D_0 D_1 x_0 + \omega_0^2 x_0 + \varepsilon D_0^2 x_1 + 2\varepsilon^2 D_0 D_1 x_1 + \varepsilon \omega_0^2 x_1 \\ = -\varepsilon c_d D_0 x_0 - k_d \tau x_0 + e\omega^2 \cos(\omega t) + \cdots \quad (5.230)$$

Letting the coefficients equal for the same order of ε on both sides of the Eq. (5.229) one has:

$$\varepsilon^0 \quad D_0^2 x_0 + \omega_0^2 x_0 = e\omega^2 \cos(\omega t) \quad (5.231)$$

$$\varepsilon^1 \quad D_0^2 x_1 + \omega_0^2 x_1 = -2 D_0 D_1 x_0 - 2 c_d D_0 x_0 - k_d \tau \, x_0 \quad (5.232)$$

From Eq. (5.230) one has:

$$x_0 = A(T_1) \exp(i\omega_0 T_0) + \Lambda \exp(i\omega T_0) + C_c \quad (5.233)$$

in which $\Lambda = \frac{1}{2} e\omega^2 (\omega_0^2 - \omega^2)^{-1}$.

Substituting Eq. (5.232) into Eq. (5.231) one gets:

$$D_0^2 x_1 + \omega_0^2 x_1 = [-2i\omega_0(A' + C_d A) - k_d t A] \exp(i\omega_0 T_0) \\ + [-2i\omega \Lambda C_d - k_d t \Lambda] \exp(i\omega T_0) + C_c \quad (5.234)$$

In order to avoid the infinite term then:

$$-2i\omega_0(A' + C_d A) - k_d t A = 0 \quad (5.235)$$

One has:

$$A = \frac{-2i\omega_0}{2i\omega_0 C_d + k_d t} A' \quad (5.236)$$

Assume:

$$A = a \exp(i\beta) \quad (5.237)$$

α, β are function of T_1. Substituting Eq. (5.236) into Eq. (5.235) and letting the real part and imaginary part equal respectively one has:

$$\begin{cases} a' = -C_d a \\ 2\omega_0 \beta' = k_d t \end{cases} \quad (5.238)$$

The solution for Eq. (5.225) is:

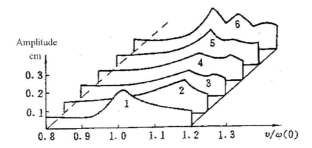

Fig. 5.35 The simulation for different stiffness coefficients

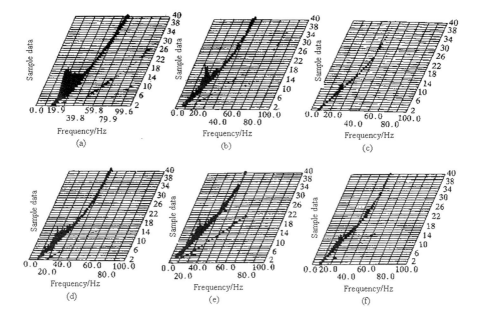

Fig. 5.36 3D spectra for different k_d

$$x = a\cos(\omega_0 t + \beta) + e\omega^2(\omega_0^2 - \omega^2)^{-1}\cos(\omega t) + O(\varepsilon) \qquad (5.239)$$

In which α, β are functions of $T_1 = \varepsilon T$ and are associated with the changing stiffness coefficient k_d. If k_d is chosen carefully it may play a role in reducing the amplitudes.

Take $\varepsilon = 0.01$, $m = 12.4$ kg, $e = 0.04$ m and conduct numerical simulation and results are shown in Fig. 5.35, while Fig. 5.36 shows the 3-D spectra from tests of different k_d values. It can be seen from the Figure that when $k_d = 10$ the amplitude is minimum. That is to say, the system vibration amplitude is well controlled by selecting different changing stiffness coefficients k_d.

5.10 Utilization of Chaos

The characters of the chaos movement have been that it has broadband power spectra and irregularity in the movement trajectory. Some people tried to apply these chaotic features to the industry applications, such as canceling the boring bar on a boring machine by generating the chaotic movements for the steel balls in the cavity of the impact vibration isolator, speeding up the fix of materials by generating chaotic movements of the multi-materials in the vibration fixing machines; some people even made attempts to design a vibration screen with chaotic movement of the materials.

For all those ideas of utilizing the chaos to accomplish some technological process, the key problem to assess its function is that the technological indices and other synthetic assessment indices of the chaotic movement equipment should be better than those of the same purposes of other equipment.

The chaos has been successfully applied to the encryption system and communications.

In addition, the harmful unstable vibration may occur during the working process of the nonlinear vibration system. In this case, it is a natural development to make the unstable vibration transit to the stable and favorite chaos movement. It is worth studying if there are some more effective measures than the chaotic movements.

To utilize the chaotic movement in engineering, we must master the condition for which chaos occurs and its computation methods. Now we will introduce the numerical methods for the chaos computation.

The nonlinear vibration problems encountered in the engineering have the features of multi-degree of freedoms, strong nonlinearity, strong couplings and slow-changed system parameters . Modern nonlinear mechanics theory and methods face some difficulty in studying the high degree of freedom systems. Hence the numerical computation and simulation become the main ways of dealing with the nonlinear dynamics. Here we integrate the frequency spectrum analysis, stability analysis, Lyapunov index analysis and global analysis.

5.10.1 *Major Methods for Studying Chaos*

The main contents for studying chaos include: (1) find out the stable responses of the system—stable solutions and judge their stability, (2) the changing patterns of the stable solutions and stability as the system parameters changes, (3) the long term tendency of system evolution for different initial conditions. To find out the stable solutions are the basic and important contents for the nonlinear dynamic analysis. The stable solutions of the systems are merely the forms of the balance point solutions, periodical solutions, pseudo periodical solutions and chaotic solutions.

The study of the changing patterns of the stable solutions and stability as the system parameters change belongs to the category of bifurcation theory and it is imperative to understand the dynamic behaviors of the dynamic system parameter changes, i.e. the bifurcation of the system movements. There are Hopf Bifurcation Theory (study of the balance point bifurcation or bifurcation of periodical solutions for the autonomous systems) and Floquet Bifurcation Theory (study of bifurcation problem of the periodical solutions).

The numerical methods are developed with the applications of the computer technology. They include: (1) initial value methods based on the numerical integrations; (2) numerical methods for finding the periodical boundary problems, such as the targeting method, difference method, harmonic balance method, PNF(Poincare-Newton-Floquet) method etc; (3) numerical methods for bifurcation curves, such as the parameter continuation algorithm, increment harmonic balance method and compound method among the continuation method, targeting and harmonic balance; (4) cell mapping methods for global characteristics.

The chaos movement is a very common nonlinear phenomenon. The essence of the chaos has not yet been fully understood and lacked of effective mathematic tools, the numerical study becomes the important methods for the chaos movement. Based on the numerical experiments, the traditional methods on studying the chaotic movements are:

(1) Observe the movement trajectory and irregularity of the strange attractor structures by numerical computation results.
(2) Check the numerical power spectra: if the power spectra are continuous, then chaos can be considered to occur.
(3) Convert the continuous dynamic systems into discrete systems using Poincare mapping and study the discrete systems. If what acquired in the Poincare section is not limited point set or simple closed curves, then the chaos may exist.
(4) Lyapunov index is the measure of sensitivity of the movement to its initial values. It's numerically computed. The number of the Lyapunov for a system is in general the dimension number of the phase space. That the maximum Lyapunov index is larger than zero can be used as an important criterion for existence of chaos.
(5) For a dissipated system, determination of the dimensional numbers of the attractors and their attracting region boundary helps judge the singularity of the attractors. The occurrence to the attractors and their attracting regions with fraction dimensional number is the important features of the chaos.
(6) The measure entropy or topology entropy is the measurement for increment of the system information in the movement process. If the measure entropy or topology entropy is larger than zero, the system can be considered to be chaotic. These entropies can be computed numerically.
(7) Cell mapping is an effective method in study of dynamic bifurcation, especially in chaos.

The numerical methods have the advantages of wide applicability high accuracy.It has become the most basic, move active and most fruit method on study of nonlinear problems. The traditional numerical integration method is the most basic numerical method.

It has played the crucial role in forming the science of the nonlinearity, exploring the complex features of the nonlinear dynamic systems. The new theoretical results or analysis results of the numerical methods need to be verified with the numerical integrations. The biggest feature it distinguishes itself from almost all other methods is that it can study either the bifurcation or the chaos phenomena. This feature is also the essential reason why it has the great vitality and been widely used up till now. Now it is the basic major numerical means to study the nonlinear basic theoretical research using the classic low dimensional dynamic systems (Duffing, Van der Pol, Mathieu Equations), is the basis or dispensable link, and the major tool for solving multi-degree of freedoms, strong couplings, strong nonlinearity and changing parameter rotor system relatived nonlinear dynamics problems.

5.10.2 Software of Studying Chaos Problems

(1) Brief introduction on major structure of the analysis software

The software is based on the Poincare mapping principles. The Runge–Kutta method is used in the numerical integration method. Its major functions include: (1) Time domain wave shape analysis for the transient, stable states, phase plane trajectory analysis, Poincare section analysis. (2) Computation of maximum Lyapunov indices. (3) Animation of initial value sensitivities. (4) Study of system bifurcation and chaotic features. The software can be used for the analysis of both the continuous system and the discrete system and for the basis for further development of nonlinear dynamic systems analysis software. The software can be installed under either DOS system or the windows system. To understand features of the systems thoroughly the software uses combination of the bifurcation analysis and global feature analysis. It is coded with C language and composed of head file (MAIN.H) and three source codes (FEFFING.D, GRAPH.C, and DEMON.C), and there are 33 functions.

The multi-initial point bifurcation analysis method for the integration and Poincare point mapping principle. The multi-initial point bifurcation analysis method is to select multi-phase points as initial conditions, and conduct bifurcation analysis from each phase point of the system. The bifurcation figures obtained in this way are called multi-initial point bifurcation figures. This is equivalent to conduct a rough global analysis by selecting a few phase points for each bifurcation parameter values. Thus it is possible to trace multi-solution branches in the multi-attractor's co-existence parameter region and thus find out the co-existence phenomenon among multi-attractors. Using the multi-initial point bifurcation analysis method we can study the co-existence phenomenon of the multi-attractors of the complex rotor systems. When only one initial point is selected, the multi-initial point bifurcation method is equivalent to a common bifurcation analysis method and getting the bifurcation figures is similar to a common bifurcation method. Figure 5.37 shows the initial interface of the software program. The figure shows that the analysis method used for bifurcation is the point mapping method, the analyzed object is the Duffing Equation $\ddot{x} + 0.3\dot{x} - \ddot{x} - x^3 = f\cos(1, 2t)$ which describes a non-autonomous system. The bifurcation vertical coordinate is the x_1 component of the fixed point, i.e. displacement component, and the horizontal coordinate is the disturbing force amplitude f. So that this is the displacement bifurcation figure with the force amplitude as the control parameter, the small line segment crossing the horizontal axis is the bifurcation icon. The little cross in the initial point region is the initial point icon which indicates the position in the phase plane of the current initial point, its sequence number is 3, the phase plane coordinates is displayed on the initial point region as "$H = 0.625\,00$" and "$V = 1.500\,00$". Other positions of the 15 initial point are denoted by the little circle points.

(2) Analysis Software Interface Introduction (Fig. 5.37)

"P-60 $f = 0.31000$" beneath the picture above indicates that for the control parameter at the bifurcation figure with little short line icon its stable solution starting from the third initial point is chaotic solution. The phase trajectory figure and attractor figure both display the phase trajectory lines and Poincare section. If the stable solution is a periodical solution at the icon position, then the attractor figure shows the fixed point of the periodical solution. A series of short lines beneath the horizontal axis indicate the positions of limit and bifurcation points identified by the software. When moving the bifurcation figure icon, the analysis software will automatically find the bifurcation or limit points nearest to the icon position and show their information. The above figure shows that the nearest bifurcation point or limit point from the bifurcation point $f = 0.31000$ is the double bifurcation point from P-8 to P-16: $f = 0.2915000$. Unless it is specially indicated, the step length for the control parameter is implied in the bifurcation horizontal divisions: if the division value precision is accurate to the tenth, then $f = 0.01$; to the hundredth, $f = 0.001$.

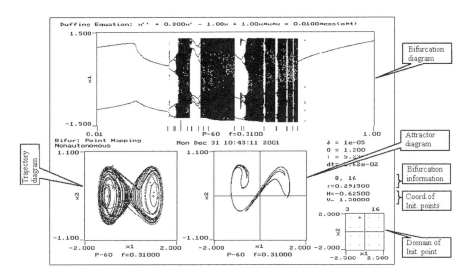

Fig. 5.37 The analysis software initial interface

5.10 Utilization of Chaos   275

The information window is located on the bottom of the screen beneath the phase trajectory figure and attractor figure and the brief computation process information and calculation results are showed.

Viewing the bifurcation records by moving the bifurcation icon and the cross icon in the initial point region the phase trajectory line figure, Poincare section figure, periods, parameters, initial point coordinates, bifurcation point information will be updated. You may increase the viewing speed by closing the phase trajectory lines. The icon can stop at any parameters, any initial point to use the function keys to observe the current stable solutions: (1) F5 key to display time-domain wave shape of the attractor. (2) F6 key to start animation of phase trajectory line sensitivity to the initial value in the region. (3) F7 key to calculate and display the fine structures of the attractors in the attractor display region. (4) F8 key to calculate the maximum Lyapunov index. (5) F9 key to remake the point mapping at the icon location. (6) F10 key to make spectrum analysis for a component of the stable solution. (7) F11 key to analyze the stability of the periodical solutions. In addition, F1 key is used to sequence review, F2 inverse sequence review of the system parameter setup (including dynamic system parameters, numerical integration parameters etc.). The animation display of the initial value sensitivity and the study process of the system bifurcation behavior can be interrupted by pressing any keys to enter into other function or re-entering the corresponding function analysis after altering the system parameter setups.

5.10.3 Application Examples of Chaos

As the rapid developments of the computer and multi-media technology, the computer network is increasingly becoming an important means of information propagation and spread, the web business transactions, remote controls etc., the information security and encryption of the computer network become a critical problem. The web information security and encryption become an issue not only for the trade secret, personal privacy but also for the national security.

One picture is worth a thousand words. The pictures are widely used as the effective means of information expression. The picture security is a big concern in the web technology. The traditional encryption techniques (such as DES encryption algorithm, RSA encryption algorithm) have the shortcomings of the low efficiency of the encryption and decryption when the picture sizes are large. Since 1989 Matthews proposed the data encryption technique based on the chaos system, the picture encryption technique based on the chaos system has become a hot spot both in domestically and internationally due to the features of the chaos such as sensitivities to initial values and the white noise statistical characteristics. The widely used figure encryption technique is the one-dimensional discrete chaos systems (such as logistic mapping). It has the features of simple form, quick generation of chaos sequence, and high efficiency of encryption and decryption. However its shortcoming is the small key space due to its simple structures. Perez and Castillo cracked the lower dimensional chaos encryption by the phase space reconstruction technique and neural network techniques.

The authors proposed a method, based on the continuous chaos system which has complex evolution patterns and high randomness, of construction of discrete chaos system by the least square method and successfully applied it to the picture encryption techniques. The results show that this method can expand the encryption key space to increase its ability to resist the un-encryption coding, and at the same time maintain the high encryption and decryption efficiency of the one-dimensional chaos system.

(1) Construction of discrete chaos system based on the continuous chaos systems

The chaos movement is an internal random process of the nonlinear deterministic system. In general the chaos movements have the following features: the long term movements have high sensitivity to initial values, i.e. long term unpredictability; the small difference in the initial values will result in the huge difference in the movement process after a certain period of time; trajectory lines have no patterns, i.e. local instability and global stability, the class-randomness of the chaos movement, the wide-band Fourier spectra, positive Lyapunov index, positive measure of entropy. As the chaos systems proved an excellent complex and class-randomness features it is suitable for the security encryption.

The simple one-dimensional discrete chaos systems (such as logistic mapping) used for encryption has security concerns due to their simple structures while the continuous chaos systems have the shortcomings of the intensive calculation, not suitable for realization of digital circuit and real time working features even though they have more complex movement patterns and high randomness for the encryptions. Thus the authors proposed the method of construction of discrete system based on the continuous chaos systems by the least square method. The results have shown that this method has much less calculation burden, and at the same time it maintains the original chaos pattern of the continuous systems and is suitable for the digital system realization.

First of all, sample the outputs of the continuous chaos system (the original system); then assume the structure of the discrete chaos system (the derived system); finally determine the structure parameters of the discrete chaos system according to the continuous chaos system by the least square method.

Take the Duffing system as an example:

$$\ddot{x} + a\dot{x} - bx + cx^3 = F\cos t \tag{5.240}$$

where $a = 0.20$, $b = 1$, $c = 1$, $F = 0.27$.

Assume the corresponding discrete chaos system has the following structure:

$$\begin{aligned} x(k) = {} & a_1 x(k-1) + a_2 x(k-2) + a_3 x(k-3) \\ & b_1 x^3(k-1) + b_2 x^3(k-2) + b_3 x^3(k-3) \\ & + f\cos(T_s(k-1)) \end{aligned} \tag{5.241}$$

5.10 Utilization of Chaos

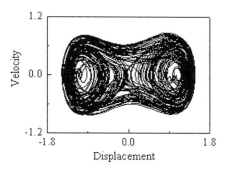

Fig. 5.38 The phase figure of the original chaos system

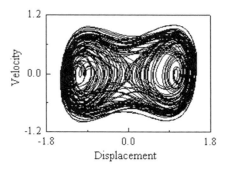

Fig. 5.39 The phase figure of the derived chaos system

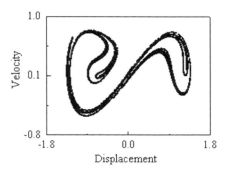

Fig. 5.40 The Poincare section figure of the original chaos system

where T_s is the sampling rate.

The coefficients are obtained by the least square method as:

$a_1 = 2.0086, a_2 = -0.9917, a_3 = 0.0077, b_1 = -0.0271, b_2 = 0.0039,$
$b_3 = -0.0015, f = 0.0075$

Figures 5.38, 5.39, 5.40 and 5.41 are the phase and Poincare section figures for the original and derived chaos systems respectively. It can be seen from the figures

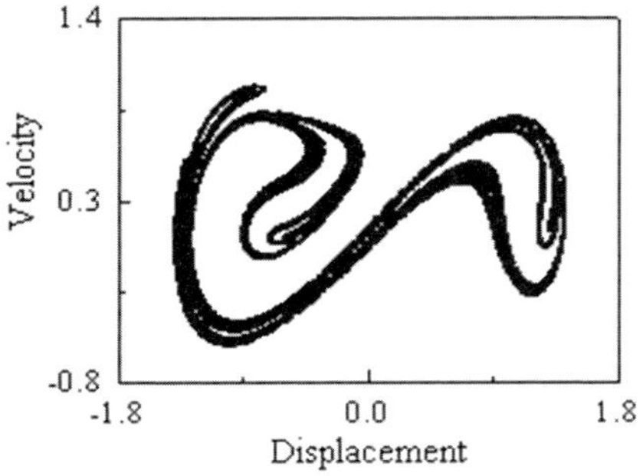

278 5 Utilization of Nonlinear Vibration

Fig. 5.41 The Poincare section figure of the derived chaos system

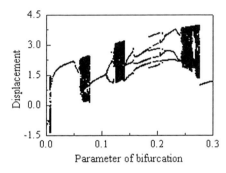

Fig. 5.42 The bifurcation of the derived chaos system with the change of the excitation amplitude

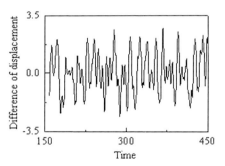

Fig. 5.43 The response difference of the derived chaos system due to the small difference in the initial values

that the derived chaos system has very similar phase figures and the Poincare section figures to those of the original chaos system and its attractors have the bifurcation features. By the further analysis we can obtain the derived chaos system's bifurcation figure changes due to the excitation amplitude changes and the response difference of the curves due to the small difference in the initial values as shown in Figs. 5.42 and 5.43.

5.10 Utilization of Chaos

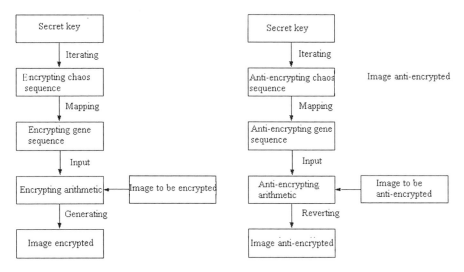

Fig. 5.44 Picture encryption and decryption model based on the discrete chaos systems

It can be seen from Fig. 5.42 that when excitation amplitude values are in the regions of [0.0068, 0.0085], [0.1230, 0.1410], [0.061, 0.078] and [0.245, 0.0260] the systems are in the chaos state.

Figure 5.43 shows the difference of two responses with slightly different initial values: [0.0068, 0.0085], [0.1230, 0.1410], [0.061, 0.078] and [0.2450, 0.0260]. It can be seen that a slight change in the initial values could result in an obvious difference in the responses.

We can know from the above analysis that based on the continuous chaos systems the derived chaos systems by the least square method have complex evolution patterns with the chaos characteristics and can be used for the information encryption.

(2) The picture encryption with the discrete chaos systems

The Fig. 5.44 shows the picture encryption and decryption based on the discrete chaos systems. We use Eq. (5.241) of the discrete chaos system to generate the chaos sequence. The encryption key is constructed by the initial values ($x(k-1)$, $x(k-2)$, $x(k-3)$) and the structure parameters (a_1, a_2, a_3, b_1, b_2, b_3, f) of the discrete chaos system.

The methods of mapping the encryption (decryption) chaos sequence into the encryption (decryption) factor sequence can be classified as bi-value (or n-value) sequence and position sequence. The bi-value sequence is to compare the real chaos sequence $\{x_k: k = 0, 1, 2, \ldots\}$ to the threshold value. It is set to 1 when it is larger than the threshold value and zero otherwise. As the randomness loss of the bi-value sequence is large, the mapping method of the n-value sequence is proposed. It is to map the real chaos sequence between its maximum and minimum linearly via 2^n

grade linear mapping to obtain n binary sequences as the encryption (decryption) factor sub-sequences.

The bit-sequence is to remove the integer part of the real chaos sequence $\{x_k: k = 0, 1, 2, \ldots\}$ and rewrite it as L-bit floating format:

$$\hat{x}_k = 0.b_1(x_k)b_2(x_k)b_3(x_k)\cdots b_L(x_k) \tag{5.242}$$

where $b_i(x_k)$ is the ith bit. The sequence encryption (decryption) factor sequence can be expressed as $\{\{b_i(x_k) : i = 0, 1, 2, \cdots, L; k = 0, 1, 2, \cdots\}\}$.

The authors use an improved bit-sequence method, taking the first w bits after the point and construct the integer:

$$\overline{x}_k = b_1(x_k)b_2(x_k)b_3(x_k)\cdots b_w(x) \tag{5.243}$$

Then take a certain bit in the binary coding to construct a bit-sequence. If the first bit of the binary expression, then it can be expressed as:

$$I_k = \overline{x}_k \ \& \ 0x0001 \tag{5.244}$$

where & is the logic "and" operator. The improved bit-sequence remains the randomness of the original real chaos sequence and is convenience for the encryption (decryption) operations afterwards.

The picture encryption process based on the discrete chaos system is described as follows:

The encryption process:

(1) Given the original picture G to be encrypted and encryption key P.
(2) The chaos sequence is derived from the discrete chaos system determined by the encryption key.
(3) The real chaos sequence is mapped into encryption factor sequence by the improved bit-sequence method.
(4) Encrypt every pixel for original picture G color components by encryption factor sequence.
(5) Transport the encrypted pictures G' in the public channel.
(6) Transport the encryption key P in the security channel.

The decryption process:

(1) Obtain the encryption key P in the security channel.
(2) Obtain the encrypted picture G' from the public channel.
(3) The chaos sequence is derived from the discrete chaos system determined by the encryption key.

5.10 Utilization of Chaos

(a)
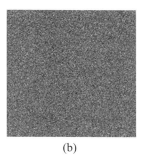
(b)

(c)

Fig. 5.45 Bmp figure encryption and decryption. **a** Original picture **b** the encrypted pictures **c** decrypted picture

(4) Use the improved bit-sequence method and map the real chaos sequence into the decryption factor sequence.

(5) Decrypt every pixel for encrypted picture G color components by encryption factor sequence.

(3) Experiment results

Using the method above to test multi-picture, Fig. 5.45 shows the results of encryption and decryption of Lena Picture and the results are satisfied.

Figures 5.45a–c are the original, encrypted and decrypted pictures respectively. The encryption key is [(−1.2877, −1.2730, −1.239), (2.0086, −0.9917, 0.0077, −0.0271, 0.0039, −0.0015, 0.0075)]. We find after analyzing the encrypted picture that distribution probability of 0.1 is 0.5:0.5, reflecting the excellent randomness of the discrete chaos system and the statistical features of the original systems is completed expanded. If the encryption key is changed to [(−1.2876, −1.2730, −1,239), (2.0086, −0.9917, 0.0077, −0.0271, 0.0039, −0.0015, 0.0075)] and then decrypt the picture and the result is shown in Fig. 5.46. It can be seen that due to the high sensitivity of the iteration to the initial values and structure parameters, even the encryption key has a slight change the results are totally different and thus cannot decrypt the original picture.

Fig. 5.46 Decrypted picture using a wrong encryption key

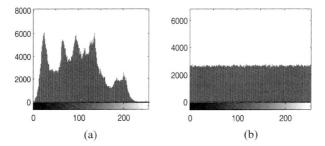

Fig. 5.47 The histograms of the pictures. **a** Histogram of the original Lena picture **b** the histogram of the encrypted Lena picture

Figure 5.47 is the histograms of the original Lena picture and the encrypted Lena picture. It can be seen from the histograms that the histogram of the encrypted picture is near the normal distribution and it coves up the statistical feature of the original Lena picture.

(4) Conclusions

The authors proposed the method of constructing discrete chaos system based on the continuous chaos system, which has the features of the complex change patterns and highly random, by the least square methods to overcome the shortcomings of small encryption key space and weak security. The multi-iteration initial value and complex structure parameters can be used to design encryption keys with the extended space and improve the encryption and decryption real chaos sequence mapping to bit-subsequence of an encryption and decryption factor sub-sequence. The test results show that the picture encryption methods can not only maintain the features of high encryption and decryption efficiency of the one-dimensional discrete chaos system,

5.10 Utilization of Chaos

but also make use sufficiently of the features of the high randomness of the chaos system and further increase the anti-decryption capability of the encrypted system.

The methods mentioned in this section can also be used to the encryption of other information.

Chapter 6
Utilization of Wave and Wave Energy

6.1 Utilization of Tidal Energy

The world energy experts extended their researches to the seas and oceans due to the global energy shortage and environment concerns. The oceans occupy 71% areas and contain 97% of water on the earth. The inexhaustible water in the oceans could have been the hopes of the new energy in the twenty-first century.

The tide is the natural phenomenon of the cyclic rising and falling of earth's ocean surface caused by the tidal forces of the moon and the sun acting on the earth. This is a wave form with the ultra-low frequencies. The changing tide produced at a given location on the earth is the result of the changing positions of the moon and sun relative to the earth coupled with the effects of the rotation of the earth and the local bathymetry. On average, high tides occur 12 h 24 min apart. The 12 h is due to the earth's rotation, and the 24 min to the moon's orbit (in the shore areas of Beibu Bay in China there is only one obvious tide). The differential between high tide and low tide is about 8.9 m. The tidal energy is directly proportional to tide height. In an area of 1 km^2 sea surface, the energy produced in the tidal movement in one second could reach to 200, 000 kW. If the tidal height is 3 m, the period is 7 s and the coverage is 10 km, then the energy generated by the tide is equivalent to the electricity generation of the Xin'an River Hydropower Station in China. It is estimated that the total energy generated by the tides is about 2.7 trillion kW·h of energy. If all the tidal energy could be converted into electricity, the annual electricity generation would be about 1.2 trillion kW • h.

900 years ago, the Ancient Chinese utilized the tides to move stones and materials when they were building a bridge over the Luoyang River in Quanzhou City in Fujian Province, China. In the fifteenth–eighteenth century, the British and French built the water-wheels driven by the tides in the Atlantis Ocean shores. The earliest tide electricity generation is the Bushm Tide power station built by Germany in 1912. In the 50's of the twentieth century the nations began developing the technology of the tide electricity generation. One of the tide power stations which was put into operation in the earliest with the largest installed capacity was the Rance Estuary in Northern France in 1968. Its installed capacity was 240,000 kW and its annual power generation

was 60 billion kW·h. In 1984, the Canadian Annapolis tide power station had the 17,000 kW installed capacity, the second largest one. In recent decadem,the US, British, India,South Korea and Russia invested heavily to develop the tide power stations.It is estimated that the total tide power electricity generation could reach 60 billion kW·h. China has more than 30 year history of studying and developing the tide energy resources. The 1982 census on tide energy resources shows that the total installed capacity for tide power generation is about 20.98 million kW and annual electric production is 58 billion kW·h. The third largest tidal power station in the world is the Jiangxia power station in Wenling City in Zhejiang Province, its installed capacity is 3,200 kW.

The basic principle for the tide electric generation is similar to that of hydraulic electro-generation, it utilizes the potentials of the water during the tide rise and fall to generate electricity, i.e. converting the energy in the tide rise and fall into mechanical energy, and then converting the mechanical energy into electricity. Specifically speaking, the tide electricity generation is to build a dam in a bay or an outlet of a river with tides, to separate the ocean with the bay or the river outlet forming a reservoir, to install water turbo generators in the dam, then to generate the electricity by utilizing the tide rise and fall through the turbo generators. The flow of the tide and flow of river are different and changing in directions, the tide electricity generation has different forms. One is single-reservoir-one-direction and it can generate electricity only when the tide is falling and the reservoir water level is always higher than average tide level as shown in Fig. 6.1. Another is single-reservoir-double-direction, it can generate electricity when the tides both rise and fall. When the tides rise the water level in a reservoir is lower than the tide level, the turbo generators operate in reverse power generation and when the tides fall, the water level in a reservoir is lower than the tide level and turbo generators operate in forward power generation. The reservoir water level relative to the tide level is shown in Fig. 6.2.

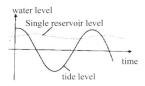

Fig. 6.1 Single-direction power generation: tide-reservoir water level relation

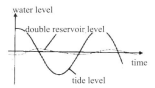

Fig. 6.2 Double-direction power generation: tide-reservoir water level relation

6.2 Utilization of Sea Wave Energy

It is estimated by marine scientists that the energy in the sea waves is as high as 9 trillion kW·h and is 500 times the world electricity production. The scientists had noticed that the sea waves can be utilized for power generation. The movement of the sea waves is a random movement. It is the technical difficulty to be resolved that how to increase the efficiency of the wave utilization. The efficiencies of the tide power stations by the US, Japan and the British have reached to above 60%.

The tide electricity generation has the advantages of low costs, high efficiency, no contamination and no adverse effects on environment. That is one of the reasons why people put their hopes on the tide electricity generations for future energy sources.

The United Kingdom built a tide power station in the isle of Islay. Its installed capacity is 40 kW and could provide electricity power to a small town. The power station uses a Wells Pneumatic turbo to convert the energy, generated by sea wave in a narrow grotto, into electricity energy. The power station uses the high efficient collecting and converting devices and everything from wave entering into the turbo, electricity generation and power supplier to the system network is controlled by the computers.

Japan developed a float type of wave device in 1995. The device is 60 m long, 30 m wide. Its silhouette looks like a whale so it is named "Whale". The 2 km long and 35 m deep "whale mouth" of the device attracts the sea waves, three pneumatic chambers are installed in its front part of the "stomach". When the waves enter the chambers the water surface in the chambers floats ups and downs and thus generates high speed air flow to drive the turbo. Then the compressed air generated by the turbo was sent to nearby sea or a bay to churn the sea water to purify the water, just like a big sea water purifying device. The rotating turbo has the ability to generate electricity. Each of the three chambers in the device is connected to a generator and each generator has the installed capacity of several thousand kilowatts.

Tom Wend of Wave Energy Company in Oklahoma in the U.S.A. designed a wave electricity generating device. It is a cylinder float of diameter 2.44 m (8 ft.) with flat bottom and dome top and looks like a flying dish floating on the sea surface. Underneath there is a pump as a spring load. The pump is tied to the sea bed. When the sea wave pushes the pump up the spring is stretched and when the sea wave pulls the pump down the spring is compressed, thus pumps the water into a turbo and generates electricity.

There are rows of strange houses in the sea near a small island in the South Wales in Australia. That is the wave electricity generation proving ground built by the Sidney University and South Wales University. The sea water is guided to parabolic-shaped low wall through tree-root-shaped low walls in the water and then all the water is reflected into a chamber. The water surfaces violently wave up and down and push the air flow in and out. The pushed up water surface compresses the air up and forces the air to inject out from a narrow channel to a turbo-generator; when the water surface is lowed, the air is attracted into the chamber to drive the turbo-generator. The air in and out drive the turbo-generator rotate in one direction and generate electricity. This system is different from others in two folds: the efficiency of the device is 5 times higher than other device and the parabolic-shaped walls walls which can reflect the scattered waves into a concentrated point in the chamber to make use of a wider range of sea waves. And the parabolic-shaped walls had not been used, the sea waves aside of the chamber would not have been made used of. The parabolic-shaped walls increase the efficiency of using the sea waves. The work state of the system is dependent upon the wavelength of the sea waves. For the electricity generation, the major factor is the height of the sea waves. In the real operation of the device, a little of sea waves could be enough, i.e. decades of center-meter of sea waves could provide the electricity for a city of population of hundreds of thousands.

Holland Teamwork Company developed the experimental equipment for wave electricity generation by making use of the water pressure variation caused by the rolling waves on the sea surface. The equipment is mainly composed of two umbrella-shaped floats submerged 15 m under the sea. The air chambers of the floats are connected by channels. Pumping air into the floats the umbrella-body of the floats will be extended and water can flow in and out. When the wave peaks pass the top of the floats, the water pressure increases and water flows into the umbrella from the bottom. The compressed air is discharged out from the tubes. At this time, the floating force decreases, and it falls into water gradually, then the air is compressed further and more air is discharged. If one float is on the wave peak and another in the valley, the pressure on the one in the valley decreases and attracts more air from the chamber of the one in the wave peak and thus increases the floating force on the float. As the waves rise and fall the floats rise and fall. This movement is used to generate electricity.

The first bank-based wave power station was built in the Pearl River Delta in China. The power station chamber area is 3 m × 4 m and a 3 kW and a 5 kW generator sets are installed. The power station is a bank-based structure with trumpet-bell-typed front port on its side. Its principle is for the chamber to convert the wave energy in the air chamber into the reciprocal movement of the air and the air flowing is made use of generating electricity.

6.3 Utilization of Stress Wave in Vibrating Oil Exploration

The elastic wave (including stress wave and seismic waves) propagates in different velocities and in different forms in different media. When the waves encounter special media and/or a discontinuity in media, it will be reflected and generate a reflecting wave. According to this principle, the stress waves can be used to detect the integrity of the posts and ship anchor state check, diagnose and monitor the defectives and early fatigue cracks on the surface and inside the parts in engineering. It can also be used to explore into the inside of the earth layers. The stress waves can also be used to disturb the oil layers in the oil fields to increase the oil flow and reduce the oil viscosity and hence increase the oil production.

6.3 Utilization of Stress Wave in Vibrating Oil Exploration

In the research on the prediction of earthquake early warnings, the effects of the natural earthquake and artificial vibration upon the oil production have caught the attentions of the scientists. An oil well in Illinois, US, is about 1,500 ft. deep. When trains pass over on the surface the oil pressures in the well fluctuate and it has some effects on the oil output. The gas and oil outputs in Shengli, Dagang and Liaohe oil fields in China, before and after The Haicheng Earthquake in 1975 and The Tangshan Earthquake in 1976, had been increased obviously. Scientists note that compress and expansion waves of the earth surface are the source of the micro additional pressures on the oil layers (mentioned already in the Introduction). Data show the waves of the surface vibration of the earth may become a source of the micro-pressure for the oil layers to increase the pressure to drive the oil and improve the fluidity of the oil. Both theory and simulation tests in labs show that vibration can increase the recovery ratio of the oil.

Former USSR was the first nation to use the controllable vibration source technology to increase the oil production. In the early 1992, a Russia expert Shelietzniev mentioned that the super low frequency controllable vibration source can be used to vibrate oil extraction in a seminar held in Changchun Geology College. The former USSR developed the first large power super-low frequency controllable vibration exciter in the world. It was used for earthquake exploration, the earthquake prediction, building vibration simulation test and vibration oil extraction. The vibration source effective radium is about 3,000~4,000 m, 500~2,000 m in depth, the production increase is about 5%~20%. However the device is complex and the cost is about $100,000/unit. China conducted the study on this and had some excellent results.

6.3.1 Mechanism and Working Principles of Controllable Super-Low Frequency Vibration Exciters

The controllable vibration source is a device of generating earth surface vibration in a human controlled way. The device is a centrifugal vibration exciter with large power, dual eccentrics. The device is composed of a motor with alternative differential adjuster, belt pulley, a transmission, an eccentric shaft with two eccentrics and a frame as shown in Fig. 6.3. The frequency and amplitude of the exciting forces of the device are altered by a distributor and a control device. The horizontal forces are canceled out and the vertical forces are superimposed due to the symmetry of the structure. The centrifugal force is above 100 kN which is about the same as the base gravity. The frequencies of the exciting force can be adjusted to 4~40 Hz. by the adjustable motor. The vibrating device is put on the surface of the earth near the oil well. When it is vibrating, the whole vibration source is like a huge hammer pounding on the earth surface, generating strong penetration elastic waves with super-low frequency and transmitting the vibration energy into the oil layer of hundreds, thousands meter beneath the earth surface and generating a wave field in

Fig. 6.3 Controllable vibration source and semi space

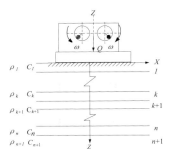

the region. The constant disturbance of the wave to the oil layer generates vibration effects on it which makes the viscosity and solidifying point lower, permeating flow accelerated, and moisture content decreased and thus the oil production increased.

The vibration oil extraction has the following advantages compared with other technology:

(1) The vibration source is on the earth surface.
(2) One point vibration can benefit large area of the oil layer for multiply of oil wells, the effecting radius can be as much as 1,000 m and the depth can be as much as 1,500 m.
(3) Applicability to all kinds of wells.
(4) There is no environment contamination.
(5) Savings on man power, less investment, quick turnaround, high efficiency, simple device.
(6) Can be repeated in one area. One device can be used to more than two regions repeatedly.

6.3.2 Effect of Stress Wave on Oil Layers

(1) The basic factors affecting the permeating flow

The real oil layer can be considered as a fissure and porous medium underneath the earth several hundred or thousand meters, and is subject to pressure and temperatures. When the pressure in the layer has pressure gradient the oil in the layer would have some permeating flow movement. For an oil well with an opening on the surface, the permeating flow will result in oil blowout.

To better describe the permeating flows, the oil layers can be simplified and assumed to be the Warren-Root Model as shown in Fig. 6.4. The model is composed of two parts: part 1 is the uniformed porosity rock system, and part 2 is fissures that separate the rocks or the fissure system. If rock has no fissure it degenerates into uniformed rock systems. Due to the existence of the fissures the above model can be called non-uniform fissure porous media model.

6.3 Utilization of Stress Wave in Vibrating Oil Exploration

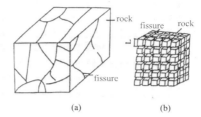

Fig. 6.4 Porous media rock.
a Fissure porous media rock,
b modulated fissure porous media rock

The basic principle to describe the permeating flow movement is Darcy Principle. Darcy Principle can be formulated as:

$$v_{in} = \frac{k_1}{\eta} \frac{\partial p_1}{\partial x_i} \tag{6.1}$$

$$v_i = \frac{k_2}{\eta} \frac{\partial p_2}{\partial x_i} \tag{6.2}$$

where subscripts 1, 2 denote the rock systems that separate the rock fissure system and porous spine system; v denotes the permeating speed; p denotes the pressure, k and η are the permeating rate and dynamic viscous damping; and i denotes the space victor.

It is obvious that the geometric structure and flow characteristics of the subsystem are given; the permeating flow speed depends on the pressure gradients. For the whole oil layer systems, the averaged permeating flow rate depends on the mutual permeating of the two subsystems:

$$q = \frac{\rho \alpha}{\eta}(p_2 - p_1) \tag{6.3}$$

where ρ is the fluid density, α is the dimensionless coefficient representing the permeating density between rock and fissure.

Obviously the mutual permeating exchange depends on the pressure differential between rock and fissure $\Delta p = p_2 - p_1$. The larger the Δp the larger the exchange rate and the faster the total exchange. Below we will use three basic equations to discuss the factors that affect the oil permeating flow in a non-uniformed fissure porous media rock system.

We know from Darcy Principle that there are three factors affecting the permeating flow no matter the rock systems are either rock system or fissure systems: pressure gradient, permeating flow rate, and dynamic viscous coefficient. The pressure gradient is dependent on the mechanics characteristics inside and outside of the oil layers. The permeating flow rate has something to do with the geometry structure, such as porosity, particle size etc. in the system. In general, the larger the particle size, the more the fissures (the smaller the rock averaged size), the higher the permeating flow rate. The dynamic viscous coefficient is associated with the oil physical properties and state and is inversely proportional to the permeating speed. The sticker the oil the slower the oil permeating.

In the fissure oil layers, the porosity of the rocks is generally large, however the permeating flow rate is relatively small; for fissures, it is just opposite: the space that the fissure takes versus the total space, i.e. the fissure is much smaller than the rock fissures, however its permeating rate is much larger than that of the rocks. The fissure systems are the major channels of the oil permeating flows in the oil layers.

When the oil is blowing out from an oil well, the oil in the fissure in the oil layer first moves the oil well which has the lowest pressure subject to the pressure gradient. While the oil in the rock systems will permeate into the fissure under the pressure differentials (Eq. 6.3). The permeating rate in rock systems is much smaller than that in the fissures; therefore, the pressure gradient will decrease in the oil extracting process. Lower production for every oil well will occur sometime of the oil extraction due to the fact that the oil in the rock systems can not compensate the oil loss in the fissure systems. In order to increase the permeating speed and intensity, we can increase the pressure gradient in the oil layer, increase the permeating rate, mutual permeating rate and decrease the viscosity.

(2) Effect of the stress waves on the oil layer permeating flow

When the disturbance force is acting on the surface in the super-low frequency, the impact load will propagate into the media underneath in the form of stress wave in its longitudinal speed and have effect on the oil layer. The most direct effect is to add a periodical dynamic disturbance force to the original oil layer pressure. This addition is to increase the pressure gradient in the disturbance source ambient and thus increase the permeating speed. According to the principles of fracture mechanics, under this impact load the vertical fractions in the porous, fissure media of the oil layers under the vibration source will tend to expand, thus the pressure in the fissures will decrease and permeating flow will increase. The pressure differential $\Delta p = p_2 - p_1$ between the rock system and the fissure system will increase according to Eq. (5.3). The result favors the exchanges of the oil in the rock system to the fissure system and increases the permeating intensity. When other properties remain unchanged, the permeating rate is dependent upon the change of the porosity: $k \approx n^3/(1-n)^2$. It is easy to see that the fissure permeating rate will definitely increase due to the fissure opening and increase of the porosity.

If we only consider fissure pressure change and neglect the gravity effect, we can derive the two-phase stationary permeating continuity equation:

$$\frac{\partial}{\partial x}\left(kx\frac{\partial p}{\partial x}\right) + \frac{\partial p}{y}\left(ky\frac{\partial p}{\partial y}\right) = S\frac{\partial p}{\partial t} + \frac{\partial \bar{e}}{\partial t} \qquad (6.4)$$

in which p is the fissure oil pressure; k is the permeating rate; e is the averaged volume strain; $S = (\varepsilon/\theta + \alpha)r/(1+\varepsilon)$ in which r is the oil density, α is the rock compression

6.3 Utilization of Stress Wave in Vibrating Oil Exploration

coefficient (a reciprocal of the volume modulus), ε is the rock porosity, θ is the oil volume modulus (effective elastic constant). The dimension of S is L^{-1}. For hard rocks, $S = (2 \times 10^{-7} \sim 1.7 \times 10^{-5})$/m, and for the water layer in depth $S = (10^{-6} \sim 10^{-5})$/m.

The left-hand side of the Eq. (6.4) denotes the change of the flow and the first term of the right-hand side denotes the reserve change due to the porous fluid elastic tension and the second term denotes the reserves change due to the porosity change. The two terms are all time variants.

The averaged effective pressure is:

$$\sigma_{\text{avr}} = \overline{\sigma}_{\text{avr}} - rp = \theta \overline{e} \tag{6.5}$$

where σ_{var} is the total averaged pressure, $\overline{\sigma}_{\text{var}} = 1/2(\sigma_x + \sigma_y)$, σ_x and σ_y are stresses in x and y directions respectively.

Taking derivative of $e = 1/2(\varepsilon_x + \varepsilon_y)$ and substituting it into Eq. (6.4) one has:

$$\frac{\partial}{\partial x}\left(kx\frac{\partial p}{\partial x}\right) + \frac{\partial}{\partial y}\left(ky\frac{\partial p}{\partial y}\right) = \left(S - \frac{\gamma}{\theta}\right)\frac{\partial p}{\partial t} + \frac{1}{\theta}\frac{\partial \overline{\sigma}_{\text{avr}}}{\partial t} \tag{6.6}$$

Eq. (6.6) describes the interaction between porous oil pressure p and the averaged total stress σ.

According to the in-door simulation testing results on media of crude oil it can be seen that the amplitude of the permeating flow change is large, it is equivalent to the intermittent output of oil in the production oil well. When the vibration is exerted the intermittent phenomenon is more obvious; the permeating flow under vibration is higher than that without vibration; it is found that the viscosity of the oil under vibration is higher obviously than that of the oil without vibration when the oil viscosity was tested under different temperatures as shown in Fig. 6.5.

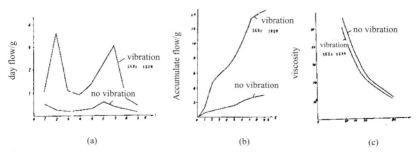

Fig. 6.5 Effects of vibration. **a** Vibration effects on day flow. **b** Vibration effects on accumulate flow. **c** Vibration effects on viscosity

(3) Effects of stress wave on the oil viscosity in the oil layer

The oil layer is a special layer in the upper shell of the earth. It differs both from a pure single solid medium and from a single viscous fluid, owing to its large porosity and full of oil in the porosity and fissures. The media in the oil layer could be an elastic solid for the higher frequency vibration while it should be a matter of transition state in between the elastic solid and viscous fluid for the 10 Hz movement. For the transition state of the viscosity the stress and strain in the media should satisfy the following relation:

$$\sigma_{ik} = 2\mu u_{ik} \tag{6.7}$$

$$\sigma_{ik} = 2\eta \dot{u}_{ik} = -2i\eta\omega u_{ik} \tag{6.8}$$

Equation (6.7) is the Hook Rules for shear stress in an elastic solid in which μ is the shear modulus; u_{ik} is the strain tensor. Equation (6.8) is the stress expression of viscous stress in the viscous fluid in which η is the dynamic viscous coefficient; $\dot{u}_{ik}$ is the strain velocity tensor. For viscous elastic media, Eq. (6.7) represents the high frequency limit situation, Eq. (6.8) represents the low frequency limit situation.

Besides, for a viscous elastic medium, the relaxing time of the character stress relax is a very important character:

$$\tau = \frac{\eta}{E} \tag{6.9}$$

In general, τ is called Maxwell relaxing time, η and E are the viscous hysteresis coefficient and Young's modulus respectively. When viscous elastic medium is subject to a force with frequency ω, if the changing period of the force $1/\omega$ is much larger than the relaxing time τ, i.e. $\omega\tau \ll 1$, then the viscous elastic medium is in the low frequency limit state just as a liquid, otherwise, $\omega\tau \gg 1$, the viscous elastic medium shows the high frequency limit state just as a elastic solid. It is obvious that for the transition state, $\omega \approx 1/\tau$. For the stresses expressed by Eqs. (6.7) and (6.8) their grade of magnitudes must be equal. Thus we have $\eta \approx \tau\mu$, or $\eta \approx \mu/\omega$.

It is known from the above results that for the viscous elastic medium which is in the transition states between the two limit states, its viscosity is frequency-dependent. When an extra-low frequency perturbation propagates to the oil layers, the viscosity of the oil when it is disturbed will be smaller than the viscosity without disturbance. Due to the decrease in viscosity, the permeating velocity and permeating exchange intensity will increase as they are inversely proportional to the viscosity coefficients as shown in Eqs. (6.1)–(6.3). Thus the oil productivity will be increased.

It can be seen that the impact forces at the disturbance sources increase the pressure gradients inside the oil layers around the source and at the same time increase the permeating rate of the oil in the rock fissures and the intensity of the rock permeating into the fissures. In addition, the low frequency disturbance happens to decrease the viscosity of the oil in the oil layers. All of these effects make the oil in the oil layers permeate quickly and increase the oil extraction. However for those wells whose positions happen to be in the regions of fissures in the rocks or stress concentration

6.3 Utilization of Stress Wave in Vibrating Oil Exploration

areas in the rock fraction, the average pressure ingredient between the well and disturbance source will decrease instead of increase, thus reduce the permeating flow speed and the oil extraction could not be increased.

(4) Hysteresis effect of the stress wave

There are two phenomena worth studying in the utilization of extra-low frequency vibration load technology: one is the lag phenomenon. After the disturbance source begins to load, the production of the oil wells would not increase immediately, depending on the distance between the vibration source and the oil wells, the near wells will show the increase in about 20 h and the far wells in 3–5 days. The second one is the delay phenomenon: after stopping the vibration loads, the oil wells remain the increased oil extraction for a rather long time period.

i. The delay phenomena of the unstable permeating flow

On the basis of the Eqs. (6.1)~(6.3) describing the basis equation for the permeating flow in the non-uniform fissure and porous media, giving the continuity equations in the fissure system and rock systems, we obtain the equations of the permeating movement satisfying both systems and further we will give the permeating equation which includes only the function variable of the gap pressure in the fissure system:

$$\frac{\partial p_1}{\partial t} - \eta_0 \frac{\partial}{\partial t} \frac{\partial^2 p_1}{\partial x_i^2} = a \frac{\partial^2 p_1}{\partial x_i^2} \qquad (6.10)$$

where $\eta_0 = k_1/\alpha$, $a = k_1/\eta n_1 \beta_1$, in which α is the mutual permeating exchange intensity between the systems, a is the hydraulic conduction coefficient, n_1 is the porosity of the fissure and β_1 is the compress coefficient in the fissures.

We could obtain the analytic solution in principle given the permeating equations and its corresponding initial and boundary conditions. However for the problems studied, the real initial and boundary conditions are unknown, they need to be solved by the equations of the displacement movements and that is obviously a complicated problem itself. For the unstable permeating flows, Liuqiao Wang etc. proposed an analytical solution. The solution contains an important delay characteristic time in its part which associates with time:

$$\tau = \frac{\eta_0}{a} = \frac{\eta n_1 \beta_1 L^2}{k^2} \qquad (6.11)$$

where L is the rock averaged size; n_1 is the porosity of the fissure and β_1 is the compress coefficient in the fissures. The meaning of τ in Eq. (6.11) is the delay time changing from unstable permeating to stable permeating flow. The larger the averaged rock size, the longer the rock fissure, the larger the delay time. It makes it easier to understand the reasons why the oil outputs do not increase immediately after the vibration forces are exerted. Of course, the delay effects would show in the process of the unstable permeating flow after the disturbance force stops exerting and the oil output would not be decreased immediately.

ii. The flow properties of the oil layers. For the extra-low frequency vibration of 10 times per second, the media in the oil layer does not display the solid elastic properties, rather, the viscous elasticity. If the impulsive vibration load sustains 10~102 days, it, in theory, would display its non-elastic yielding properties for a medium in the oil layers for such movement in such a large scale of time.

To illustrate the flow properties, let's assume first that it is a typical Maxwell viscous matter. For such a viscous elastic medium, its constitutive relation among stress, strain and strain rate is as follows:

$$\frac{\eta}{E}\frac{d\sigma}{dt} = -\sigma + \eta\frac{d\varepsilon}{dt} \quad (6.12)$$

where σ and ε are the stress and the strain respectively; E is the Young's modulus and η is the dynamic viscosity coefficient.

The properties of the flow materials are described by its creeping and relaxing characteristics. The creep characteristic is meant the stress changes with the time under a certain stress while relaxation characteristics is the stress change with the time under some unchanged strain. Figure 6.6 shows Maxwell material creep characteristics. Subject to the continuous stress, the strain of the material linearly increases with the time and the material strains "slowly flow" just like a viscous fluid. Figure 6.7 shows the relaxation characteristics. When the strain remains unchanged, the stress attenuates gradually to zero with the time.

iii. the flow properties of the material subject the extra-low impulse stress. When the Maxwell material is subjected to the extra-low impulse stress, assume its rectangular impulse stress with a period of $2T$ to be:

Fig. 6.6 Maxwell material creep characteristics

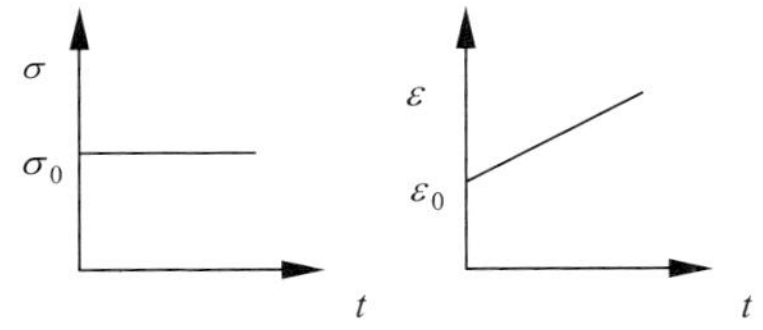

Fig. 6.7 Maxwell material relaxation characteristics

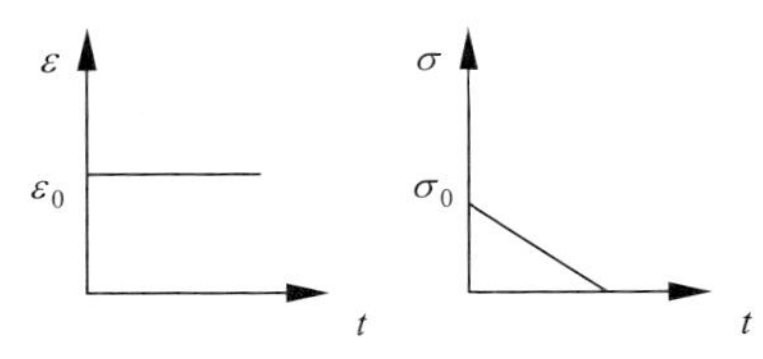

6.3 Utilization of Stress Wave in Vibrating Oil Exploration

$$\sigma(t) = \begin{cases} 0 & t < 0 \\ \sigma_0 & 0 \leq t \leq T \\ 0 & T < t < 2T \\ \sigma_0 & 2T \leq t \leq 3T \\ \cdots & \cdots \end{cases} \quad (6.13)$$

Under such impulsive stress, the strain situation of the Maxwell material can be solved by Eq. (6.11). Note that when integrating Eq. (6.12) for different time segments, the jump time instance $t = 0, T, 2T, \ldots$, its integrals are respectively $t(0^-) = 0 - \delta$, $t(0^+) = 0 + \delta$, $t(T^-) = T - \delta$, $t(T^+) = T + \delta$, $t(2T^-) = 2T - \delta$, $t(2T^+) = 2T + \delta \cdots$, and $\delta \to 0$, we have:

$$\int_{0^-}^{0^+} \sigma(t)dt = 0, \quad \int_{T^-}^{T^+} \sigma(t)dt = 0$$

Thus we could find stress solutions for different time segments:

$$\sigma = 0, \quad \varepsilon = 0, \quad \text{when} \quad t < 0$$

For the first period:

$$\sigma = \sigma_0, \quad \varepsilon(t) = \frac{\sigma_0}{\eta}\left(\frac{\eta}{E} + t\right), \varepsilon(0^+) = \varepsilon^0$$

$$\varepsilon(T^-) = \frac{\sigma_0}{\eta}\left(\frac{\eta}{E} + t\right), \quad \text{when} \quad 0 \leq t \leq T$$

$$\sigma = 0, \quad \frac{d\varepsilon}{dt} = 0,$$

$$\varepsilon(T^+) = \varepsilon(T^-) - \varepsilon_0 = \frac{\sigma_0}{\eta}T, \quad \text{when} \quad T \leq t \leq 2T$$

For the second period:

$$\sigma = \sigma_0, \quad \varepsilon(t) = \varepsilon(T^-) + \frac{\sigma_0}{\eta}\left(\frac{\eta}{E} + t\right)$$

$$= 2\varepsilon_0 + \frac{\sigma_0}{\eta}(T + t), \quad \text{when} \quad 2T \leq t \leq 3T$$

$$\sigma = 0, \quad \frac{d\varepsilon}{dt} = o, \varepsilon(3T^+) = \frac{4T_0}{\eta}T, \quad \text{when} \quad 3T \leq t \leq 4T$$

It is not difficult to see that for an impulsive stress in the $n+1$ period, i.e. $2nT \leq t \leq (2n+1)T$, the impulsive stress makes the accumulated strain of the Maxwell material to be:

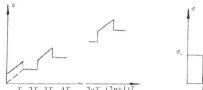 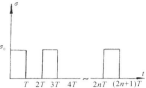

Fig. 6.8 The strain accumulation under the action of the periodical rectangular impulsive stress

$$\varepsilon(t) = (n+1)\varepsilon_0 + \frac{\sigma_0}{\eta}(n^2 T + t) \qquad (6.14)$$

For a real medium, its strains are not increasing linearly when subjected to stress, but decreasing with time. Although each impulsive stress action could be small the strain accumulation could not be ignored when it's subjected to the periodical impulses. Figure 6.8 shows the strain accumulation process under the action of the periodical rectangular impulsive stress. It is not difficult to prove that for each period the strain jump and fall should be a constant accompanying on each stress impulse jump and fall instance: $\Delta\varepsilon = \varepsilon_0 \sigma_0 / E$.

Of course for a real medium with oil, the strain accumulation could not last forever, it will approach to a saturated state. When the extra-low frequency disturbance source stops action, the oil layer has been maintained a large strain values due to the constant strain accumulation. According to Maxwell material relaxation characteristics, the stress corresponding to the strain would attenuate exponentially. It is worth noting that this stress value is not σ_0 when the impulsive stress is exerted initially. If the symbol $\sum \Delta\varepsilon(t)$ replaces the accumulated strain in Eq. (6.14), then the corresponding stress value should be $E\sum \Delta\varepsilon(t)$. As $\sum \Delta\varepsilon(t)$ is a large value, the stress inside the oil layer after stop of the disturbing source is a large and un-ignorable value. It is this kind of stress that makes the oil layer maintain the oil output just before the disturbance stops. For the oil layer media, their relaxation time is in general very large, thus even the vibration impact stops, the above stress value could maintain for a rather long time period inside the oil layer and the oil well output could sustain its increased oil productivity momentum for a period of time. Making use of lag and delay effects, people could stop the vibration impact after the oil extraction shows a stable increase tendency, the oil output as much as that during the disturbance period can be sustained due to the stress by the strain accumulated inside the media. By the same token, when the oil output data show the decrease without impulsive vibration impact the new round of extra-low vibration load should be exerted.

6.3.3 Experiment Results and Analysis

The experiment is conducted via "well measurement execution plan". An oil well factory is divided into two sections: Northeast (NE) and Northwest (NW), segmenting 9 wells (see Fig. 6.9 and Table 6.1) into layers and taking vibration sweep on the layers and collecting data.

When measuring, send the cables of a 3D seismic detector into the well bottom, automatically pushed onto the well wall and then conduct frequency sweeping vibration excitations. The data were collected in the frequency range of 5~13.5 Hz with 0.5 Hz increment. After the well bottom test and measurements, the detector was raised at layer 940 m, 900 m, 700 m, 500 m, 300 m and 100 m. Repeat the above sweeping and measurement process and the data were collected.

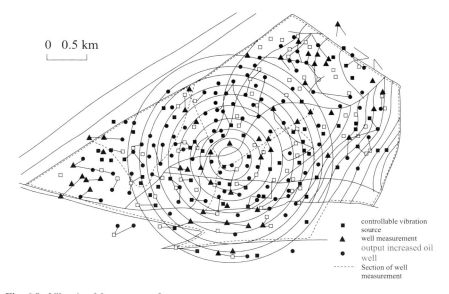

Fig. 6.9 Vibration Measurement Layout

Table 6.1 Well measurement distribution table

NW direction	40,044#	40,045#	42,047#	D161#	4348#
Focal Distance Δ/m	25	150	480	620	750
Oil layer depth/m	915–1075	920–1065	812–952.5	818–1032.5	831.5–952
NE direction		4040#	4139#	4236#	43,035#
Focal distance Δ/m		400	530	820	1000
Oil layer depth/m			888–975	902–955	905–1007.5

300   6   Utilization of Wave and Wave Energy

(1) The attenuation characteristics of the vibration response in the vertical layers of the wells

Each well response includes the vibration amplitude attenuation relations both in the vertical and horizontal directions, the horizontal vibration amplitude is resultant by two mutually perpendicular vibration amplitudes, as shown in Figs. 6.10, 6.11, 6.12, 6.13, 6.14, 6.15, 6.16, 6.17 and 6.18.

It can be seen from Figs. 6.10, 6.11, 6.12, 6.13, 6.14, 6.15, 6.17 and 6.18 that the energies (Amplitudes AH, AV) from the earth surface to the well bottom are attenuating exponentially in three wells (#40,044, #40,045 and #4040) whose source distance is less than 500 m and the attenuation patterns are very simple. The attenuation patterns are complex for those wells whose source distance is larger than 500 m. For those large source distance wells the attenuation patterns are exponential from the earth

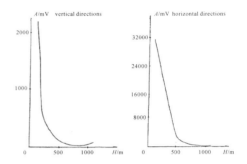

Fig. 6.10 Response attenuation curves in the vertical layer on the well #40,044

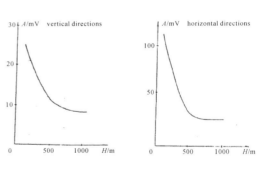

Fig. 6.11 Response attenuation curves in the vertical layer on the well #40,045

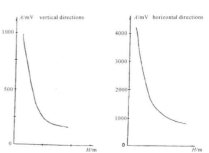

Fig. 6.12 Response attenuation curves in the vertical layer on the well #4040

6.3 Utilization of Stress Wave in Vibrating Oil Exploration

Fig. 6.13 Response attenuation curves in the vertical layer on the well #42,047

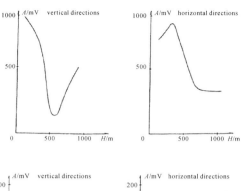

Fig. 6.14 Response attenuation curves in the vertical layer on the well #4139

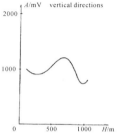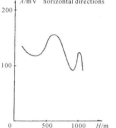

Fig. 6.15 Response attenuation curves in the vertical layer on the well #D161

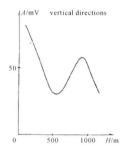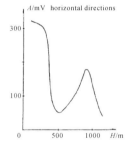

Fig. 6.16 Response attenuation curves in the vertical layer on the well #4348

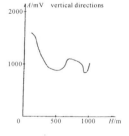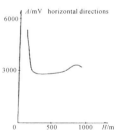

surface to the 500 m depth, when the well depth is over 500 m, the in-depth energy displays a sudden jump phenomenon. It indicates that the vibration source generated elastic waves have complex propagation pattern in different layers.

Fig. 6.17 Response attenuation curves in the vertical layer on the well #4236

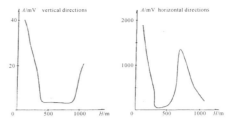

Fig. 6.18 Response attenuation curves in the vertical layer on the well #43,035

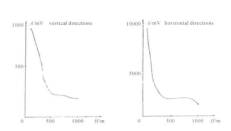

(2) Vibration response characteristics of the oil layers

It can be seen from AH/AV in Table 6.2 that the horizontal amplitudes are larger than the vertical amplitudes and the ratio could be as large as 9.85 (well #4236). It indicates that due to the surface vibration by the vibration sources, the oil layers are subject to larger horizontal actions than the vertical ones.

It can be seen from Table 6.2 that the principal frequencies of the responses are in the range of 10.0~11.5 Hz for those wells whose source distance is less than $\Delta = 500$ m while the principal frequencies vary between 9.0 and 10 Hz for the wells whose source distance is larger than 500 m. It means that changing the principal frequencies could have different effects on wells even in the same well regions.

We can see from the oil layer frequency-amplitude response curve in Fig. 6.19 for well #4139 that even though the response attenuates quickly as the frequencies increase when the frequency is below 9 Hz, its energy is still large. When the frequencies are larger than 9 Hz the resonant principal frequency is the largest near 10 Hz. Since the vibration below 9 Hz has big effects on nearby buildings we should not choose the controllable vibration source below 9 Hz as the vibration principal frequency.

Table 6.2 The oil layer principal frequencies and wave amplitude ratios

NW direction	40,044#	40,045#	42,047#	D161#	4348#
AH/AV	1.38	1.41	4.04	1.33	3.3
f_0	0.0~11.5	10.2~11.5	9~10.5	8.5~10.5	8.0~10.2
NE direction		4040#	4139#	4236#	43,035#
AH/AV		1.4	1.25	9.85	2.09
f_0		10.2	9.5~10.2	8.5~9.0	9.5

6.3 Utilization of Stress Wave in Vibrating Oil Exploration

Fig. 6.19 The oil layer frequency-amplitude response curve

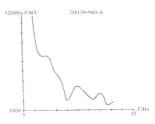

(3) The response characteristics of the oil layers for different source distances

To study the response energy distribution situation of the well oil layers for different vibration source distances, we rearranged the 9 wells by their vibration source distances and plotted to find some relations between the oil layer response amplitude and the vibration source distance in the well NE and NW directions (Fig. 6.20). It can be concluded that with the well depth being within 400 m the oil layer responses are similar and amplitudes are comparable. With well depth being larger than 400 m the maximum changes happen for increments 150~200 m. Well #42,047, #4139 and #43,035 responses are the largest while the wells #D161, #4236 responses are small. We can see from the vibration amplitude changes that the vibrations radiated from the vibration sources do not attenuate by relation of $1/r^2$ but change in the patterns similar to the sinusoidal patterns.

Fig. 6.20 The relation between the oil layer vibration response and the vibration source distance

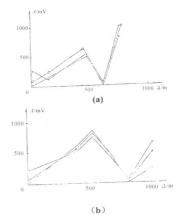

(4) The spectrum analysis of the vibration responses

The principal frequency can be determined by the oil layer vibration amplitude response curves. Comparing with the averaged vibration amplitudes, it is determined that 9~11.5 Hz are the best excitation frequencies. For each single frequency vibration excitation response curve, the best frequency of the maximum amplitude is not a single frequency rather a frequency band forming an energy group of the extra-low frequency envelope.

To further understand the distribution characteristics of the amplitude-frequency responses we have to conduct the spectrum analysis on the response curves. Take well #4236 as an example, the well has the vibration source distance 820 m, the depth of the oil layer is 920~955 m. The measurements are taken at the top, middle and bottom, they are 900 m, 940 m and 1030 m respectively. Figure 6.21 shows the vibration response frequency spectrum curves in 3 coordinate directions. It can be seen from the oil layer response spectrum curves at 900 m depth that the vibration amplitude at the vertical direction is smaller than those in two horizontal directions. The amplitudes in the horizontal directions display a wave-like shape changes. The peak is basically an integer multiple of the surface vibration frequency and the resultant amplitude frequency is higher and the amplitude values tend to be larger than desired. It can be seen from the oil layer response spectrum curves at 940 m depth that the vertical amplitude is smaller than those in two horizontal, the vertical vibration amplitude is stable in the vibration process, the horizontal vibration amplitudes change in a wave-like pattern, the peak frequency is an integer multiple of the surface vibration source frequency. The peak frequency with the obvious pattern has the characteristics

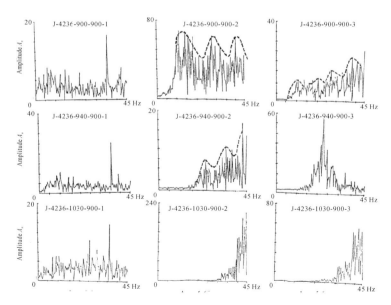

Fig. 6.21 Vibration measurement spectrum analyses for Well No.4236 oil layer

of large amplitude (60 mV), narrow bandwidth (half-power point bandwidth 2 Hz) and it's 2 times of the surface vibration frequency. It is considered to be the major energy source on oil output. At the 1030 m depth, the vertical vibration amplitude is relatively stable, the peak value is not obvious, the horizontal amplitude becomes obviously large at the 4 times of the surface vibration frequency, while the low frequency amplitude is very weak.

6.3.4 Elastic Stress Wave Propagation When a Controllable Vibration Source is Working

In a general case, the earth media can be considered as an elastic body when subject to a short period of force. The earth radius is very large compared with the seismic wave lengths, the earth can be considered as a half-infinite elastic body. In reality, the wave propagations in the oil, due to the internal viscosity and nonlinear connections the internal friction would lead to energy losses. Thus the earth media should be considered precisely to be non-complete elastic media. The plastic strains make the loss of the vibration wave energy and it is different from the diffusion of the vibration wave energy in a complete elastic medium due to the wave-sphere surface increases. It does not depend upon the propagation distance but mainly has something to do with the medium properties and vibration frequencies etc.

Assuming the oil to be viscous elastic medium, in addition to λ, G express the elastic expansion and shear characteristics, λ_1 and G_1 denote the expansion and shear characteristics in a plastic medium, then the wave equation in the viscous elastic medium is:

$$
\begin{aligned}
\rho \frac{\partial^2 u}{\partial t^2} &= (\lambda + G) \frac{\partial \Theta}{\partial x} + G \nabla^2 u + (\lambda_1 + G_1) \frac{\partial}{\partial x}\left(\frac{\partial \Theta}{\partial t}\right) + G_1 \nabla^2 \frac{\partial u}{\partial t} \\
\rho \frac{\partial^2 w}{\partial t^2} &= (\lambda + G) \frac{\partial \Theta}{\partial z} + G \nabla^2 w + (\lambda_1 + G_1) \frac{\partial}{\partial z}\left(\frac{\partial \Theta}{\partial t}\right) + G_1 \nabla^2 \frac{\partial w}{\partial t}
\end{aligned}
\quad (6.15)
$$

Adding them up after taking derivatives of Eq. (6.15) respect to x and z one has:

$$
\rho \frac{\partial^2 \Theta}{\partial t^2} = (\lambda + 2G) \nabla^2 \Theta + (\lambda_1 + 2G_1) \nabla^2 \frac{\partial \Theta}{\partial t} \quad (6.16)
$$

in which

$$
\Theta = \frac{\partial u}{\partial x} + \frac{\partial w}{\partial z}, \quad \nabla^2 = \frac{\partial^2}{\partial x^2} + \frac{\partial^2}{\partial z^2}
$$

Taking derivative of the first equation with respect to z and the second with respect to x and subtracting the first from the second one has:

$$\rho \frac{\partial^2 \overline{\omega}_y}{\partial t} = G \nabla^2 \overline{\omega}_y + \omega_1 \nabla^2 \frac{\partial^2 \overline{\omega}_y}{\partial t} \tag{6.17}$$

in which

$$2\overline{\omega}_y = \frac{\partial u}{\partial z} - \frac{\partial w}{\partial x}$$

If the movement is harmonic then one has:

$$\frac{\partial \Theta}{\partial t} = i\omega\Theta, \quad \frac{\partial \overline{\omega}_y}{\partial t} = i\omega\overline{\omega}_y \tag{6.18}$$

Substituting Eq. (6.16) into Eq. (6.17) one has:

$$\rho \frac{\partial^2 \Theta}{\partial t^2} = [(\lambda + 2G) + (\lambda_1 + 2G_1)i\omega]\nabla^2 \Theta \tag{6.19}$$

$$\rho \frac{\partial^2 \overline{\omega}_y}{\partial t^2} = (G + i\omega G_1)\nabla^2 \overline{\omega}_Y$$

It can be seen that the wave speeds of P wave (longitudinal wave) and S wave (transverse wave) are:

$$c_{P1}^2 = \frac{(\lambda + 2G) + i\omega(\lambda_1 + 2G_1)}{\rho}, \quad c_{S1}^2 = \frac{G + i\omega G_1}{\rho} \tag{6.20}$$

Assuming $h_1 = \frac{\omega}{c_{P1}}, k_1 = \frac{\omega}{c_{S1}}$, then:

$$h_1^2 = \frac{\rho\omega^2}{(\lambda + 2G) + i\omega(\lambda_1 + 2G_1)}, \quad k_1^2 = \frac{\rho\omega^2}{G + i\omega G_1} \tag{6.21}$$

Letting $\delta_1 = \frac{\lambda_1 + 2G_1}{\lambda + 2G}, \quad \delta_2 = \frac{G_1}{G}$, then:

$$h_1 = \sqrt{\frac{\rho}{\lambda + 2G}} \cdot \frac{\omega}{\sqrt{1 + i\omega\delta_1}} = \sqrt{\frac{\rho}{\lambda + 2G}} \cdot \frac{\omega\sqrt{1 - i\omega\delta_1}}{\sqrt{1 + \omega^2\delta_1^2}}$$

$$k_1 = \sqrt{\frac{\rho}{G}} \cdot \frac{\omega}{\sqrt{1 + i\omega\delta_2}} = \sqrt{\frac{\rho}{G}} \cdot \frac{\omega\sqrt{1 - i\omega\delta_2}}{\sqrt{1 + \omega^2\delta_2^2}} \tag{6.22}$$

When $\omega\delta_1$ and $\omega\delta_2$ are very small, one has:

$$h_1 = \sqrt{\frac{\rho}{\lambda + 2G}} \cdot \omega\left(1 - \frac{i\omega\delta_1}{2}\right), k_1 = \sqrt{\frac{\rho}{G}} \cdot \omega\left(1 - \frac{i\omega\delta_2}{2}\right) \tag{6.23}$$

6.3 Utilization of Stress Wave in Vibrating Oil Exploration

In the complete elastic medium:

$$h = \frac{\omega}{c_P} = \frac{2\pi}{Tc_P} = \frac{2\pi}{L_P} \quad k = \frac{\omega}{c_S} = \frac{2\pi}{Tc_S} = \frac{2\pi}{L_S} \quad (6.24)$$

in which L_p and L_s are the P and S wave length respectively.

In the viscous elastic medium, under the influence of the viscous damping, wave number h_1 and k_1 are smaller than those in the elastic medium, how much small depends on the damping. If denoting η_P and η_s to be the P wave and S wave attenuation coefficients, then we have:

$$h_1 = \frac{2\pi}{L_P} - i\eta_P, \quad k_1 = \frac{2\pi}{L_s} - i\eta_S \quad (6.25)$$

Comparing Eqs. (6.25) and (6.23) one obtains:

$$\begin{aligned}\eta_P &= \sqrt{\frac{\rho}{\lambda+2G}} \cdot \frac{\omega^2 \delta_1}{2} = \frac{\omega^2}{2c_P^3} \cdot \frac{\lambda_1 + 2G}{\rho} \\ \eta_S &= \sqrt{\frac{\rho}{G}} \cdot \frac{\omega^2 \delta_2}{2} = \frac{\omega^2}{2c_S^3} \cdot \frac{G_1}{\rho}\end{aligned} \quad (6.26)$$

We know from the above equations that when the medium viscosity characteristics coefficients λ_1 and G_1 are very small, for the P wave and S wave attenuation coefficients are directly proportional to the vibration frequency squared and have very close relationship to the medium physical properties. The relationship is in generally expressed as:

$$\eta = f(\omega)m \quad (6.27)$$

where $f(\omega)$ is the function of source vibration frequency; m is the coefficient associated with the medium constants.

The vibration propagation pattern can be expressed as:

$$R(r)\mathrm{e}^{-\eta t} = R(r)\mathrm{e}^{-f(\omega)mr} \quad (6.28)$$

where $R(r)$ expresses the propagation pattern of wave in the ideal elastic medium and is a function of distance r and has something to do with wave types; the second factor denotes the effect of medium damping.

The energy transmitted to the base from the vibration source is carried out by both P wave and S wave. The basic characteristics of the wave fields are shown in Fig. 6.22. Volume wave propagates outward along a radial half-sphere wave array, while Rayleigh wave propagates along a radial cylinder outward. Every wave propagates outward, the volume of the matter would increase gradually, thus,

Fig. 6.22 The basic characteristics of Rayleigh wave fields

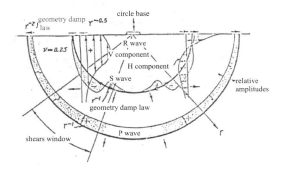

the energy density of every wave would be decreasing as the propagation distance away from its source is increasing. The amplitudes of the volume waves are directly proportional to $1/r$. But it is exceptional to the spatial surface waves as its amplitudes are directly proportional to $1/r^2$. Rayleigh wave amplitudes are directly proportional to $1/\sqrt{r}$. For a circular source oscillating vertically on the surface of an isotropic elastic half space in uniform medium, the three elastic waves have the following energy ratios: Rayleigh wave 67%, shear wave 26% and compress wave 7%. Because the Rayleigh wave contends 2/3 of total energy, and Rayleigh wave attenuates energy with distance much slower than volume wave, for the media near or approaching the surface, Rayleigh wave has an important significance. It can be seen from Fig. 6.22 that the shear waves display the wave-type changes with the source distances and therefore is the segment which has the maximum effect on the oil layers. As Rayleigh wave and shear wave contend 93% of the total energy, the major energy affecting the oil well output is from Rayleigh wave and transverse wave.

Chapter 7
Utilization of Vibrating Phenomena and Patterns in Nature and Society

7.1 Utilization of Vibration Phenomena and Patterns in Meteorology

There exist a variety of dimensions of the vibration phenomena no matter it is in time domain evolution or in spatial distribution. For example, in space perspective, it can be seen that there are a variety of fluctuations of contours in the high attitude synoptic charts. The fluctuations include the fluctuation with wavelength over thousands of kilometers. There exist fluctuations with wavelength over several hundreds kilometers in the atmosphere turbulence flows in the weather phenomena. In the time evolutions of the weather elements, there exist the periodical change phenomena in the time scales. In addition, the international meteorologists find other scales of the periodical changes, such as the micro-physical change in weathers may take just several seconds. The changes of the circulation index in the synoptic charts may take several days or tens of days, the changes of some weather elements may take several years or decades, even several hundreds, several thousands of years. The study on the fluctuation phenomena existed in time and in space is very important in the weather analysis and forecast.

In the process of the atmosphere evolutions, not only does it exist the fluctuations in space scale and time scale, but also there are complex mutual interactions among them. The fluctuations in different space scales have often the periodical changes in the different time scales. The 4-dimensional fluctuation phenomenon, including the time, is a very important perspective of the study on the weather.

The weather elements are a changing random process function (or random function) in the whole physical process, the weather time sequence analysis is the nonlinear dynamic analysis on the basis of random process theory. It is well known that the weather systems include a variety of time–space scales and are a multi-layer complex structure. In another word, the weather data are multi-scales. Therefore, the weather data seem disorganized. The weather system should be, in essence, a random, nonlinear system, and also a random nonlinear vibration system. The analytical method of the time sequence should have common significance. The specific problems to be resolved in the analysis of the weather are: (1) infer the inherent

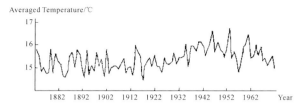

Fig. 7.1 Annual average temperature 1873–1972 in Shanghai

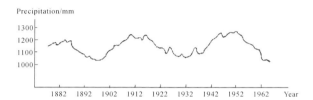

Fig. 7.2 Annual average precipitation 1873–1972 in Shanghai

relations of the weather physical process generating the time sequences; (2) forecast the specific numbers (grades) for future weather and atmosphere changes.

The forecast on precipitation trends has been emphasized. To make precipitation trend prediction, according to the precipitation patterns, could reduce the disaster damage. The precipitation has obvious periodicity. It can be divided into long period, mid period and short period.

The weather trends could be either the large scale weather changes, or the small scale regional weather changes due to human activities. In either cases, there exist the obvious vibration characteristics. Figures 7.1 and 7.2 show the averaged annual temperature (°C) and precipitation (mm), they have obvious vibration patterns.

How to analyze the atmosphere movement patterns existed in a variety of complex fluctuation phenomena in time and in space? As a matter of a fact the spectrum analysis method based on the Fourier analysis had been used on the periodicity of the weather element time sequence. In the atmosphere element evolution process in time, even though it is complex in time domain, we can distinguish it by different frequency fluctuation in the frequency domain by Fourier expansion, effectively analyze the inherent periodicity of the annual changes and effects, discover their change patterns, and further forecast the weather trends. The harmonic wave analytical method is used to study the weather changing patterns.

Assume that a time sequence of a weather element of a given period T is expressed by n observations in equal interval:

$$y_0, y_1, y_2, y_3, \ldots, y_{n-1}$$

where $y_i (i = 0, 1, 2, \ldots, n-1)$ are the observations in an equal interval within a period T, and can represent the annual or month average temperature sequence, annual, quarter precipitation sequence, or weather element sequence observed within 24 h.

7.1 Utilization of Vibration Phenomena and Patterns in Meteorology

We know that the observations are highly nonlinear random vibration. We should add other periods of vibration and random perturbation in addition to the basic period T. The purpose of the harmonic analysis is to separate the harmonic waves with amplitude and phase from the irregular vibration curves and study their individual patterns and characteristics. By the Fourier Series Theory, when $f(t)$ is a function in interval of a basic period T, if some conditions are met, it can always be expressed in a sum of a series of harmonic waves and the harmonic wave has double frequency than its previous one:

$$f(t) = c_0 + \sum_{j=1}^{\infty} c_j \sin(j\omega t + \phi_j) = \frac{a_0}{2} + \sum_{j=1}^{\infty} (a_j \cos j\omega t + b_j \sin j\omega t) \quad (7.1)$$

where frequency $= \omega 2\pi/T$, which is the basic frequency, the integer multiple of the basic frequency $j\omega$ is the harmonic frequency, c_0 is the arithmetic average of all terms; c_j ($j = 1, 2, 3, \ldots$) is the amplitude of harmonic; φ_j ($j = 1, 2, 3, \ldots$) is the initial phase angle of the harmonics.

Equation (7.1) indicates that a complex periodical function can be decomposed into superposition of a series of harmonic functions. Fourier coefficient $a_0 = 2c_0$, $a_j = c_j \sin\varphi_j$ and $b_j = c_j \cos\varphi_j$:

$$a_j = \frac{2}{T} \int_{-T/2}^{T/2} F(t) \cos j\omega t \, dt \quad (j = 0, 1, 2, 3, \ldots) \quad (7.2)$$

$$b_j = \frac{2}{T} \int_{-T/2}^{T/2} F(t) \sin j\omega t \, dt \quad (j = 0, 1, 2, 3, \ldots) \quad (7.3)$$

They represent the participation of the harmonic components $\cos j\omega t$ and $\sin j\omega t$ in function $f(t)$. Notice that $a_0/2$ is the mean values of $f(t)$. As long as the integrations exist for defined a_j and b_j, it is possible to express $f(t)$ in terms of Fourier Series. If $f(t)$ can not be expressed by a function, it may be estimated. For a discrete time sequence $\{y_i\}$ ($i = 0, 1, 2, \ldots, n-1$), the above expression can be approximated to be in a form of a limited sum. In general the coefficients are estimated by n equal interval in the period interval $(0, T)$:

$$a_j = \frac{2}{n} \sum_{i=1}^{n-1} y_i \cos j\omega t \quad (j = 0, 1, 2, 3, \ldots) \quad (7.4)$$

$$b_j = \frac{2}{n} \sum_{i=1}^{n-1} y_i \sin j\omega t \quad (j = 0, 1, 2, 3, \ldots) \quad (7.5)$$

In general, the number of the harmonic waves is maximum $n/2$, thus the right side of the Eq. (7.1) can only be limited terms.

$$\hat{y}_i = \frac{a_0}{2} + \sum_{j=1}^{J}(a_j \cos j\omega t + b_j \sin j\omega t) \tag{7.6}$$

where y_i represents the estimate taken J harmonic superposition, while period T_j and wave number j and the length of the sequence have the following relation:

Wave number j	1	2	3	...	J
Period	T_j $n/1$	$n/2$	$n/3$	...	n/J
Frequency	ω_j ω	2ω	3ω	...	$J\omega$

where $\omega_j = \frac{2\pi}{T_j} = \frac{2\pi j}{n}$. Eqs. (7.4) and (7.5) can also be rewritten as:

$$a_j = \frac{2}{n}\sum_{i=1}^{n-1} y_i \cos \frac{2\pi j}{n} t \quad (j = 0, 1, 2, 3, \ldots, J) \tag{7.7}$$

$$b_j = \frac{2}{n}\sum_{i=1}^{n-1} y_i \sin \frac{2\pi j}{n} t \quad (j = 1, 2, 3, \ldots, J) \tag{7.8}$$

where the variable j is called the wave number, because it really means how many circular vibrations there are in the range of the basic period, $j = 2$ means there are two circular vibration of length $n/2$ in the n equal time intervals.

By similar regression analysis, it can be proved that the total variance of the original sequence $\{y_i\}$ and the harmonic component amplitude c_j have the following relation:

$$\sigma_y^2 = \frac{1}{n}\sum_{i=1}^{n-1}(y_i - \overline{y})^2 = \sum_{j=1}^{J}\frac{1}{2}(a_j^2 + b_j^2) = \sum_{j=1}^{J}\frac{1}{2}c_j^2 \tag{7.9}$$

The contribution from the jth harmonic to the original sequence is $c_j^2/2$, the ratio of it to the total variance is $c_j^2/2\sigma_y^2$. Due to the orthogonality of the trigonometry the harmonics are independent of each other. We can determine if we will keep a term by its contribution to the variance. Actually the variance contribution is just another expression of c_j.

It is worth noticing that the more terms used the closer to the solution. For a weather sequence with a year as its basic period, two to three terms would be good enough to get a decent approximation. By the harmonic analysis, the known basic frequency phenomena can be separated out to compare their coefficients c_j. This method is

7.1 Utilization of Vibration Phenomena and Patterns in Meteorology

commonly used in the weather analysis. It can be used to study the specific situation of the weather elements in annual and daily changes, and compare characteristics in different regions. It can also be used to filter the annual and daily change to study the shorter period of the weather changes, such as the relative weather periods of dry-wet period and cool-warm period. In addition, it can be used to study the space distribution and seasonal changes of high attitude circulation field to reveal the information of the atmosphere super long waves or long waves.

It includes a variety of circular vibration forms in the time sequence. The harmonic analysis is to analyze the circulation vibrations with the known basic period (sequence length). However, it can obtain only double frequency components which have $1/n$ as its basic frequency. For a real sequence, there are other non-double frequency components. This fact brings some difficulty for discussion and utilization. For example, given an observation sequence length, $n = 100$, the period lengths for the harmonics obtained by the harmonic analysis are only 100, 50, 33.3, 25, 20, 16.6, 14.3. The main wave periods existing in the real sequence, however, may not be the same as above, especially in the low frequency fluctuation, it is very possible to miss some of the main periods. For this purpose, the double frequency components may not be used to estimate the real sequences, rather, based on the real sequences it allows the frequencies change in interval $(0, \pi)$, i.e. using the real sequences and the tested periods, looking for the implicit real periods in the sequence within the length of n, $n-1, n-2, \ldots, 2$, the sequence $\{y_i\}$ is considered the results of the superposition of the real periods. We can express the periodical vibrations as the sine functions:

$$y_i \approx C_1(t) + C_2(t) + \cdots = \sum_{j=1}^{\infty} c_j \sin\left(\frac{2\pi}{T_j} t + \phi_j\right) \qquad (7.10)$$

where T_j is the vibration of the circular fluctuation; φ_j is the initial phase. Analyzing the periodical periods is to find the implicit periods T_j, amplitudes C_j and phase φ_j. The method of using the tested periods with the real sequence to find out the implicit periods is called the periodical graphics method. We will exam the character of a periodical function before we introduce the method.

Assuming that the periodical function in an equal time interval is:

$$z = c \sin\left(\frac{2\pi}{T} t + \phi\right) \qquad (7.11)$$

and that the observed time sequences are:

$$z_1, z_2, \ldots, z_n$$

if we take different τ as the tested periods and expand z_i into Fourier Series, then the Fourier Series Coefficients are

$$A_\tau = \frac{2}{J_\tau} \sum_{j=1}^{J_\tau} z_j \cos \frac{2\pi}{\tau} t \\ B_\tau = \frac{2}{J_\tau} \sum_{j=1}^{J_\tau} z_j \sin \frac{2\pi}{\tau} t \right\} \quad (7.12)$$

where J_τ is the maximum multiple of τ in the sequence length including τ. The first order harmonic amplitude and phase are:

$$c_\tau = \sqrt{A_\tau + B_\tau} \\ \phi_\tau = \arctan \frac{A_\tau}{B_\tau} \right\} \quad (7.13)$$

Since τ is not necessarily the real period, c_τ may not be its amplitude neither Denoting the amplitude as S_τ and calling it as the intensity of the tested period we have:

$$S_\tau^2 = A_\tau^2 + B_\tau^2 \quad (7.14)$$

It is obvious that S_τ^2 is a function of τ and it can be proved that S_τ^2 reaches its maximum when $\tau = T$, then S_τ^2 is the square of the real harmonic amplitude c_τ^2. The graph of S_τ^2 is called period figure. These characteristics can be used to determine the real existed periods T, and calculate their corresponding amplitude and phase. However, the real sequence is mixed with the random components, even the original sequences do not include the implicit periods, the sampled S_τ^2 may have several extreme values, i.e., not every extreme value corresponds to a real periodical vibration. However it can be proved that if a sequence is a normal distributed white noise sequence, then the mathematical expectations of A_τ and B_τ are zero. Generally speaking, the computed S_τ^2 should be a small value in a small neighborhood of zero. Although the sequence contains a periodical vibration and random components, the change patterns of the periodical function S_τ^2 are "disturbed", the effect of the periodical vibration is far more than that of the random components when t is taken a value near the T. The change of S_τ^2 is determined mainly by the periodical function, approximately we have: $S_\tau^2 = \max$, $S_\tau = c$, $\arctan A_\tau/B_\tau = \phi$.

To distinguish the extreme value caused by the periodical vibration and the extreme value caused by the random fluctuation, the significance test is needed.

It can be proved that if the original sequence does not have the periodical vibrations, for arbitrary tested periods, the computed S_τ^2 is the random variable of the mathematic expectation $E(S_\tau^2) = 4\sigma^2/J_\tau$ in which σ^2 is the variance of the original sequence. For the statistic variable

$$K = \frac{S_\tau^2 J_\tau}{4\sigma^2} \quad (7.15)$$

7.1 Utilization of Vibration Phenomena and Patterns in Meteorology

Table 7.1 Anomaly sequence of April–May averaged precipitation of the 16 observation stations

Years	Jan.	Feb.	Mar.	Apr.	May	Jun.	July	Aug.	Sep.	Oct.
1951–1960	0.6	21.8	−10.7	−9.7	−22.6	−4.8	19.8	24.4	−23.7	0.4
1961–1970	−12.6	−42.1	50.0	88.3	1.8	−21.0	33.8	−20.8	18.8	22.8
1971–1980	−20.1	−13.7	−9.7	−18.6	−5.6	−11.1	13.4	−13.3	−30.9	−10.6
1981–1990	−31.4	−16.1	39.6	20.9	34.5	−16.8	−8.8	8.7	−15.6	21.9
1991–1999	35.3	−7.9	−21.3	−5.7	−42.8	−15.4	−21.0	53.2	−5.8	

we have

$$P(K \geq k_\alpha) = e^{-i\alpha} = \alpha \tag{7.16}$$

For a given confidence level α, we can obtain the corresponding lower limit k_α. If we obtain $K \geq k_\alpha$, then we reject the null "hypothesis" that sequence has no periodical vibration, i.e. the periodical vibration is significant.

The Hetao region in China lies in the mid of the Yellow River, North of Qin Mountain and Huai River, West of Taihang Mountain. It is a drought and half drought region, especially in Spring seasons. We will use the Periodical Graph method to make a preliminary analysis on the precipitation characteristics and development trends in this region in April–May period. The data are from the records of 160 observation stations from January 1951–May 1999 by the China Central Weather Bureau. First we compute the precipitation in April–May by 160 stations, then use the Yinchuan Station as a correlation point, there are 16 stations whose correlation coefficient ≥ 0.5 in Huhehaote, Baotou, Xingtai, Anyang, Hezhe, Minxian, Taiyuan, Linfen, Yulin, Yanan, Xifengzhen, Lanzhou, Zhongning, Yinchuan, Linxia, Wuwei. The 16 stations' anomaly sequences of the April–May averaged precipitation are shown in Table 7.1.

We know from the Table 7.1 that the years of the more precipitation (larger than 20 mm) are 11 years (1952, 1958, 1963, 1964, 1970, 1983, 1984, 1985, 1990, 1991, 1998) and the years of the less precipitation (less than −20 mm) are 11 yeas (1955, 1959, 1962, 1966, 1968, 1971, 1979, 1981, 1993, 1995, 1997).

Using 3 periods, the computed April–May averaged precipitation expected value in the Hetao region $y(t)$ ($t = 1, 2, 3, 4$) is:

$$\hat{y}(t) = c_1(t) + c_2(t) + c_3(t)$$
$$= 16.37 \sin\left(\frac{2\pi t}{7} + \frac{122.9\pi}{180}\right) + 8.64 \sin\left(\frac{2\pi t}{24} + \frac{243.8\pi}{180}\right)$$
$$+ 10.67 \sin\left(\frac{2\pi t}{3} + \frac{238.1\pi}{180}\right)$$

Fig. 7.3 The curves of the measured and estimated averaged precipitation in April and May by 16 observation stations in Hetao Region

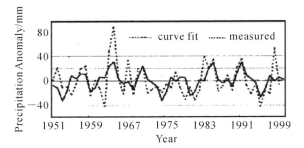

The comparison of the computed estimates and the real measurements are shown in Fig. 7.3. We can see from Fig. 7.3 that the estimated values reflect the characteristics of the measured value and the two have the same trends. However the two have differences because the factor determining the annual precipitation change can not be only one pseudo period. The computed values basically correlate well to the major characteristics of the real value changes, thus the superposed value of the components can be used as the predicted value for the future time $y(t)$. It should be noticed that using the Period Graph method to extrapolate the prediction, there is an issue of extrapolate time limit, otherwise the superposition could be added infinitely.

In addition to the Period Graph method, there are other methods such as Spectrum Analysis method, Variance Analysis method, Maximum Entropy method, Sub-Harmonic Analysis to analyze the implicit periodical vibration problems for the weather time sequences.

7.2 Periodical Vibration and Utilization of the Tide

Ocean areas are about 360,000,000 km^2, about 71% of the earth areas and about 2.5 times of the land areas. The world oceans are a huge treasure house, it has the foods for human needs and rich resources for industry. We have to know and explore the mysteries of the oceans in order to develop and utilize them. The ocean waters move every times, while the periodical rises and falls of the ocean waters are the main patterns of the water movements. This movement phenomenon is the tide. The tide phenomena are very complex. In some places, it is hard to observe the change of the tides, and in other places the tides could be as high as several meters. In Keelong in Taiwan, the sea surface difference between the tide flowing and ebbing is only about 0.5 m while in Hangzhou bay it is about 8.93 m. According to observation, the tide flowing and ebbing has an averaged period of 24 h and 50 min (i.e. a solar day in Astronomy, that is the time for the moon passes the upper culmination the second time in a day). The most common phenomena are two tide rises and two tide falls. In some places, two tide rises and two tide falls are almost equal, i.e. the tidal difference between the first rise and fall is almost equal to the tidal difference between the second rise and fall. This kind of tide whose rise time and fall time are almost

7.2 Periodical Vibration and Utilization of the Tide

equal (6 h 12.5 min) is called "half day tide".In other places, the rises and falls in a day are much different and the rise and fall times are much different, and the tides are called "mixed tides". In a period of a solar day, there is only one rise and one fall in some places, and the time interval between tide rise and tide fall is 12 h 25 min, and the tide is called "full day tide".

On the causes and patterns of the tides, there were many understanding opinions in ancient China. Even though the causes for tides are complex, it is generally acknowledged that the moon's "Tide Generating Force" is one of the major causes. There are two factors for the "Tide Generating Force": (1) The attraction of the moon. The Universal Gravitation tells us that there is a mutual attraction between two substances. The earth and moon constitute a gravitational system with a common center of mass which is located on the earth radius of 0.73 time. (2) A centrifugal force is generated when the earth rotates around the common center of mass. The resultant force of the earth centrifugal force and the moon attraction force is the "Moon Tide Generating Force". In addition, the sun's "Tide Generating Force" is the second force. The mass of the sun is about 2,700 times of the Moon's and the averaged distance between the sun and the earth is about 389 times of that between the sun and the moon. It can be deduced that the radio of the sun's "Tide Generating Force" to that of the moon's is 1:2.18. The tide caused by sun is hard to be observed, however, it affects the moon's "Tide Generating Force".

When the sun, the moon and the earth is on a straight line, the sun and the moon's "Tide Generating Force" has obvious resonance decompress effect. The ancient Mayan, Egyptian, Indian and Chinese all emphasized the peculiar astronomical phenomenon of the "3-point-on-one-line" and thought that the phenomenon indicated some severe disasters would happen. Some people had used the prediction theory to verify the severe earthquakes happened in the last 100 years and found that its accuracy was 99%! Chinese scientists verified from typhoon data in the past and they found that in the recent decades there were 5 typhoons landed at the Chinese continents with rainstorms of 1,000 mm precipitations. Each of them encountered the double triggering of the sun's and moon's "Tide Generating Force" which caused the sudden decompresses in the low atmosphere pressure weather system and thus strengthened the typhoons. The "FeiFei" typhoon in Central America in September, 1975, "Andrew" in Florida, US, in August, 1992, "9809" typhoon in the Philippines on October 14, 2002, "Miche" Typhoon in Central America on October 30, 2002, all happened to have the sun's and moon's "Tide Generating Force" resonance effect within 36 h the typhoon landed and suddenly became more disastrous and more severe consequences. This was the result of the couplings between the rainstorm situation and the resonance decompression. If one of the factors was missing the severe rainstorm like that would not have happened.

The regular patterns of the celestial body movements mean that there exist the periods of the tides. To study the periodical is the first important thing for the tide analysis. The tide study is associated closely with the defense constructions, sea transportation, ocean resource development, energy utilization, environment protection, and port construction etc. There are many irrigation gates in Chinese coastal areas. These constructions are subject to the effects of the tides and the tide factors must be considered for the designs of the projects. In constructing the ports, the tides must be considered: when the tide rises the water won't submerge the ports and when the tide falls the boats won't be ashore.

When the tides rise, the high salt density sea water is pushed ashore, this is a good opportunity to extract the salt, it is time to open the gate and let the water in for the future salt extraction.

The tidal differentials contain a huge energy which can be utilized for electricity generation. It is roughly estimated that the tidal energy in the world could generate 1.24 trillion kW·h electricity. The principle of the tidal electricity generation is the same as a general generator. It is to utilize the tidal differentials to drive a hydraulic turbine and the turbine drives a generator to generate electricity. There are three types of installations: single-way type (the water is introduced to the pool when the tides rise, then use the water differential outside and inside the pool to generate electricity); two-way type (generate electricity when the retaining and discharging); continuous type (using multiple of pools to generate electricity).

7.3 Vibration Patterns and Utilization in Other Natural Phenomena

7.3.1 Periodical Phenomenon of Tree Year-Ring

The growth of the trees is not only controlled by their own DNA but also affected by the environments in which they grow. The two factors are mutually connected. There are different types of trees at different elevations, and the same tree grows differently at different elevation and regions. That are the results affected by the weather factors during their growth process. The tree year-ring meteorology is a branch of science which induces the weather in the past by studying the changes of the year-rings.

The study of the tree year-ring meteorology is suitable to the dry and half-dry and even soaking regions. It reflects the most sensitive information and most of the information needed and stored. Except the tropical areas, the tree grows a year-ring in a region with obvious weather changes. The year-ring is not only a time record, but also a treasure house for scientific information. In each of its year- ring, it contains the information chain on the natural environment and human activities and condensed them into the combination of the reproduction cells of the year-rings. The width of the year- rings has associated strongly with the weather conditions. The wide or narrow and uniformities of the year- rings reflect not only the tree growth speed, the quality of the tree growth, but the record of the atmosphere changes. In warm and wet years, the trees grow quickly, the year- rings are wide; in cold and dry years, the trees grow slowly and the year- rings are narrow. By analyzing the year- rings, the information on past weather changes could be obtained and the change situation of some weather elements could be inferred to compensate the lack of the historical weather data, the weather change patterns for hundreds, even thousands years could be obtained. The data can be used to forecast the weathers.

The first American tree-ring meteorologist is the Astronomer A.E. Douglass. He was initially interested in the relation of the sunspots to the precipitation and thus thought that the tree growth might be associated with the precipitation. In 1900s he devoted himself to the study of year- rings in the Southwestern US drought region and collected the samples and the year determination. He was the first person to expound the relation between the changes of the tree year- ring width and the real precipitation in more than 500 years and established a lab dedicated for the study of the tree year rings 1930's. After him many tree year-ring meteorologists conducted the studies on relations of the physiology process of the tree growth to the weather changes, and acknowledged on the selections of the sample trees and year-ring sequences and established the basic principles and methodologies for the year-ring meteorology. Douglass found that the changes of the year-ring width have a period of 11 years, and he analyzed the trees collected from the Britain, Norway, Sweden, France and Austria and draw the same conclusion. This finding brought the joy to him while he was puzzled. He suddenly thought the period would be the same as the period of the sunspot occurrence. He compared the changes of the year- rings and the changes of the sunspots, they were the same! It turned out that the changes of the sun radiation have effects on the tree growth. When the sunspots occur and affect the weather and typhoons and rainstorms occur. It also indicates that when the sunspot activities are strong, the sun radiates more lights and thermal than others. The effects make the trees grown faster and the year-ring widths become wider. In later 1970's, the analysis on prairie tree year-rings in Tibet showed that there were two large cold weathers in Tibet in last century and the period of the weather changes is 200 year.

The modern hydrology records are similar to the weather observation records, it is a short time and has no enough data to reveal the history of the atmosphere and hydrology changes. It limits the effects of the hydrology statistical results and hinders the accuracy of the statistics application to the hydrology forecasts. To compensate the limitation the scientists use the year- ring analysis to extract information on the weather and hydrology from the tree year- rings to restore the historical process and form the long sequence of the hundreds even thousands years. This method reduces the errors by the unstable factors in the sequences.

The year- rings can reflect the pollution situations in the atmosphere. When the air is polluted, the year-rings contain the pollutants. The contents of the pollutants in the year- rings can be used to monitor the pollution history.

7.3.2 Bee's Communications Using Vibrations

Scientist found that bees use a special "dance" to spread the food position information to their partners. But people do not know how the bees see the "dance" in the dark honeycomb cells.

A study group in Germany conducted a study on the bee behavior patterns. They found that when the bees dance to pass some information on to their partners, their honeycomb will generate weak, low frequency vibration. If the honeycomb vibration is held back, the number of the bees which can understand the "dance" information is reduced about a quarter of the bees who understand the information without holding the vibration.

The study also revealed that for any of six-sided honeycombs, the vibration directions of the opposite honeycomb walls are the same while at a place where is about 2–3 honeycomb distances, the opposite honeycomb walls have the same vibration direction. The scientists think that the bees sense the vibrations via their six legs and acquire the information from their peers in the dark and noisy honeycombs. Other experts think that the vibration of the honeycomb is similar to an earthquake, the study on the bee vibration would help the earthquake resistance of the buildings.

7.4 Utilization of Vibration Phenomena and Patterns in Some Economy Systems

7.4.1 Fluctuation and Nonlinear Characteristics in Social Economy Systems

(1) The economy systems in constant fluctuation processes

In the economic fields, there exist vibrations (fluctuations) everywhere and we can sense it intuitively. For example, the economical situations in a country or a region are good for some years and bad for other years. The produce prices rise for a while and then fall down; bond rates, stock price, fix investment, inventories and electricity consumed are changing and fluctuating with time. The data show that there are periodical fluctuations with different frequencies in the economical operation processes.

The occurrence of the economy crisis and the financial crisis are a presentation of vibration. The economical fluctuation phenomenon has its own patterns. The economists can forecast the economic growth or developments in a region or a country by the domestic and international economic, political and social factors, i.e. the vibration patterns in economic development process. They might propose the corresponding effective measures to reduce the social and economic losses due to some factors.

Economy systems have fluctuations, and some times have terrifying waves. The perfect forms of economic theories and cruel economy reality display a huge contrast. Great Depression in world economy in 1930s and financial crisis in Southeast Asia in 1997 all proved it. When Financial Crisis erupted, fortunes of Southeastern people were devalued half over night, while financial opportunists made a big fortune. The reason the opportunists made a big money was that they had in depth understanding for the nonlinear essence and fluctuations of world economy. They applied the Uncertainty Principle in Quantum Mechanics and created the famous "Reflexology". Based on the theory, they formed their own recognition on violent ups and downs of world economy, proposed a unique risky venture method. It is found after studying their ideas that what they used in their thinking patterns was the wave and nonlinear patterns. Mastering the habitual practice the opportunists used some measures could be taken under the permit of the economy power strength to prevent the occurrence of the economy crises. It is very difficult to resist this kind of risk and crises for those countries whose economy strength is weak. The possible methods could be to seek international association and assistance.

Economy fluctuation phenomena have its own patterns. Economists could predict the rises and falls of a stock price by the internal, external influential factors, the general rules of the stocks, i.e. mastering the vibration patterns of the stock rise and fall.

From the angle of ecology, in the human reproduction process the reproduction speed is high in some time and low in other times. This phenomenon can be described by the vibration theory.

(2) Complex nonlinear economy systems

As the progress of the science and technology and development of the modern industry, especially the progress in the nonlinear science, traditional theories of the sciences are impacted strongly. The scholars in the economy fields acknowledged that the economy systems in some senses are more complex than the physical and chemical systems. The economy systems are the large scale one with human participations. The complicities of the economy systems are in the following aspects:
(1) Multiple variants. The economy systems contain many variables and the changes of each variables affect the changes of the systems in some extents.(2) Multiple levels. The main components composed of the system play different roles in the evolution process of the systems and their patterns may be drastically different.(3) Strong couplings. The variables in different levels or the variables in the same levels in the economy systems are mutually associated and constrained, each variable's change affects other variables and at the same time is affected by other variables. (4) Nonlinearity. The relationships among the variable are nonlinear. The action patterns and extents of the variables upon the systems are all associated with the environments in which the variables lie and the conditions upon which the variables are dependent and the changes of the systems display a variety of forms. That is the important sources of the system complexity. Many economists found that the objects of the economic studies are extremely complex systems of multiple factors which are mutually and nonlinearly correlated, and the simple and linear systems are very few. The behaviors of those complex systems are dynamic, unstable, random and irreversible with multiple development tendencies. The stable and balanced systems are few. The existence of the irreversibility and randomness makes the economy systems themselves to possess very complex nonlinear characteristics. The asymmetric supply and demand, asymmetric economic periodical fluctuation and the time delay in the economical variable iteration processes are all the specific presentations of the nonlinear characteristics. Besides, the complete deterministic and simple ideal behaviors would also possibly result in the random behaviors of the economy systems.

Let's retrospect the development process of the economy theories. Traditional economists thought that long term economy change tendency had its own unique reasons, while short term irregularity was caused by random factors. The corresponding economy models are linear equations (or logarithm linear) plus random terms. Economist Samuelson described the macro economy movement by multiplier-acceleration linear difference. Due to periodic solutions of the linear difference equations are unstable and harshness on parameter value requirements, it is hard to find solutions which satisfy real situations. Later on economists Hicks and Goodwin introduced nonlinear mechanism in economic models and obtained "limit ring" solutions (i.e. periodic solutions of the equations). Although the solutions were stable, they could not explain the phenomena of coexistence of multiple periods. In 1959, taking logarithm on stock prices people found that the stock prices followed Brown Movement. That promoted the economy world replace the linear determined model with the linear random equation. They attributed the irregular rise and fall to "white noise" which is beyond the economy systems and they assumed that economy systems themselves were within the stable region. Those view points put traditional economy theories into crisis of methodology. At the same time the economy theories could not be subject to the real world verification. The predictions from those models were so much different from the reality they could not predict the severe economy retreats in 1981–1982 and the US strong economy recovery and inflation decrease, thus resulted in a mistake in economy policies. That is because they attributed the nonlinearity to the external factors. They could not predict the real situation accurately and could not find a reason for it.

The facts prove that we need a new theory and method to recognize the economy systems again. For traditional economy theories we could not put some patches on them. The fundamental problem for the traditional theories is their assumptions are not correct.

The important purpose of the studies in the economy fields is to make accurate predictions for the further economy postures. It is no doubt that the traditional economy theories and prediction methodologies play an important role in the social and economic developments. Using the nonlinear system theory to study the social and economic phenomena was emphasized by the sociologists and economists. Prigogine pointed out that we were at the apex the linear science could ever reach and at the same time we were at the beginning of the new science: the nonlinearity science. The latter opens a new window for us to look forward for the futures of the world.

(3) Chaos in the operation process of the economy systems

The economy issues, such as the economy growth, enterprise competition, stocks, exchange rates and precious metal prices, display a variety of complex phenomena and behaviors including the chaos due to their natures of the nonlinearity. The study of the economy phenomena needs urgently to combine the results of modern science development in nonlinear system theory to develop new theories and methodologies for the economy forecasts.

The US economist Stutzer in 1980 discovered the chaos phenomena using Harvey economy growth equations and it is acknowledged that there are inherent randomness in the economic models based on the traditional economics theories, which generated a great impact on the mainstream economics in the West. It indicates that even though the governments interfere the economy using measures such as budgetary finances, the randomness and complexity of the economy behaviors put a great constraint to accurate economy forecast and even reasonable prediction behaviors become complex. The inherent randomness in the economy systems shakes the hypothesis of new classicism that the source of the economy fluctuation come from the factors outside the economy.

The US economists Benhabid and Day in 1983 studied the consumer tendency function and the results showed that the consumption behaviors for different incomes are different and the poor people's consumption behaviors are stable and the rich are possibly periodical or chaos.

Baumol and Wolff in 1983 found when they were studying the relation between the enterprise expense on R&D and the enterprise production growth rates that when the ratio of R&D expense to the enterprise revenue reaches to a certain range the enterprise production growth rates display periodical changes and even chaos; this indicated that the rationale of the expending on R&D depends on the absorption ability of the enterprise on the new technology innovation. Baumol later found after studying the relation between enterprise advertisement expending and the enterprise sales revenue that when the consumer favoritism on market demand happens to obey a pattern, the advertisement expending level reaching to a certain proportional limit would lead to the periodical fluctuation in sales profits or chaos.

In 1985 Baumol etc. studied the predictableness of economy by considering the situation in which one-dimensional iteration function could generates chaos and economy characteristics.

The dynamic characteristics displayed in large amount of the time sequence in the real economy life showed similarity with the chaos results in the physical system digital simulation. The economists were looking for the empirical evidence of the chaos in the real economy data. The real economy data were processed mathematically and the test method in determining the nonlinear characteristics was used to find a strange attractor to judge the existence of the chaos.

In 1988 economist Ramsey checked the data of quantity of money in the US monetary market from January 1959 to November 1987 and obtained evidence of chaos existence. Chinese scholar Ping Chen in 1987 provided the empirical evidence to show the existence of chaos in the real economy data, and obtained the monetary strange attractor of 1.47 dimension with basic period of above 4.7 year by Poincare and Lyapunov index criteria.

On the fractional dimension of the Chinese stock markets, Longbing Xu etc. in 1990 studied the time sequences of the Shanghai Stock Market index (sampled between December 1990 and October 1998) and the component index of Shenzhen Stock Market (sampled from April 1991 to September 1998) and obtained that fractional dimensions are 1.629 and 1.555 to reflect the inherent nonlinear characteristics of the Chinese stock markets.

In 1994 Rosser described the model of the multiplier-two stage accelerator economy of the capitalist market economy and long investment wave model for socialist planning market and proved that necessity of existence of market fluctuation and planning systems of socialist economy would have long term of fluctuation and chaos.

People predict that there exist more complex dynamic behaviors than chaos- super chaos movement. Therefore it could be beneficial to understand the economy system development and reveal the real phenomena by studying the nonlinear problems in economic areas systematically.

7.4.2 Growth and Decline Period in Social Economy Development Process

The specific contents of the social economy formations are in the constant changes and fluctuation. The fluctuations in general display the periodical properties. There is no complete repeatability, but in most cases there are some periodicities. For example, the prices of merchandise are high sometimes and low other time; the development in some countries or regions is quick or slow in different time frame; in the stock fluctuations, the bear markets and bull markets appear alternatively. The averaged values of many economy systems often display gradually increasing tendency but the periodicity of the fluctuation process is very obvious.

(1) The growth period in the economy development process and rare chances

The economy formations are complex, their models can be abstracted and the ideal models can be established. A system may have many periods, long periods, mid periods and short periods. We can expand the models in Fourier series, and the periodical functions can be expressed by a sum of sub-harmonics with a basic frequency and the integer multiple frequencies of the basis frequency. Those harmonics may have different effect on the sum. In Fourier series it takes a few terms as approximation. Sometimes only one term is used, i.e. the system is simplified by a sinusoidal function (Fig. 7.4).

If we idealize an economy formation as a sine curve, then there are peaks, also called apex, top and the lowest point, also called valley point. And the curve can be divided into

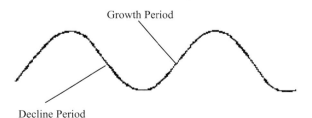

Fig. 7.4 Growth and decline period of an idealized economy system

7.4 Utilization of Vibration Phenomena and Patterns …

segments of increase and decrease. The special growth points can be chosen in the increase segments. The so called growth points represent the quick development periods. Those periods are important if they can be identified correctly. The leaders and decision makers of the enterprises should grab those growth points, the success chances.

In order to succeed in a cause, the first thing is to choose a time to start an investment, i.e. the growth period in the curve, not the decline period. Secondly select the optimal point on the curve, i.e. ideal growth point. The optimal point or ideal growth point should be after the decline period, i.e. after the valley point. The ideal point is at 1/4–1/2 positions.

(2) The recession period and measures to be taken in the economy development process

Any matters are in state of change, they have growth period and decline period. There are no such things which have only growth period and no decline period. For any fluctuation matters, there are growth stages and decline stages. If an ideal model for an economy system is a sine curve, then in the decline stage, we can find the accelerating decline period. To avoid losses, a variety of measures could be taken:

(1) Macro-adjustment measure: to reduce the period of the decline or reduce the amplitude of the decline period.
(2) Quick withdraw the investment or stop production before the decline period is coming.
(3) Reduce investment and reduce the production to the minimum.
(4) Change directions, i.e. transfer the enterprise to another enterprise.

7.4.3 Active Role of Macro-adjustment in Preventing Big Economy Fluctuations

The economy developments in any country have different forms of fluctuations. The fluctuations are big sometimes and are small other times. When the fluctuation scales are large the necessary measures must be taken to prevent over fluctuations. This is the most important task for decision makers.

In the last half year of 2004, there was some over-heating in China economy, such as over investment in steal, concrete industry, real estates. Some economists predicted that if no effective measures would be taken, the economy in some aspects would be misadjusted and it would bring an adverse effect on the future of the economy. The Chinese government took macro-adjustment policies to limit the investments and construction on those projects which should not be developed or low level repeats and also limit the investments scale on the fixed properties. It was proved that the effects of the measures are showed.

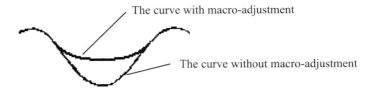

Fig. 7.5 The active effect of the macro-adjustment on economy fluctuation

We would compare two situations showed graphically in schematic Fig. 7.5 (only fluctuation was showed): (1) the vibration curve with no limits, no macro-adjustment and let it go ; (2) the curve with macro-adjustment.
The macro-adjustment can alleviate, even eliminate the adverse effects (figure shows only one situation); it is obvious that the curve without macro-adjustment has a valley and if it is severe it could lead to the collapse.

7.5 Utilization of Vibration Phenomena and Patterns in Stock Market

7.5.1 Stock Fluctuation is One of Typical Types of Economy Change Form in Social Economy Fields

The change and volatility of the stock markets are one of the most typical change forms of the social economy. It has the following characteristics:

(1) Volatility

The stock traders observe the stock prices are changing constantly, sometimes they are higher and sometimes they are lower; the exchange volumes are experiencing the same thing.

The Bull Markets and Bear Markets appear alternatively in the stock market oscillations if they are observed from big rise-fall trends. The alternative periods sometimes sustain several years and sometimes just a couple of months. Of course the stock traders would expect the Bull Market last as long as possible, which is not realistic as we know. According to the stock volatile patterns the total amount of the stock price increase during the Bull Market is roughly equal to the total amount of the stock price decrease during the Bear Market if the averaged stock prices were to keep the same. If the stock market maintains its rising trends, the total amount of the stock price increase in a Bull Market should be larger than that of the stock price decrease; while the stock market maintains its falling trends, the total amount of the stock price increase in a Bull Market should be smaller than that of the stock price decrease. This volatility always exists in stock markets and is an inevitable pattern so

7.5 Utilization of Vibration Phenomena and Patterns in Stock Market

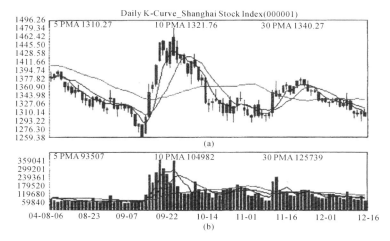

Fig. 7.6 Fluctuation curves in stock market **a** Shanghai Index daily-K curve **b** The fluctuation of a stock price

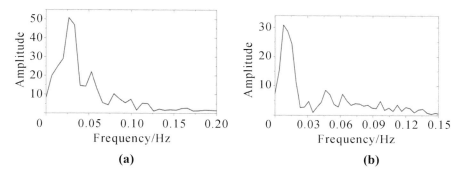

Fig. 7.7 The spectrum analysis on curves in Fig. 7.6 **a** Shanghai Index daily-K curve **b** The fluctuation of a stock price

called the volatility rule. Figures 7.6a, b show the daily-K curve and time-price curve of the Shanghai Stock Index respectively. Figure 7.7 shows the spectrum analysis results of Fig. 7.6. The period of the peak value in Fig. 7.6a is 37.7 days while the period of the peak value in Fig. 7.6b is 128 min. However both spectra of the curves are broadband spectra.

(2) Non-repeatability

The stock price changes lack of repeatability and are random. The stock price changes are affected by a variety of unpredictable internal and external factors. It is difficult to find the same situation happened even in some very closed situations. The non-repeatability brings difficulty in predicting the stock oscillations. There are some patterns by which can be followed even though the non-repeatability exists.

(3) Pseudo periodicity

The patterns could not be classified or calculated exactly as periods of repeatability. It is called the pseudo periodicity. In long term of stock price fluctuation, you may find more than one pseudo periodicity, the big pseudo periods may be as long as 1–5 years, mid pseudo periods could be calculated as several months while the small pseudo periodicity may be defined as several days even as short as hours or minutes.

In the long term fluctuation process, it is found that there are many periods, big ones are 1–5 years, mid ones are several months and small ones are only several days, the smallest ones count only hours or even minutes.

In the alternative happening process between Bull and Bear Markets, there are some of small irregular fluctuations. Those small fluctuations may have their own pseudo periods.

Generally speaking, it is difficult to calculate the exact values of these pseudo periods. Their approximations can be calculated.

7.5.2 Stock Market Characteristics and General Patterns of Oscillation

Stock oscillation is one of the most typical vibration phenomena in the social economy fields. Stock market is in any countries, and any stocks are always in the process of constant fluctuations with their prices and exchange volumes. The stock oscillations have their own intrinsic patterns. If stock traders know these patterns and apply them in their trading strategy the winning probability will be much higher than the losing probability.

The oscillation patterns of stock markets can be roughly summarized as follows:

(1) Situation driven and Policy Guidance

The Bear Market and the Bull Market are all associated with the current political and economic situations. The political stability, economy high speed growth, important international expositions or approval of the exposition application all have the tendency to push the stock prices up; while unstable society, political party disputes, regional war crisis, economy recession, disasters, all have the tendency the pull the stock prices down.

Changes or adjustments of national policies could affect stock markets. For example, a nation sets up or declares policies which could benefit the stock market developments, those policies could drive stock prices up. Vice versa.

(2) Stock Quality is a Basic Factor

There are a variety of types of stocks. Like a merchandise, the stocks have better quality and bad quality. It is important that the stock investors should distinguish them. Investors should understand: types of business, their volumes, profits, PE ratios, investment directions, and development potentials, and their background, their major shareholders, major funding, and performances. Those things can be used to judge if the stock is a quality one.

(3) Inertia Principle

The so called stock inertia principle, also called "trend", once established, will continue to develop in its direction.

i. In physics, the inertial is a measure of ability of a substance to maintain its movement status. Similar to the physics, the larger the total shares are, the bigger the inertial and by the same token, the smaller the volume is, the smaller the inertial. In order for the "large inertial" stock price to changes, large volume of stocks must be changed hands.
ii. When the whole market goes up almost to its peak, it still goes up and up; while the whole market is down almost to its bottom, it still goes down and again. This is the inertial. Take the Shanghai Stock Index as an example. In September 1996 after the Index reached to 1,000, it went up till 1,258. From February to May 1997, the Index reached to 1,510 from 870 after more than 80 days increases in a row with more than 22 yellow card warnings. The Index 1,047 in January 1999 and 1,341 in new year of 2000 were all completed in continuous decreases. The most classical one happened on the second half of 1999 from 1,756 to 1,341. During this period, even though a series of good news was issued and it could not stop the declines for a half year. This is the results of the inertial force.

    The inertial forces generally push the increasing market excessive up and pull the decreasing market excessive down. The stock inertial forces are sometimes spontaneous and sometimes artificial, or both. But it is mainly artificial. First it is the main force behavior driven by the wills and benefits. The main forces and individual main forces, from their own benefits, complete their purchases by forcing stock price down and they did it again and again to force the middle and small investors out. This generated a down turn inertial. Once they collected enough chips, they would try to push the stocks, even the whole market, up. The middle and small individual investors will escape the push up. However as the main forces have too many chips and dump the chips, they either benefit less or avoid large losses therefore, the market has to be pushed up and displays the pushing up inertial. The second reason is the investors' thinking pattern and "obeying the majority" psychology. The majority of the investors have a linear thinking pattern , i.e. thinking a problem in one direction and reluctanting to change directions. The "obeying the majority" psychology makes it even worse. In the weak market, the inertial forces of the bad news are large; while in the strong market, the inertial forces of the good news are tremendous. As a matter of a fact, the inertial of the large increase or decrease of the market is caused by the investor's psychological thinking—for a Bear Market they are afraid of declining and for a Bull Market they wish it would continue, not by the company's fundamental changes. The market inertial is coming from investor's thinking inertial. The psychological factors of the investors play an important role in the continuous developments of the market situation.

iii. Take the individuals as an example. A continuously increasing stock will remain increase for a while and a continuously declining stock will remain decline for a while. This obeys the inertial rule. If stock turns from decline to increase, or from increase to decline it takes a while to release its energy or cumulate its energy. This is so called the upside down dome.

(4) Energy Principle

When a matter moves, its energy is the half of the product of its mass and velocity squared:

$$T = \frac{1}{2}mv^2 \tag{7.17}$$

Referring to the physical energy, a stock energy is considered to be similar to Eq. (7.17). For a stock, its mass is the trade transaction, the velocity C is the stock price change per unit time (or stock price changes in a day, increase takes $+$ and decrease takes $-$). For the ith day the stock energy is:

$$T_i = \frac{1}{2}\left[m_i(C_i)^2\right] \tag{7.18}$$

The total energy in n days is:

$$\sum_{i=1}^{n} T_i = \frac{1}{2}\sum_{i=1}^{n}\left[m_i(C_i)^2\right] \tag{7.19}$$

It is obvious that the total energy for a stock indicates the directions of the stock. The larger the energy is, the larger the possibility for the stock goes up or down; vice versa. We can compare different stocks to judge if a stock could go up or down. The above formula must be modified and the relative quantity should be used instead of the absolute quantity:

$$\sum_{i=1}^{n} \Delta T_i = \frac{1}{2}\sum_{i=1}^{n}\left[\Delta m_i(\Delta C_i)^2\right] \tag{7.20}$$

where Δm_i is the stock change hand rate; ΔC_i is the ratio of the stock price change to the stock price in unit time (or the price change in a day).

If we express the energy of a stock simply we can use the momentum analogue in physics:

$$E = mv \tag{7.21}$$

where E is the momentum of the matter, m is the mass and v is the velocity.

7.5 Utilization of Vibration Phenomena and Patterns in Stock Market

In stock market, the energy can be expressed by the product of the change hand rate and a relative stock price change(the ratio of the stock price change to the stock price in unit time):

$$\sum_{i=1}^{n} \Delta V_i = \sum_{i=1}^{n} [\Delta m_i \Delta C_i] \tag{7.22}$$

The positive sign is for stock increase and negative for decrease. The momentums of different stocks can be computed and compared.

(5) Probability Principle

The probability principle is the most common principle in stock market.

i. The advanced stocks in a stock market are different in different days. If the advanced stocks are 70%, the unchanged stocks are 5% and declined stocks are 25%, then the probability of the advanced stocks is 70. 5% for the unchanged, and 20% for the declined. If an investor purchases a stock in the higher probability of stock increase, the investor has higher chance to profit, and vice versa.

ii. In the stock market, individual stock price change rates are different, for example, a stock price changes up 2% and another changes down 8%, then their advance or decline probability are 2% and 8% respectively. If two stocks have the same advanced level in several days, the two are good stocks. The stock, which increased higher than the another one in the first day, may have the lower increase probability, than the other one on the second day, and so on and so forth.

iii. For different stocks, the fundamentals for some stocks are very positive and they have high profits while others have negative profits. In general the stocks with high profits would have higher probability to increase than the negative profit stocks.

(6) Block Effect

Many stocks in Chinese markets have some special characteristics of important economy intensions. When the common characteristics are recognized in the markets, the stocks form a block structure. The blocks are divided into three types: industry, region and other special concepts.

The block effects occur in the stock markets. For example, when some good news is issued, the block of the stocks in the markets would increase collectively, while when bad news appears in an industry the block of stocks in that industry would decrease collectively. Or some stocks in one block increase, the other stocks in the same block would increase in the following days, or vice versa. When one or two stocks in one block leads to the increases, other stocks in the same block would follow their leads. This is the block effect. How to judge the stocks in a block to follow the lead stocks in that block? According to the Grey System Theory, when the geometric shapes between two data sequences are very close, their change patterns are very close and their correlation is large. We can analyze the correlation to judge their mutual interaction.

(7) Main Force Behavior

In the stock operations, the phenomena, which do not obey the stock oscillation patterns, are the main forces to push down and pull up, which exist in any countries and any regions. The phenomena are described in section of Inertial (3) above.

7.5.3 Some Principles in Stock Operations

There are about 80 million individual stock investors in China. The stock markets are relatively new in China and many stock investors lack the understanding on the stock oscillation, much less applying the vibration to describe the stock oscillation and applying those rules in the daily stock trade transactions. Now we might as well talk a little bit about the principles in the stock trade transactions.

7.6 Obey the General Rules in the Stock Operations

(1) Emphasizing on Policies and Fundamentals

The rises and falls of the stocks are closely associated with the political and economic situations and the stock quality themselves. The stock traders must concern the situation, predict the future developments, and concern the announcements, implementations and the adjustments of the national policies and regulations and their effect on stock markets.

The selections of the stocks for purchases are one of the most important things for individual stock traders. The stock quality is the first important thing for consideration. Analysis should be conducted for stocks' development futures and possible problems, and whether the introduction information is accurate or not.

(2) Obey the Inertial Principle

There is inertial in the stocks themselves or the operation process, the inertial principle should be applied in the operation process.

    i. The stocks with large volume in transactions have large inertial, and vice versa. To purchase with quick price changes, the low volume stocks are the selections. The large volume stock prices remain stable and have small fluctuation. Just like a bid ship, when it is sailing at sea, even the waves are large the ship is not easy to fluctuating.

7.6 Obey the General Rules in the Stock Operations

ii. In the process of the stock rise and fall, if it is continuously falling till 90% or even 100% shareholders are losing money, the stocks would continue to fall for a while until it reaches to the upside down dome and then it would go up slowly. This is similar to the bottom segment of the sine curves and the stocks will continue to increase up to the top of the dome. The investors should master the patterns and operate their transactions accordingly. For different situations they should take different measures.

iii. The stock prices are fluctuating constantly, the time of buying stocks should not be in the quick decline segments, according to the inertial principle, the stocks would continue to decline. The best time would be the time when the velocity is going to be zero; while for selling the stocks, the selling time should not be the time when the stocks are quickly increase instead it should be the time when the increasing velocity is about zero.

(3) Obey the Energy Principle

Here are some of the strategies:

i. Investors should notice the change of the stock energy and should avoid to buy the low energy stocks because it takes a long time for the low energy stocks to gain the energy to go up.

ii. If a stock change hand rate does not have big changes, even it is going up the increase tendency does not sustain because of the small energy. When a stock price is going up and so is its volume, that is to say the stock energy is high and the stock price could go up in a sustainable way.

iii. The stock holders should compare the energy computation results for different stocks and purchase the stocks with large energy.

(4) Obey the Probability Principle

According to the probability principle, we have the following strategies:

i. According to the probability principle, the investors should buy the stocks in the early stage of the Bull Market. At this time the majority of the stocks are in the advance stage process. For example, the stock prices increase 60%–80% a day, the increase probability is in the range of 60%–80%.

ii. When a stock price is going up and it would go down eventually someday. If it goes up several days, then the probability of its going down would be larger for the next day, and vice versa.

iii. Block rotation.The block rotation has its probability. In a general case, when a block remains unchanged for a long time, the block rotation may happen. The block probability should be made use of buying stocks.

When a block has no increase for a long time with any special situation, then the block rotation effect would be in action,and it could be a time to buy it on the block increase probability.

(5) Note the Block Effect

Here are some strategies for the block effect:

i. Making use of the mutual correlation effect. When a stock becomes an attractive object for other people to purchase, the stocks within the same block should be traded in, this is an important way of making short term profits. When 1 or 2 stocks in the block are leading to the large amplitude increase (or decrease), or a stock becomes an attractive object for other investors to push up (or pull down), the purchase (or sell) should be made on the well correlated stocks in the block to achieve the short term gains or avoid risks.

ii. Making use of the lag effects of the block. When a good news is issued, it would induce some stocks in the block to increase, then the stocks in the block would start to increase. However the stock response speeds to the news are different, some starts to increase immediately while others may response to the news slowly or in other days. The lag effect should be made use of.

iii. Making use of the block rotation effects. The block effects are often changing alternatively. The stock blocks will not remain unchanged. The stock block rotation prediction would create some chances. The prediction may be in the policies and the technology perspectives.

(6) Follow the Main Force

The market push-up and pull-down by the main forces happen from times to times. In some cases they determine the market directions. How to identify the main force actions is up to individual's ability. To follow them may be a choice in an abnormal situation.

7.7 The Entering Point and Withdrawing Points in the Stock Operations

Investors buy stocks at times and sell them at other times. When to buy and when to sell are two of the most important issues. This is the crucial points we will discuss: the selection of the entering point and withdrawing point on the fluctuation curves in the stock operations.

The stock markets are always in the change and fluctuations. In most cases, they show some pseudo periodical. The Bear and Bull markets change alternatively. The pseudo periodicity sometimes would be obvious.

The investors should first understand or pre-estimate the fluctuation patterns of the stocks, master the entering point when the stocks increase and the withdrawing point when the stocks decline. This practice would help the investors take corresponding measures to succeed in the stock operations and gain profits. The patterns of the fluctuations are very complex in general, the simplest or the ideal form for the fluctuation is the sine curves. If we simplify the stock fluctuation as a sine curve, there

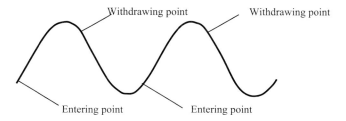

Fig. 7.8 The entering and withdrawing points in the stock investment

is a top point, also called the apex point and a lowest point, also called the valley point. In additions, there are a segment of increase and a segment of decline on the curves. On the increase segment there are the entering regions in which the appropriate entering point could be chosen. The so called entering point is the time when the stock fluctuation process is in the rapid development. The investors should master the entering points which are the keys for success.

(1) The entering point in the stock oscillation process

Figure 7.8 shows the segment of the sine curve. The first thing is to study if the stocks are in the increase regions and if the entering points in the increase regions are ideal. The rules of the thumb are to select the position of 1/4–1/2 amplitude of the top point as the entering point.

(2) The withdrawing point in the stock oscillation process

After the stock investors buy the stocks, they will choose a proper time to sell them to get profits. They will select a withdrawing point. Selecting a withdrawing point is as important as selecting the entering point. The ideal withdrawing point should be somewhere on the curves after the apex point. The rules of the thumb are the withdrawing point should not be more than 10%–20% from the apex point.

7.8 Utilization of Vibration Phenomena and Pattern in Human Body

7.8.1 *Vibration is a Basic Existing Form of Many Human Organs*

There is no exception that the every organ in human body is in vibration all the times. For example, heart beats, blood circulation, lung breath, stomach peristalsis, brain cell thinking and ear drum vibration and vocal cord vibration, they are all effective utilization of the vibration of our body and those vibrations have certain patterns. The human organs accomplish their functions in the constant vibration states. If the required vibration patterns are not in kept with, the abnormality would occur in human organs, i.e. a symptom for a disease. If the phenomena are severe, that is to say the disease is severe, could lead to the end of a life. Therefore it is imperative that human organ vibrations have important significance.

To study human organ vibration patterns, we could manufacture the artificial organs to replace the malfunctioned human organs.

7.8.2 Some Diseases Make Abnormal Fluctuations (Vibration) in Human Organs Physical Parameters

For a healthy person, his/her heart beats, blood pressure, body temperature, electrocardiography, electroencephalography, white blood cell contents, etc. are all in the normal changing ranges. When a person is ill these physical parameters would change and are in some abnormal ranges. For the different diseases these parameter changes are different. Doctors use those changes and fluctuation patterns to diagnose some diseases.

The human disease diagnostics and machine fault diagnostics have something in common. But they have an essential distinguish: the machines have no life but human does. The two have differences in their internal structure, life mechanism and movement mechanism. Therefore their diagnostics are different. The diagnostic methods have an essential distinguishes. One common point should be noticed: they all use the vibrations theory or vibration patterns for diagnostics.

People have life, everyone has a life once, therefore significance and importance of treating diseases are obvious. Scientists propose a variety of check and diagnostics methods. Many of the diagnostics methods are associated with the parameter changes, i.e. their fluctuation patterns.

(1) The nonlinearity in heart beat variation

The human hearts not always beat in a constant speed. The heart beat variation (HBV) signals describe the micro-rise and fall in the normal heart beats. The cause for the HBV is very complex, either physiological or psychological or even pathological. Thus the HBV signals are very irregular. The HBV signals contain the nonlinear, chaos and bifurcation characteristics.

Here we chose the Electrocardiography (ECG) in the MIT-BIH physiology data bank and code the HBV extraction program and analyze the HBV signals. Figure 7.9a shows the 10 s HBV signal process extracted from the normal ECG sample 16,265, while Fig. 7.9b shows the 10 s HBV signal process extracted from the over beat ECG sample cu01.

It is necessary to analyze the movement trace of the signals in the multi-dimensional space for the HBV scattered plot. Figure 7.10a is the scattered plot for a healthy HRV signal in Fig. 7.9a while Fig. 7.10b is the scattered plot for an unhealthy HRV signal in Fig. 7.9b. The scattered plot for a healthy person's physiological signals is different from those from an unhealthy person. We could conclude that the

7.8 Utilization of Vibration Phenomena and Pattern in Human Body

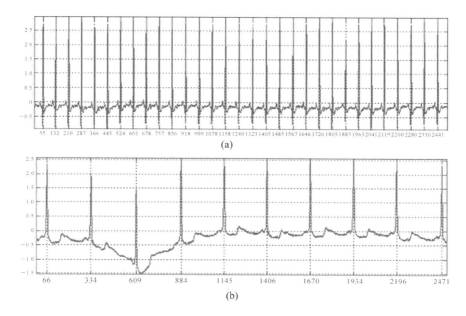

Fig. 7.9 ECG signals. **a** Normal ECG signal **b** over beat ECG signal

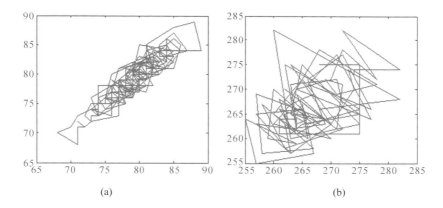

Fig. 7.10 HBV scattered plots. **a** Healthy HBV plot **b** unhealthy HBV Plot

healthy human's HBV scattered plots are meteor-shaped, while the unhealthy HBV scattered plots are in complex forms, such as fish-shaped, fan-shaped etc.

(2) Nonlinear analysis on electroencephalograph

Human's electroencephalograph (EEG) is the total reflections on the head skin of physiological activities of the brain nerve cell groups and it can directly reflect the activities of the brain central nervous system. Many studies have proven that EEG signal can reflect human's physiological state and spirit states as well. Some of the typical rhythms in EEG signals could change as the human spirit state. According to reference, 4 rhythms on EEG signals have the following relations with human spirits: δ-wave (1–4 Hz.), can be observed when a healthy person is in sleep; θ-wave (4–8 Hz.), it occurs when human body is in tired state; α-wave (8–13 Hz.), it occurs when human is sober, tranquil and with eyes closed. This band is associated with whether human moods are quiet, amiable or not, whether the feelings are pleasant. Worry and nervousness would restrain the α-wave. When human opens eyes, thinking problems, or excited the α-wave disappears; β-wave (14–30 Hz.), it appears when the cerebral cortex is excited. It is caused generally by worry and nervousness. The further studies showed that when human body is subject to a static force, α-wave rhythm decreases or disappears, β rhythm is reinforced; as the time increases, the load increases, δ and θ waves appear indicating the cerebral cortex is constrained; after the load is released gradually, the slow waves disappear gradually, β rhythm decreases gradually and α rhythm is restored. It is seen that if the δ and θ rhythms increase, β rhythm decreases, which means that human is in the tired and sleepy states.

The 10 different sets of the EEG signals from MIT > BIH physiological data bank are analyzed and calculated. The repeat sampling rate is 128 Hz. The conclusions by the wavelet cell analysis are as follows:

It can be seen from Fig. 7.11 that the δ-wave energy ratio is large in the EEG signal Fig. 7.11a, it contains almost all the energy indicating the signal is the EEG for a healthy sleeping person; Fig. 7.11b shows the δ-wave contains most of the energy however some θ-wave is contained as well, which means the person is feeling very tired, and in the half awake and half sleeping dreaming state, and his spirit is a little unstable; while in Fig. 7.11c δ-wave energy is almost zero, the person is not sleeping yet, the large α-wave indicates the person is in eye-closed state, the large amount of the β-wave indicates the person is worrying and nervous we can judge that the person is in mood fluctuation, awaking state; in Fig. 7.11d it contains large amount of δ-wave indicating the person is sleeping, but at the same time it contains large amount of α-wave and β-wave which indicate the person is in worrying state, the small amount of θ-wave indicates the person is tired and not in a good mood.

As a reference, Fig. 7.11 also gives out the amplitude-frequency spectra corresponding to different frequency bands based on the FFT computations. Even though person is in different mood states (such as different sleeping states) the wave shapes have obvious qualitative difference but can not be described and determined quantitatively. Or from the view of point of diagnostics, the time domain wave shapes can not be used as the diagnostic characteristics. However based on the FFP spectrum it contains wide brand characteristics and the spectrum characteristics are not obvious.

Table 7.2 shows 10 sets of the computed bandwidth energy ratio of the sleeping EEG in different spirit states.

To realize the quantitative evaluation for the human spiritual state, the relationship between the rhythm characteristics in human EEG signals and human spirit states can be used and the decomposition method on the wavelet cell is used to analyze emphatically the human EEG in different spirit states to extract the 4 typical rhythms mentioned above and compute the bandwidth energy ratio among the rhythms. The

7.8 Utilization of Vibration Phenomena and Pattern in Human Body

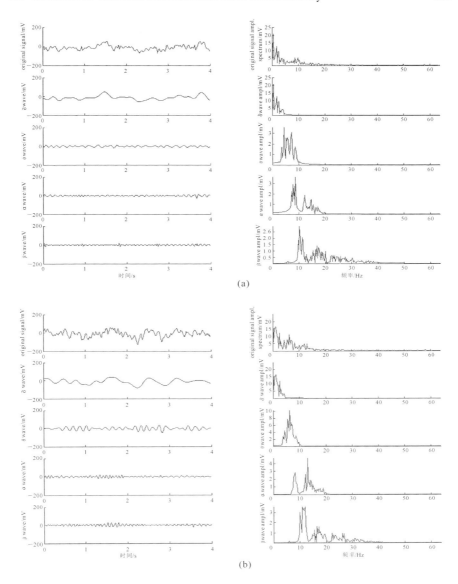

Fig. 7.11 4 typical EEG signal wavelet cell decomposition and amplitude-frequency spectra in different spirit states. **a** healthy human in deep sleeping state, **b** human in tired sleeping state, **c** human in mood fluctuating and awaking state, **d** human in worrying and nervous sleeping state

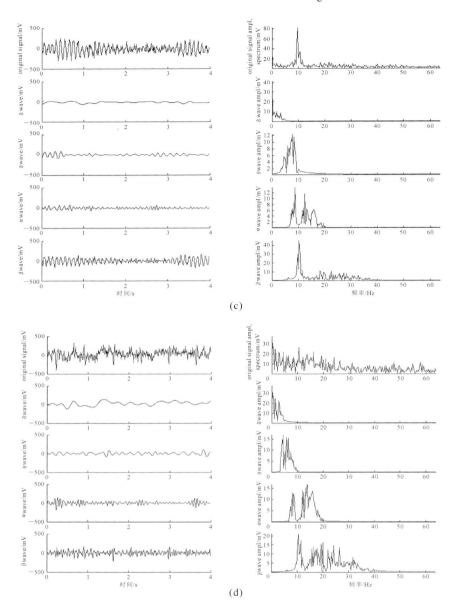

Fig. 7.11 (continued)

7.8 Utilization of Vibration Phenomena and Pattern in Human Body

Table 7.2 The computed values and analysis results on typical EEG signal bandwidth energy ratios

Record name	δ wave band energy ratio	θ wave band energy ratio	α wave band Energy ratio	β wave band Energy ratio	Spirit state assessment
1	0.9982	0.0014	0.0002	0.0002	Health person in deep sleep
2	0.9991	0.0004	0.0002	0.0003	Health person in deep sleep
3	0.9985	0.0012	0.0001	0.0003	Health person in deep sleep
4	0.9921	0.0063	0.0010	0.0006	Health person in deep sleep
5	0.8563	0.1377	0.0031	0.0029	Person in tired sleep state
6	0.0066	0.0189	0.5181	0.4565	Person in emotional fluctuation and waking state
7	0.2194	0.0079	0.2492	0.5235	Person in worrying and nervous sleep state
8	0.9995	0.0000	0.0002	0.0003	Health person in deep sleep
9	0.9699	0.0027	0.0010	0.0264	Health person in deep sleep
10	0.1903	0.0242	0.3647	0.4209	Person in worrying and nervous sleep state

rhythm bandwidth energy ratios effectively reflect the information on human spirit state and can be used quantitatively to assess the characteristics of human spirit states.

7.8.3 Medical Devices and Equipment Based on Vibration Principles

In order to accurately diagnose human diseases, and effectively treat the diseases, scientists and researchers develop a variety of medical devices and equipment, many of which use the vibration and wave principles.

(1) B-ultrasonic Applications

Normal and abnormal human organs absorb and reflect the sound waves differently. The ultrasonic diagnostic device is working on this principle. The ultrasonic medical devices radiate the ultrasound into human body and the sound propagates in human body. When the sound encounters the organs and tissues it will be reflected and refracted. The medical devices receive those signals, process them and show them in wave curves, or figures for doctors to make decisions if the organs are healthy or not.

(2) Stone Crash by Ultrasound

The ultrasound can be used to crash the stones such as kidney, gallbladder stones.

(3) Electrocardiograph Technology

The current in the EEG device is used to detect the heart and blood circulation in human body.

(4) Electroencephalograph Technology

The device can detect the brain blood circulation.

(5) X-Ray Machine

X-Ray can be used in the fluoroscopy of lung and chest etc.

(6) X-Ray CT

Medical CT uses X-ray as its light source and scans sections of human body to be examined.

(7) Magnetic Resonance Imaging CT

The MRI scanning has much less harms to human body than the X-Ray.

7.8.4 Artificial Organs and Devices Using Vibration Principles

(1) Artificial Heart

When the heart is abnormal and can not be cured the artificial heart may be the substitute. The artificial heart is in essence an self-exciting vibration generator.

(2) Heart Pacer

When human heart beat is not normal, a heart pacer is placed in human body to regulate human heart beat in its normal pace. The pulse frequency generated by the pacer is the same as that of a normal human heart and is continuously triggered to keep the pace.

(3) Hearing Aid

Hearing aid is actually a sound amplifier. When it receives a sound signal, it amplifies the sound signal to broadcast so the ears could hear the sound.

7.9 Prospect

(1) It has very important and practical significance to study the vibration problems in nature and human society. As a matter of a fact a leader in a country, or a manager in an enterprise encounters a variety of vibration problems in his/her territory and has seen a variety of vibration phenomena. Handling the growth and recession periods in the fluctuation process and selecting a proper entering point and withdrawing point, i.e. mastering the vibration patterns of systems, and taking a proper measure and strategies, promoting the social and economic quick, stable and sustainable development and avoiding the losses in the recession period in the economy process are the major important tasks for a quality leader and those are also the characters of the leader's intelligence and skills.

(2) It is not enough currently to study the vibration problems in nature and in the society using the vibration theories in two folds: the vibration problems in nature and in society have a variety of forms, the vibration problems in nature and in society are much more complex than those in structures and mechanical systems. This book only considers the vibration problems as a branch of utilization of vibration and introduces some application examples in this branch. The special efforts should be devoted to further study and explore in the perspective of predicting the associated problems in nature and human society by further developing the vibration theory and methodology and combining the theory and methodology with the problems.

(3) The phenomena in nature and social society such as sub-harmonic, super-harmonic, jumping, leg delay, slow-changing, sudden changing, bifurcation, chaos, and utilization of the patterns of their random characteristics should be further studied by applying the vibration theory and nonlinear dynamics.

(4) Authors proposed in this book a theoretical framework on the branch of "Utilization of Vibration Engineering", conducted some studies on its contents, applied some of the theoretical results to the engineering practices and obtained important social and economic benefits and social effects. However some fields, such as the vibration phenomena and patterns in nature and human society, have not yet been studied thoroughly. It can be predicted that there are wide and promising development futures. We expect more scientists and researchers to participate the studies in these fields and in the near future they would make great achievements in this area and contribute to their countries and the world.

References

1. Wen BC, Liu FQ (1982) Theory and applications of vibrating machines. China Machine Press, Beijing
2. Wen BC, Gu JL, Xia SB et al (2000) Advanced rotor dynamics. Machine Industry Press
3. Wen BC, Zhao CY, Su DH et al (2002) Vibration and control synchronization for mechanical systems. Science Press, Beijing
4. Wen BC (1987) Proceedings of international conference on mechanical dynamics. Press of Northeast University of Technology, Shenyang
5. Wen BC (1989) Proceedings of Asia vibration conference. Press of Northeast University of Technology, Shenyang
6. Wen BC (1998) Proceedings of international conference on vibration engineering. Press of Northeastern University, Shenyang
7. Wen BC (2001) Proceedings of Asia-Pacific vibration conference 2001. Press of Jilin Science and Technology, Changchun
8. Wen BC, Li Y (1997) The new research and development of vibration utilization engineering. In: Proceedings of 97' Asia-Pacific vibration conference, Kyongju, Korea, vol 11, pp 19–24
9. Wen BC (1981) Research concerning frequency entrainment of nonlinear self-synchronous vibrating machines. In: Proceedings of 9th international conference on nonlinear oscillations, Kiev, USSR, vol 3, pp 54–58
10. Wen BC, Ji SQ (1981) Forced oscillations of nonlinear system with impact dry friction and sectional masses. In: Proceedings of 9th international conference on nonlinear oscillations, Kiev, USSR, vol 3, pp 58–62
11. Wen BC (1984) Study on vibratory synchronization transition. In: Lecture at third international conference on vibrations in rotating machinery, New York, UK
12. Wen BC, Duan ZS (1984) Study on the nonlinear dynamics of self-propelled vibrating machines with rocked impact. In: Proceedings of international conference on nonlinear oscillations, Valna, Bulgaria, pp 798–801
13. Wen BC (1985) The probability thick layer screening method and its industrial applications. In: Proceedings of international symposium on mining science and technology, Xuzhou, China
14. Wen BC, Wu WY (1986) Dynamic characteristics of the rotors with axial vibration. In: Proceedings of international conference on rotor dynamics, Tokyo, Japan, pp 577–582
15. Wen BC (1986) Frequency entrainment of two eccentric rotors in nonlinear near resonant system. In: Proceedings of international conference on vibration problems in engineering, Xian, China
16. Wen BC, Ji SQ (1986) The theory and experiment about the resonant probability screen with inertial exciter. In: Proceedings of international conference on vibration problems in engineering, Xian, China
17. Wen BC (1986) Theory experiments and application of probability screen to coal classification. In: Proceedings of international conference on modern techniques of coal mining, Fuxing, China

18. Wen BC, Liu J (1986) Analysis and removal steps of bending vibration break down of long distance vibrating conveyer. In: Proceedings of international conference on diagnosis and monitoring of machine faults, Shenyang, China
19. Wen BC (1986) Some nonlinear vibration problems in the field of vibration utilization. Lecture at Tokyo University, Tokyo, China
20. Wen BC (1987) The qualitative method of the nonlinear systems with varying mass. In: Proceedings of international conference on nonlinear oscillations, Budapest, Hungary
21. Wen BC, Guan LC (1987) Synchronization theory of self-synchronous vibrating machine with two asymmetrical vibrators. In: Proceedings of international conference on mechanical dynamics, Shenyang, China, pp 434–439
22. Wen BC (1987) Synchronization theory of self-synchronous vibrating machines with ellipse motion locus. In: Proceedings of ASME vibration and noise conference, Boston, USA
23. Wen BC (1987) Equivalent mass and damping coefficient of nonlinear system with varying mass. In: Proceedings of ASME vibration and noise conference, Boston, USA
24. Wen BC (1987) Theoretical research and experiments of vibratory synchronization transmission. In: Proceedings of the 7th world congress on machine design and mechanisms, Severlia, Spain
25. Wen BC (1987) The qualitative method of the nonlinear systems with varying mass. In: Proceedings of international conference on nonlinear oscillations, Budapst, Hungry
26. Wen BC, Zhang H, Li DS, Lou MF, Yu SD (1991) Some nonlinear vibration problems in the mechanical engineering. In: Proceedings of the Asia-Pacific vibration conference, Melbourne, Australia
27. Wen BC (1993) Some recent situations on vibration investigations in China. In: Proceedings of A-PVC, Kitakyushu, Japan
28. Wen BC (1993) Nonlinear responses and stability of electromagnetic exciter. In: Proceedings of A-PVC, Kitakyushu, Japan, pp 1674–1679
29. Wen BC, Zhang TX, Van J, Zhao CY, Li YN (1996) Controlled synchronization of mechanical system. In: Proceedings of the 3rd international conference on vibration and motion control, Tokyo, Japan, vol 2, pp 352–357
30. Wen BC, Li DS (1984) Research concerning the nonlinear vibration characteristics of the longitudinal vibrating machinery with asymmetrical spring system. In: Proceedings of international conference on nonlinear oscillations, Varna, Bulgaria, pp 794–797
31. Wen BC, Zhang TX, Zhao CY, Li YN (1995) Vibration synchronization and controlled synchronization of mechanical system. In: Proceedings of Asia-Pacific vibration conference '95, Kuala Lumper, Malaysia, pp 30–39
32. Wen BC, Zhang TX, Fan J, Zhao CY, Li YN (1996) Controlled synchronization of mechanical system. In: Proceedings of the 3rd international conference on vibration and motion control, Tokyo, Japan
33. Wen B, Li Y (1997) The new research and development of vibration utilization engineering. In: Proceedings of 97' Asia-Pacific vibration conference, Kyongju, Korea, vol 11, pp 19–24
34. Wen BC, Zhang YM, Liu QL (1998) Response of uncertain nonlinear vibration system with 2d matrix function. Int J Nonlinear Dyn 15(2):179–190
35. Wen BC, Li YN, Hen QK, Zhang H (1999) Some nonlinear dynamic problems in process of vibrating compaction. In: The proceedings of 10th world congress on Tomm, Oulu, Finland
36. Wen BC, Li YN (2003) Recent developments and prospect of vibration utilization engineering. In: 10th Asia-Pacific vibration conference, Royal Pines Resort Gold Coast, Australia, pp 759–764
37. Wen B, Li Y (1997) The new research and development of vibration utilization engineering. In: Proceedings of 97' Asia-Pacific vibration conference, Kyongju, Korea, vol 11, pp 19–24
38. Wen BC (2003) The probability thick-layer screening method and its application. Coal Prep 23(1–2):77–88
39. Liu J, Xu XN, Ji SQ, Wu L, Wen BC (1995) Response characteristics of a subsystem of the series connected mass-spring system in origin anti-resonance. In: Proceedings of Asia-Pacific vibration conference '95, Kuala Lumper, Malaysia

40. Liu J, Li XJ, Wu L, Wen BC (1995) Partial differential linearization on mathematical model of electromagnetic vibrating machine. In: Proceedings of Asia-Pacific vibration conference '95, Kuala Lumper, Malaysia, pp 400–405
41. Liu J, Lin XY, Gong ZM, Wen BC (1988) Diagnosis and preventive measures of noise sources in vibrating machines. In: Proceedings of ICTD '88, Shenyang, China
42. Liu J, Wen BC (1987) The flexible vibration of vibrating conveyers and its application in design. In: Proceedings of international conference on mechanical dynamics, Shenyang, China
43. Liu J, Wen BC (1993) Fault diagnosis and monitoring of electromagnetic vibratory machines. In: Proceedings of A-PVC, Kitakyushu, Japan
44. Zhang TX, Wen BC, Fan J (1997) Study on synchronization of two eccentric rotors by hydraulic motors in one vibrating system. Shock Vib 4(5–6):305–310
45. Zhang TX, Wen BC (1995) Dynamical analysis on synchronous vibrating machine with two hydraulic motors. In: Proceedings of international conference on structural dynamics, vibration, noise and control, Hong Kong
46. Zhang TX, Wen BC, Fan J (1994) Study for special feature of rotating speed and synchronization of hydraulic eccentric rotors. In: Proceedings of the international conference on vibration engineering, Beijing, China, vol 833–836
47. Zhang TX, Wen BC (1993) Dynamic analysis and computer simulation study for self-synchronous rotation state of hydraulic motor-eccentric rotor system. In: Proceedings of Asia-Pacific vibration conference, Japan
48. Duan ZS, Wen BC (1986) An analysis of the drift of an operating point of vibrating machines and the correct measures. In: Proceedings of international conference on diagnosis and monitoring of machine faults, Shenyang, China
49. Duan ZS, Wen BC (1988) An application of the theory of grey system in estimating residual of AR model with high order, Shenyang, China
50. Duan ZS, Lou MF, Wen BC (1991) A grey diagnosis system with self-learn function for machinery faults. Proceedings of the Asia-Pacific Vibration Conference, Melbourne, Australia 5:91–95
51. Ji SQ, Wen BC (1990) The determination of the exciter positions of the self-synchronous vibrating screen with the exciter deflection. In: Proceedings of international conference on vibration problems in engineering, Wuhan, China
52. Ji SQ, Wen BC, Zhang JM (1991) The study on the occurrence condition of the swing oscillation in the secondary isolation system. In: Proceedings of 2nd international conference on technical diagnosis and seminar, Guilin
53. Ji SQ, Wen BC (1993) The study on transitional processes of the inertial vibrating machinery. In: Proceedings of A-PVC, Kitakyushu, Japan, pp 1261–1265
54. Ji SQ, Hao XL, Liu J, Liu SY, Wen BC (1995) The research on nonlinear theory of the material shock. In: Proceedings of Asia-Pacific vibration conference '95, Kuala Lumper, Malaysia, pp 105–110
55. Guan LZ, Wen BC (1987) Research on synchronization of self-synchronous vibrating machine with dual mass within resonant region. In: Proceedings of international conference on mechanical dynamics, Shenyang, China, pp 445–449
56. Liu SY, Yuan Y, Ji SQ, Wen BC (1995) Dynamic analysis of the vibrating conveyer with large inclination when considering material forces. In: Proceedings of Asia-Pacific vibration conference '95, Kuala Lumper, Malaysia, pp 337–342
57. Wen CX, Wen BC (1998) Bifurcation in nonlinear mechanical system with piece-wise linearity. In: Proceedings of international conference on vibration engineering, Dalian, China, pp 302–304
58. Wen CX, Wen BC (1989) Numerical method of nonlinear system with piece-wise linear springs. In: Proceedings of ASME vibrations and noise conference, Montreal, Canada
59. Wen CX, Wen BC (1987) Global stability of the nonlinear system with piece-wise linearity using cell mapping method. In: Proceedings of international conference on nonlinear oscillations, Budapest, Hungry

60. Wen CX, Wen BC (1989) Numerical calculation of nonlinear system with piece-wise linearity. In: Proceedings of Asia vibration conference, Shenzhen, China, pp 107–112
61. Wen CX, Wen BC (1995) Point mapping and cell mapping synthesis method for dynamical system. In: Proceedings of international conference on structural dynamics vibration noise and control, Hong Kong
62. Fan J, Wen BC (1992) Microcomputer controlled synchronization of two induction motors. In: Proceedings of 1st international conference on motion and vibration control, Yokohama, Japan
63. Fan J, Wen BC (1993) Track control of synchronous vibrating machine using nonlinear control. In: Proceedings of A-PVC, Kitakyushu, Japan
64. Li Y, Zheng L, Wen BC (2002) Experimental research on evaluating structure damage with piezoelectric dynamic impedance. Chin J Mech Eng 15(3):204–208
65. Li Y, Wen BC et al (1997) Study of the mechanism of using the controllable hypocentre with super low-frequency to extract oil with vibration. In: Proceedings of 97'Asia-Pacific vibration conference, Kyongju, Korea, vol 11, pp 983–986
66. Fu Y, Wen B, Li Y (1998) The development of controllable epicenter and study of its mechanism. In: Proceedings of international conference on vibration engineering, Dalian, China, vol 8, pp 411–414
67. Li Y, Zheng L, Wen BC (2000) Study on structural damage identification with wave information. In: Proceedings of the first international conference on mechanical engineering, ICME2000, Shanghai, vol 11, pp 19–22
68. Li Y, Jing Z, Seij C, Wen BC (1998) Impedance-based technique and wave propagation measurement for non-destructive evaluation. In: Proceedings of international conference on vibration engineering, Dalian, China, pp 476–481
69. Li Y, Zheng L, Wen BC (2001) Experimental research on evaluating structure damage with piezoelectric dynamic impedance. In: ICSV8, 2001, Hong Kong
70. Li YN, Zheng L, Wen BC (2000) Study of dynamic characteristics of slow-changing process. Shock Vib 7(2):113–117
71. Zhang YM, Liu QL, Wen BC (2002) Dynamic research of a nonlinear stochastic vibratory machine. Shock Vib 9(6):277–281
72. Zhang Y, Chen S, Liu Q (1994) The sensitivity of multibody systems with respect to a design variable matrix. Mech Res Commun 21(3):223–230
73. Zhang Y, Chen S, Liu Q, Liu T (1996) Stochastic perturbation finite elements. Comput Struct 59(3):425–429
74. Zhang Y, Wen BC, Chen S (1996) PFEM formalism in Kronecker notation. Math Mech Solids 1(4):445–461
75. Zhang Y, Wen BC, Chen S (1997) Eigenvalue problem of constrained flexible multibody systems. Mech Res Commun 24(1):11–16
76. Zhang Y, Wen BC, Liu Q (1998) First passage of uncertain single degree-of-freedom nonlinear oscillators. Comput Methods Appl Mech Eng 165(4):223–231
77. Zhang Y, Wen BC, Leung AYT (2002) Reliability analysis for rotor rubbing. J Vib Acoust-Trans ASME 124(1):58–62
78. Zhang Y, Liu Q (2002) Practical reliability-based analysis of coil tube-spring. Proc Inst Mech Eng Part C-J Mech Eng Sci 216(C2):179–182
79. Zhang Y, Wen BC, Liu Q (2002) Sensitivity of rotor-stator systems with rubbing. Mech Struct Mach 30(2):203–211
80. Zhang Y, Liu Q, Wen BC (2002) Quasi-failure analysis on resonant demolition of random structural systems. AIAA J 40(3):585–586
81. Zhang Y, Liu Q (2002) Reliability-based design of automobile components. Proc Inst Mech Eng Part D-J Autom Eng 216(D6):455–471
82. Zhang Y, Liu Q, Wen BC (2002) Dynamic research of a nonlinear stochastic vibratory machine. Shock Vib 9(6):277–281
83. Zhang Y, Wen BC, Liu Q (2003) Reliability sensitivity for rotor-stator systems with rubbing. J Sound Vib 259(5):1095–1107

84. Zhang Y, Wen BC, Liu Q (2003) Uncertain responses of rotor-stator systems with rubbing. JSME Int J Ser C-Mech Syst Mach Elements Manufact 46(1):150–154
85. Zhang Y, Wang S, Liu Q, Wen BC (2003) Reliability analysis of multi-degree-of-freedom nonlinear random structure vibration systems with correlation failure modes. Sci China Ser E-Technol Sci 46(5):498–508
86. Zhang Y, Liu Q, Wen B (2003) Practical reliability-based design of gear pairs. Mech Mach Theory 38(12):1363–1370
87. Zhang Y, He X, Liu Q, Wen B (2005) An approach of robust reliability design for mechanical components. Proc Inst Mech Eng Part E-J Process Mech Eng
88. Zhang Y, He X, Liu Q, Wen BC, Zheng J (2005) Reliability sensitivity of automobile components with arbitrary distribution parameters. Proc Inst Mech Eng Part D-J Autom Eng
89. Song ZW, Fang XD, Wen BC (2001) Synchronal control of multi-cylinder hydraulic system. Proceeding of Asia-Pacific Vibration Conference, Hangzhou 10:1102–1105
90. Qi JY, Zhang, GZ, Wen BC (1986) The numerical solution of machine body movements of rock drilling machines 8:113–119
91. Qi ZY, Zhang GZ, Wen BC (1991) An analysis on the working condition of rock drill body. In: Proceedings of 2nd international conference on technical diagnosis and seminar, Guilin, China
92. Han QK, Wen BC (1998) Nonlinear vibration of mechanical system with asymmetrical Picec-Wisely curvature hysteresis. In: Proceedings of international conference on vibration engineering, Dalian, China, pp 296–301
93. Han QK, Zhang ZM, Zhang H, Wen BC (1995) Identification of the nonlinear system of the vibration compacting machine. In: Proceedings of international conference on structural dynamics, vibration, noise and control, Hong Kong, pp 906–911
94. Zhao CY, Wen BC (1995) Dynamic characteristics of the asymmetrically piece-wise linear vibrating machines. In: Proceedings of Asia-Pacific vibration conference '95, Kuala Lumper, Malaysia, pp 371–376
95. Zhao CY, Wen BC (1997) The optimal design of the fuzzy controller. In: International conference on artificial intelligence and soft computing, Banff, Canada, vol 8, pp 61–64
96. Yan SR, Wen BC (2001) Study on nonlinear dynamics of pressing mechanism of a road-paver. Proceeding of Asia-Pacific Vibration Conference, Hangzhou 10:93–95
97. Xiong WL, Duan ZS, Wen BC (2001) Characteristics of electromechanical coupling self-synchronization of a multi-motor vibration transmission system. Chin J Mech Eng (English Edition) 14(3):275–278
98. Xiong WL, Wen BC, Duan ZS (2004) Engineering characteristics and its mechanism explanation of vibratory synchronization transmission. Chin J Mech Eng 17(2):185–188
99. He Q, Wen BC (1998) Advances and trends of metal plastic deformation processes with vibration. In: Proceedings of international conference on vibration engineering, Dalian, China, pp 175–179
100. He Q, Jin P, Wen BC (2001) A class of hysteretic systems in oscillatory plastic working. Proceeding of Asia-Pacific Vibration Conference, Hangzhou 10:188–191
101. Li HG, Zhang JW, Wen BC (2002) Chaotic behaviors of a bilinear hysteretic oscillator. Mech Res Commun 29(5):283–289
102. Zeng HQ, Luo YG, Wen BC (2001) Studies on ultrasonic vulcanization of natural rubber. Proceeding of Asia-Pacific Vibration Conference, Hangzhou 10:1177–1179
103. Xu PM, Cao ZJ, Wen BC (2001) A strategy of numerical integral for bifurcation analysis of nonlinear system and its application. Proceeding of Asia-Pacific Vibration Conference, Hangzhou 10:77–84
104. Liu YX, Wen BC, Yang JD (2001) Applying the Lyapunov exponent to research chaos of economic time series. In: Proceeding of Asia-Pacific vibration conference, Hangzhou, vol 10, pp 121–123
105. Zhang H, Wen BC, Tong BX, Zhang RM (1994) An on line method for measuring the soil compaction by the vibration response signals. In: Proceedings of the international conference on vibration engineering, Beijing China, pp 897–900

106. Tani J, Chonan S, Zhang WP, Wen BC (1988) Vibration failures in the mining and metallurgical machines and their dynamic design. In: Proceedings of ICTD '88, Shenyang, China (1988)
107. Fu Y, Wen BC, Li Y (1998) The development of controllable epicenter and study of its mechanism. In: Proceedings of international conference on vibration engineering, Dalian, China, vol 8, pp 411–414
108. Fu Y, Wen BC, Li Y (1997) Study of the mechanism of using the controllable hypocentre with super low-frequency to extract oil with vibration. In: Proceedings of 97' Asia-Pacific vibration conference, Kyongju, Korea, vol 11, pp 983–986
109. Rui YN, Li Y, Wen B (1997) Study on the mechanism and application of filter with supersonic vibration. In: Proceedings of '97 Asia-Pacific vibration conference, Kyongju, Korea, vol 11, pp 979–982
110. Ma ZG, Yan YH, Wen BC (1997) Optimal design of intelligent material for the active vibration control. In: Proceedings of '97 Asia-Pacific vibration conference, Kyongju, Korea, vol 11, pp 1194–1198
111. Zou GS, Jin JD, Wen BC (2001) Nonlinear and chaotic oscillations of pipes conveying fluid under harmonic excitation. Proceeding of Asia-Pacific Vibration Conference, Hangzhou 10:132–135
112. Cao WH, Wen BC (1989) Dynamic characteristics of asymmetrically piecewise linear vibrating mechanism. In: Proceedings of Asia vibration conference, Shenzhen, China, pp 86–89
113. Ge SP, Wen BC (1989) Dynamic experiments and analysis for 2ZSM-2065A vibrating screen. In: Proceedings of Asia Vibration conference, Shenzhen, China, pp 457–461
114. Blehmann EE (1971) Synchronization of dynamical system. Science Press (in Russian)
115. Blehmann EE (1981) Synchronization of nature and engineering. Science Press (in Russian)
116. Gopalakrishnan S, Martin M, Ddoyle JF (1992) A matrix methodology for spectral analysis of wave propagation in multiple connected Timoshenko beams. J Sound Vib 158(1):11–24
117. Stutzer MJ (1980) Chaotic dynamics and bifurcation in a macro model. J Econ Dyn Control 2:253–276
118. Benhabib J, Day RH (1981) Rational choice and erratic behavior. Rev Econ Stud 459–471
119. Baumol WJ, Wolff EN (1983) Feedback from productivity growth to R&D. Scand J Econ 1:147–157
120. Baumol WJ, Benhabib J (1989) Chaos significance mechanism and economic application. J Econ Perspect 1:77–105
121. Baumol WJ, Quandt RE (1985) Chaos models and their implications for forecasting. East Econ J 11:3–15
122. Scheinkman J, Lebaron B (1989) Nonlinear dynamics and stock returns. J Bus 62:311–337
123. Sayers CL (1988) Statistical inference based upon nonlinear science. Eur Econ Rev 35:306–312
124. Frank M, Stengos T (1988) Some evidence concerning macroeconomic chaos. J Monet Econ 22:423–438
125. Rosser JB, Rosser MV (1994) Long waves chaos and systemic economic transformation. World Future 99:197–207
126. Feichtinger G, Hommes CH, Herold W (1994) Chaos in a simple deterministic queuing system. Math Methods Oper Res 40:109–119
127. Bültmanna U, Kanta I, Kaslb SV, Beurskensa AJHM, van den Brandta PA (2002) Fatigue and psychological distress in the working population psychometrics, prevalence, and correlates. J Psychosom Res 52:445–452
128. Nickel P, Nachreiner F (2001) The suitability of the 0.1 Hz component of heart rate variability for the assessment of mental workload in real and simulated work situations. HFES/EUR
129. Shamsuddin KM, Tatsuo T (1998) Continuous monitoring of sweating by electrical conductivity measurement. Physio Meas (19):375–382
130. Lenneman JK. The evolution of autonomic space as a method of mental workload assessment for driving. In: Proceedings of the second international driving symposium on human factors in driver assessment, training and vehicle design, pp 125–129

131. Gogolitsin YL, Kroptov YD (1981) Application of the stimulus-time histogram to the analysis of discharge post patterns in neuronal populations of the human brain during intellectual activity. In: Bechtrerva NP (ed) Psychophysiology: today and tomorrow. Pergamon Press, New York, pp 23–31
132. Okogbaa G, Richard L, Filipusic D (1994) On the investigation of the neurophysiological correlates of knowledge mental fatigue using the EEG signal. Appl Ergon 25(6):355
133. Lal SKL, Craig A (2001) Electroencephalography activity associated with driver fatigue: implications for a fatigue countermeasure device. J Psychophysiol 15:183–189
134. Fisch BJ. Spehlmann's EEG primer, second revised and enlarged edition. Elsevier Science B.V., Amsterdam
135. Schmidt P, Peltzer P (1976) Das Synchronisieren Zweier Unwuchtruttler an Schwingmaschinen. Aufbereitungs Technik 3:108–114
136. Mariop (1970) Self-synchronization of two eccentric rotors in plane motion. Shock Vib Bull 41(10):159–162
137. Meirovitch L (1975) Elements of vibration analysis. McGraw-Hill, Inc
138. Melver P, Clever M (1995) Water-power absorption by a line of submerged horizontal cylinders. Appl Ocean Res 17:117–126
139. Chonan S, Jiang ZW et al (1998) Development of a palpation sensor for detection of prostatic cancer and hypertrophy. Int J Appl Electromagn Mech 16(9):25–39
140. Lalande F, Rogers CA (1996) Qualitative nondestructive evaluation research at CIMSS. In: Proceedings of the 3rd international conference on intelligent materials, Lyon, France
141. Liang C, Sun FP, Rogers CA (1997) Coupled electro-mechanical analysis of adaptive material systems. J Intell Mater Syst Struct 8:335–343
142. Wang XM, Ehlers C, Neitzel M (1996) Investigation on the dynamic actuation of piezoelectric patches bonded on a beam. In: Proceedings of the seventh international conference on adaptive structures. Rome, Italy, pp 425–434
143. Liang C, Sun FP, Rogers CA (1996) Electro-mechanical impedance modeling of active material systems. Smart Mater Struct 5:171–186
144. Doyle JF (1997) Wave propagation in structures. Springer, New York, Inc
145. Harrington PF (1961) Time-harmonic electromagnetic field. McGraw-Hill
146. Kong JA (1986) Electromagnetic wave theory. Willy
147. Jackson JD (1979) Classical electromagnetic theory. Willy
148. Johnson CC (1965) Field and wave electrodynamics. McGraw-Hill
149. Stratton JA (1941) Electromagnetic theory. McGraw-Hill
150. Collin RE (1966) Foundations for microwave engineering. McGraw-Hill
151. Winick H (1994) Synchrotron radiation source a primer. World Scientific, Singapore
152. Collin RE (1960) Field theory of guided waves. McGraw-Hill
153. Marcus D (1991) Theory of dielectric optional waveguides, 2nd edn
154. Gupta KC et al (1981) Computer-aided design of microwave circuits. Artech House
155. Kaifez D, Guill P (1986) Dielectric resonators. Artech House
156. Jones DS (1986) Acoustic and electromagnetic waves. Clevendon Press, Oxford
157. Liu F, Wang WB (1989) An analysis of transient field of linear antenna. IEEE Trans EMC 39(4):405–409
158. Felsen LB (1976) Transient electromagnetic field. Spring, Berlin, Heidelberg, New York
159. Baum CE (1989) Radiation of impulse-like transient field. Sensor and simulation notes. Note 321
160. Seseler M (1988) New particle acceleration techniques. Phys Today 6:26–34
161. Dawson JM (1989) Plasma particle accelerators. Sci Am 3:34–41
162. Hubner K (1998) Future accelerators. In: XXIX international conference on high energy physics, p 6
163. Seseler M (1998) Gamma ray colliders and muon colliders. Phys Today 3:48–53
164. Tigner CM (1999) Handbook of accelerator physics and engineering. World Scientific